Nanodust in the Solar System: Discoveries and Interpretations

For further volumes:
http://www.springer.com/series/5664

Astrophysics and Space Science Library

Ingrid Mann • Nicole Meyer-Vernet
Andrzej Czechowski
Editors

Nanodust in the Solar System: Discoveries and Interpretations

Springer

Editors
Ingrid Mann
EISCAT Scientific Association
Kiruna
Sweden

Nicole Meyer-Vernet
LESIA
Observatoire de Paris
Meudon
France

Andrzej Czechowski
Space Research Centre
Polish Academy of Sciences
Warsaw
Poland

ISSN 0067-0057
ISBN 978-3-642-27542-5 ISBN 978-3-642-27543-2 (eBook)
DOI 10.1007/978-3-642-27543-2
Springer Heidelberg New York Dordrecht London

Library of Congress Control Number: 2012938964

Cover figure : Transmission electron micrograph of nanometer sized cosmic dust analogs obtained at JEOL 3200FS instrument with 300 kV electron beam at the Institute for Molecular Science in Japan. The analogs were produced in a vapor phase condensation experiment at Tohoku University in Japan (Figure by Y. Kimura).

Printed on acid-free paper

Springer is part of Springer Science+Business Media (www.springer.com)

Preface

When the authors of this book gathered for the first time in September 2009, it was with the goal to investigate how nanodust can be detected and studied in the solar system and how this knowledge can contribute to the understanding of nanodust in other cosmic environments. It quickly became apparent that each of us had a different understanding of the term nanodust and a different research approach. Coming from different fields of expertise, the discussions were full of questions to the others and our views of the topic diverse. Meanwhile, the interest in nanodust was growing, as well as the amount of observational and experimental results. In May 2010, we decided that the best approach to assemble and communicate our current knowledge would be in the form of a book addressing a specific aspect of the subject in each chapter. The book in front of you, finalized in August 2011, consists of an introduction and nine self-contained chapters. We hope it will attract readers working in space research, as well as advanced students and researchers working in related or other fields.

We thank all the contributors of the individual chapters who have made this book possible with their individual work and joined discussions. We also thank a number of scientists for acting as reviewers of the manuscripts and/or for discussing with us during the preparation of this book, in particular Joseph A. Burns, Konrad Dennerl, Bruce Draine, Dieter Gehrlich, Doug Hamilton, Louis d'Hendecourt, Mihaly Horanyi, Daniel Jontof-Hutter, Michiko Morooka, Tadashi Mukai, Edmond Murad, Joseph A. Nuth, and Pawel Oberc. Their comments contributed a lot to our project and helped in assembling a whole out of the different views on the subject. Our group discussions were held as an International Team formed at the International Space Science Institute, ISSI in Bern, Switzerland. We thank the ISSI staff and the editorial staff at Springer for their support.

Kiruna, Sweden — *Ingrid Mann*
Meudon, France — *Nicole Meyer-Vernet*
Warsaw, Poland — *Andrzej Czechowski*

Contents

Contributors

Andrzej Czechowski Space Research Centre, Polish Academy of Sciences, Warsaw, Poland

Hsiang-Wen Hsu Laboratory of Atmospheric and Space Physics, University of Colorado at Boulder, Boulder, CO, USA

Geraint H. Jones Mullard Space Science Laboratory, University College London, Holmbury St. Mary, Dorking, Surrey, UK

Vasili Kharchenko Department of Physics, University of Connecticut, 2152 Hillside Road, Storrs, CT 06269, USA

Harvard-Smithsonian Center for Astrophysics, 60 Garden Street, Cambridge, MA02138, USA

Yuki Kimura Department of Earth and Planetary Materials Science, Graduate School of Science, Tohoku University, Sendai, Japan

Harald Küger Max-Planck-Institut für Sonnensystemforschung, Katlenburg-Lindau, Germany

Max-Planck-Institut für Kernphysik, Saufercheckweg 1, Heidelberg, Germany

Nicholas Lewkow Department of Physics, University of Connecticut, Storrs, CT, USA

Aigen Li Department of Physics and Astronomy, University of Missouri, Columbia, MO, USA

Ingrid Mann Belgian Institute for Space Aeronomy, Brussels, Belgium

Laboratoire d'Etudes Spatiales et d'Instrumentation en Astrophysique, Observatoire de Paris, Universite Pierre et Marie Curie, Universite de Paris Diderot, CNRS, Meudon, France

EISCAT Scientific Association, Kiruna, Sweden

Nicole Meyer-Vernet LESIA - Observatoire de Paris, CNRS, UPMC, Université Paris Diderot, Meudon, France

Frank Postberg Institut für Geowissenschaften, Ruprecht-Karls Universität, Heidelberg, Germany

Max-Planck-Institut für Kernphysik, Saufercheckweg 1, Heidelberg, Germany

Peter Wurz Physikalisches Institut, Universität Bern, Bern, Switzerland

Arnaud Zaslavsky LESIA, Observatoire de Paris, CNRS, UPMC, Université Paris Diderot, Meudon, France

Introduction

Ingrid Mann and Nicole Meyer-Vernet

1 Introduction

The presence of cosmic dust particles induces and strongly influences many astrophysical processes ranging from planet formation and stellar evolution to galaxy evolution. Formation, destruction, and interactions of cosmic dust particles are observed in many different cosmic environments and can be studied at close range in the solar system.

Nanodust often interacts in different ways and often is detected in different ways than the larger cosmic dust particles. Due to the large surface area compared to the small mass it is an important agent for interactions with particles and fields. Nanodust can also be a nuisance: in near-Earth orbit it produced a large number of events in field measurements aboard the STEREO spacecraft that were unexplained at the beginning of the mission. The title of our book implies addressing three major questions: What is nanodust? How was it discovered in the solar system? And how do we interpret the observations? One might consider the chapters of this book as being divided into three groups, each addressing primarily one of these questions.

The term nanodust is used with many different meanings, even within one field of research, as in astrophysics or in solar system research. The reason is that nanodust has different properties than larger dust and the first three chapters illustrate these

I. Mann (✉)
Belgium Institute for Space Aeronomie, 3 Avenue Circulaire, 1180 Brussels, Belgium

Laboratoire d'Etudes Spatiales et d'Instrumentation en Astrophysique, Observatoire de Paris, Universite Pierre et Marie Curie, Universite de Paris Diderot, CNRS, Meudon, France

EISCAT Scientific Association, P. O. Box 812, SE-981 28 Kiruna, Sweden
e-mail: ingrid.mann@eiscat.se

N. Meyer-Vernet
Laboratoire d'Etudes Spatiales et d'Instrumentation en Astrophysique, Observatoire de Paris, Universite Pierre et Marie Curie, Universite de Paris Diderot, CNRS, Meudon, France

I. Mann et al. (eds.), *Nanodust in the Solar System: Discoveries and Interpretations*, Astrophysics and Space Science Library 385, DOI 10.1007/978-3-642-27543-2_1,

differences from various angles. The first chapter, entitled: "Nanodust in the Interstellar Medium in Comparison to the Solar System" defines nanodust as the dust particles that undergo stochastic heating. It deals with the astrophysics of dust in the interstellar medium (ISM), where the presence of nanodust has been inferred from observations since long time. The major interactions of nanodust in the ISM are described and it is shown that these processes are less important in the solar system. Nanodust can also be defined as the dust particles whose properties are predominantly determined by surface effects. This is done in the subsequent chapter "Phenomena of Nanoparticles in Relation to the Solar System", which also reviews the definitions of nanodust in solid-state physics. It presents laboratory experiments on nanodust that reveal their special properties like melting point decrease and enhanced diffusion within the solid. The specific properties of nanodust influence the growth of cosmic dust, for instance in the early solar system as well as in the late stage of stellar evolution. Nanodust in the interplanetary space can be described as those particles whose dynamics, due to the predominance of electromagnetic forces, is more similar to ions than to larger solid objects, the latter being mainly influenced by gravity and radiation pressure. The chapter "Nanodust Dynamics in Interplanetary Space" presents trajectory calculations in the interplanetary magnetic field. It shows that the nanodust reaches velocities of the order of the solar wind velocity and under certain conditions can be trapped along magnetic field lines near the Sun.

The subsequent three chapters are devoted to the discoveries of nanodust in the solar system, where space missions provide the unique opportunity to detect nanodust in situ in cosmic environments. The chapter "Dynamics, Composition, and Origin of Jovian and Saturnian Dust-Stream Particles" describes the streams of nanodust observed with the in situ dust detectors on board the spacecraft Ulysses, Galileo, and Cassini. The trajectories of the stream particles are shaped in the magnetospheres of these giant planets and finally in the solar wind. Compositional information suggests that the particles form as condensates or by sputtering of larger particles. These stream particles were observed with dedicated dust instruments, however, beyond their range of calibration. Impacts of nanodust were also observed near Saturn with a plasma spectrometer instrument aboard the Cassini orbiter. This is discussed in the following chapter "Nanodust Measurements by the Cassini Plasma Spectrometer". The observations point to the presence of charged nanodust in the upper atmosphere of Saturn's largest moon Titan, and in the plume of material ejected from the south pole of the icy moon Enceladus. All nanodust detections from spacecraft came by surprise. The chapter "In Situ Detection of Interplanetary and Jovian Nanodust with Radio and Plasma Wave Instruments", describes the probably greatest surprise of observing nanodust in the interplanetary medium with STEREO near the orbit of the Earth. The chapter also discusses why the nanodust can be detected despite its small mass and refers to the results of a similar plasma instrument onboard Cassini. The Cassini instrument observed the Jovian stream particles in Jupiter' s magnetosphere at the same time as the (dedicated) dust instrument onboard. The discussed observations offer perspectives for using plasma wave instruments as dust detectors in future space missions.

The last three chapters look into the question of how the observations can be interpreted. The chapter "Erosion Processes Affecting Interplanetary Dust Grains" discusses the lifetime of nanodust in the solar system. Nanodust in the inner solar system is destroyed by sublimation and sputtering. Grains that are composed partly or fully out of ice are also subject to photon-stimulated desorption, so that erosion of icy grains is highly effective even in the outer solar system where water ice does not sublimate. To this date, there is no brightness observation of nanodust in the interplanetary medium. Since the scattering of light is usually most efficient for wavelengths similar to the size of the particle, one may expect to observe nanodust by X-rays. The chapter on "Charge-Exchange and X-ray Processes with Nanodust Particles" considers these interactions as possible processes to observe nanodust. Scattering of X-ray photons is calculated and the intensities are compared to other diffuse X-ray emissions in the solar system. The final chapter discusses "Causes and Consequences of the Existence of Nanodust in Interplanetary Space", with particular emphasis on the formation process of the nanodust during the fragmentation of larger particles. Though being a component of the interplanetary dust cloud, its formation during high velocity collisions makes it quite possible that the nanodust has a different composition from the larger particles. Its formation is likened to the collision evolution in the ISM. Some processes that possibly determine the lower size limit are discussed, but the current observations of nanodust in the solar system do not indicate the size of the smaller dust particles and this remains an open question for future studies.

Nanodust in the Interstellar Medium in Comparison to the Solar System

Aigen Li and Ingrid Mann

Abstract Nanodust, which undergoes stochastic heating by single starlight photons and dominates the near- and mid-infrared emission in the interstellar medium, ranges from angstrom-sized large molecules containing tens to thousands of atoms to grains of a couple of tens of nanometers. The presence of nanodust in astrophysical environments has been revealed by a variety of phenomena: the optical luminescence, the near- and mid-infrared continuum emission and spectral bands, the Galactic foreground microwave emission, and the ultraviolet extinction. These phenomena are ubiquitously seen in the interstellar medium of the Milky Way and beyond.

Nanometre-sized grains have also been identified as presolar in primitive meteorites based on their isotopically anomalous composition. Considering the very processes that reveal the presence of nanodust in the ISM for the nanodust in the solar system shows that observing solar system nanodust through these processes is less likely.

1 Introduction: The Interstellar Medium and Nanodust

The stars in our Galaxy, the Milky Way, are far apart and, for instance the nearest star, Proxima Centauri, is at distance roughly 1.3 parsec (pc[1]) from the Sun. The space between stars contains gaseous ions, atoms, molecules and solid dust grains:

[1]With 1 pc = 3.086×10^{16} m this is $\sim 2.67 \times 10^{5}$ AU, where 1 AU = 1.496×10^{11} m is the average distance between Sun and Earth.

A. Li (✉)
Department of Physics and Astronomy, University of Missouri, Columbia, MO 65211, USA
e-mail: lia@missouri.edu

I. Mann
Belgium Institute for Space Aeronomie, 3 Avenue Circulaire, 1180 Brussels, Belgium
e-mail: ingrid.mann@aeronomie.be

I. Mann et al. (eds.), *Nanodust in the Solar System: Discoveries and Interpretations*, Astrophysics and Space Science Library 385, DOI 10.1007/978-3-642-27543-2_2,

the interstellar medium (ISM). With a mean number density of ∼1 H-atom/cm^3 it is more empty than the best vacuum (which has a density of ∼10^3 molecules/cm^3) that can be created on Earth. The gas and dust of the ISM contain ∼10% of the total mass of the visible matter in the present-day Milky Way.[2] The bulk of the heavy elements, including most of the interstellar silicon, magnesium, iron and a large fraction of the interstellar carbon, are depleted from the gas phase and form submicron-sized grains, which make up ∼1% of the total mass of the ISM. In a spiral galaxy like the Milky Way, most of the interstellar dust and gas are concentrated in its spiral arms and a relatively thin gaseous disk of a thickness of a few hundred pc.

The ISM plays a crucial role in galaxy evolution: New stars form out of dusty molecular clouds which present a dense phase of the ISM while stars in late stage of evolution return gas and newly formed dust to the ISM (either through stellar winds or supernova explosions). The astrophysics of the ISM, from the thermodynamics and chemistry of the gas to the dynamics of star formation, is strongly influenced by the presence of the dust. Dust particles within in the ISM are subject to various processes of destruction (e.g., sublimation, collisional fragmentation, sputtering) and formation (e.g., agglomeration, condensation, accretion, fragmentation of larger dust), their respective importance depending on the specific environment.

Already in 1956 John Platt suggested that very small grains or large molecules of less than 1 nm in radius could grow in the ISM by random accretion from the gas (Platt 1956). Today the presence of nanodust in the ISM is generally accepted, but there are still uncertainties in interpreting observations. The discussion in this chapter will focus on the ultrasmall, nanometre-sized interstellar grains. By nanometre-sized grains, or nanodust, we mean grains with a spherical radius of $a \lesssim 10$–20 nm which undergo stochastic heating and exhibit near-IR and mid-IR emission features in the Galactic ISM (Draine and Li 2001; Li 2004).

For a long time most of our knowledge about interstellar dust was derived from interstellar extinction and reddening, and to a lesser degree from interstellar polarization (which is caused by preferential extinction of one linear polarization over another by aligned nonspherical dust). We will discuss the ISM extinction in the following section. Infrared observations from satellites started in the 1980s and allowed for observing the emission from interstellar dust. The observations indicate the presence of stochastically heated nanodust and emitting polycyclic aromatic hydrocarbon molecules (PAHs) discussed in Sect. 3. Other observational results are also explained with the presence of nanodust: the microwave emission of rotationally excited nanodust (Sect. 4), the photoluminescence of nanodust (Sect. 5), and indirect evidence comes from the photoelectric heating of interstellar gas (Sect. 6). Another population of nanodust are presolar nanograins that are identified in primitive meteorites and collected interplanetary dust and were present in the ISM at the time of the formation of the solar system (Sect. 7). Based on the different

[2]This mass fraction is much higher for galaxies at early times (since the ISM is gradually consumed by star formation as galaxies evolve) and much lower (∼0.1%) for elliptical galaxies. Note that here we do not discuss the observations of nanodust in other galaxies.

processes that provide evidence for the existence of nanodust in the ISM we then make a comparison to the nanodust in the solar system (Sect. 8) and end with a conclusion. The interstellar extinction and the stochastic heating process are elaborated in Appendixes 1 and 2.

2 The Interstellar Extinction

Small solid dust grains in the ISM absorb and scatter the starlight, so that it appears fainter and redder as expected, this is denoted as interstellar extinction. The Galactic interstellar extinction curves are measured today for various sightlines over a wide wavelength range ($0.1\,\mu m \leq \lambda \leq 20\,\mu m$). Although the extinction curves vary in shape from one line of sight to another, they do exhibit some common appearance (see Fig. 1). The extinction curves are plotted as A_λ/A_V, where A_λ is the extinction, measured in astronomical magnitudes, at wavelength λ and typically given relative to the extinction A_V in the visible wavelength band. The extinction curves shown vs. inverse wavelength λ^{-1} rise almost linearly from the near-infrared (IR) to the near-ultraviolet (UV), with a broad absorption bump at about $\lambda^{-1} \approx 4.6\,\mu m^{-1}$ ($\lambda \approx$2,175 Å) followed by a steep rise into the far-UV at $\lambda^{-1} \approx 10\,\mu m^{-1}$, the shortest wavelength at which the extinction has been measured.

In the wavelength range of $0.125\,\mu m \lesssim \lambda \lesssim 3.5\,\mu m$, the Galactic extinction curves can be approximated by an analytical formula involving only one free parameter: the total-to-selective extinction ratio, R_V (Cardelli et al. 1989, see Appendix 1). The sightlines through diffuse gas in the Milky Way have $R_V \approx 3.1$ as an average value, but there are considerable regional variations and also the strength and width of the 2,175 Å extinction bump vary markedly in the ISM (see Xiang et al. 2011 and

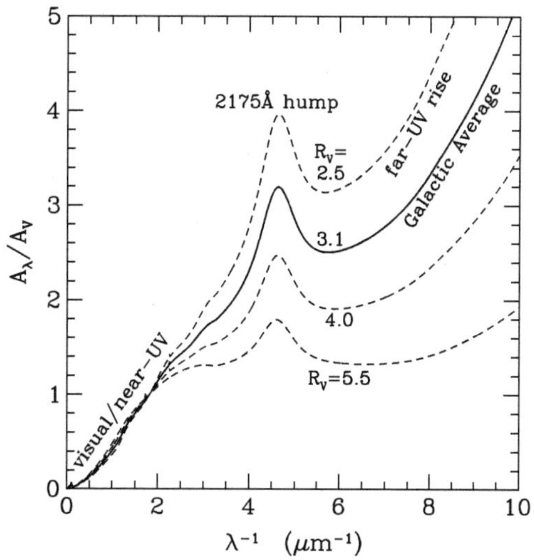

Fig. 1 Interstellar extinction curves of the Milky Way ($R_V = 2.5, 3.1, 4.0, 5.5$). There exist considerable regional variations in the Galactic optical/UV extinction curves, as characterized by the total-to-selective extinction ratio R_V, indicating that dust grains on different sightlines have different size distributions

references therein). Lower-density regions have a smaller R_V, a stronger 2,175 Å bump and a steeper far-UV rise at $\lambda^{-1} > 4\,\mu m^{-1}$, while denser regions have a larger R_V, a weaker 2,175 Å bump and a flatter far-UV rise.

The exact nature of the carrier of this bump remains unknown since its first discovery nearly half a century ago (Stecher 1965). It has been postulated to be nano carbon particles (e.g., Duley and Seahra 1998) or PAHs (Joblin et al. 1992; Li and Draine 2001a; Cecchi-Pestellini et al. 2008; Steglich et al. 2010). Laboratory experiments even suggest that the bump is possibly caused by particles consisting of interplanetary materials (Bradley et al. 2005).

The far-UV part ($\lambda \gtrsim 6\,\mu m^{-1}$) of the Galactic interstellar extinction continues to rise up with shorter wavelength to $\lambda = 0.1\,\mu m$ and there does not appear to be any evidence of saturation even at this wavelength.[3] Since it is generally true that a grain absorbs and scatters light most effectively at wavelengths comparable to its size $\lambda \approx 2\pi a$, we can therefore conclude that there must be appreciable numbers of ultrasmall grains with $a \lesssim 0.1\,\mu m/2\pi \approx 16$ nm. In the far-UV wavelength range, the grains of a couple of nanometres are in the Rayleigh regime (i.e., $2\pi a/\lambda \ll 1$) and their extinction cross sections per unit volume, $C_{ext}(a, \lambda)/V$, are independent of size. Hence the far-UV extinction indicates the presence of nanodust, but does not allow to constrain the sizes of the nanodust in the ISM.

3 Emission Brightness from Vibrationally Excited Nanodust and Molecules

The near- and mid-IR[4] emission of the ISM constrains the size and composition of nanodust, because of its stochastic heating and its characteristic emission in the near-IR. A dust particle in space is subject to substantial temporal fluctuations in temperature, if (1) its heat content is smaller than or comparable to the energy of a single stellar photon (Greenberg 1968) and (2) the photon absorption rate is smaller than the radiative cooling rate (Li 2004). In the diffuse ISM, nanodust is stochastically heated by single photons to temperatures much higher than its "equilibrium" temperature (even though an "equilibrium" temperature is not physical for a stochastically heated nanograin, it can still be *mathematically* determined from the energy balance between absorption and emission).

Figure 2 illustrates the time evolution of grain temperature within a day for PAH/graphite grains exposed to the solar neighbourhood interstellar radiation field (Draine 2003). Grain size decreases from the top to the bottom panel. We see that

[3] However, the Kramers–Kronig dispersion relation requires that the far-UV extinction rise with inverse wavelengths must turn over at some smaller wavelengths as the wavelength-integrated extinction must be a finite number (see Purcell 1969).

[4] We here use "near-IR" for wavelengths $1\,\mu m \lesssim \lambda \lesssim 12\,\mu m$, "mid-IR" for $12\,\mu m \lesssim \lambda \lesssim 60\,\mu m$ and "far-IR" for $60\,\mu m \lesssim \lambda \lesssim 1{,}000\,\mu m$.

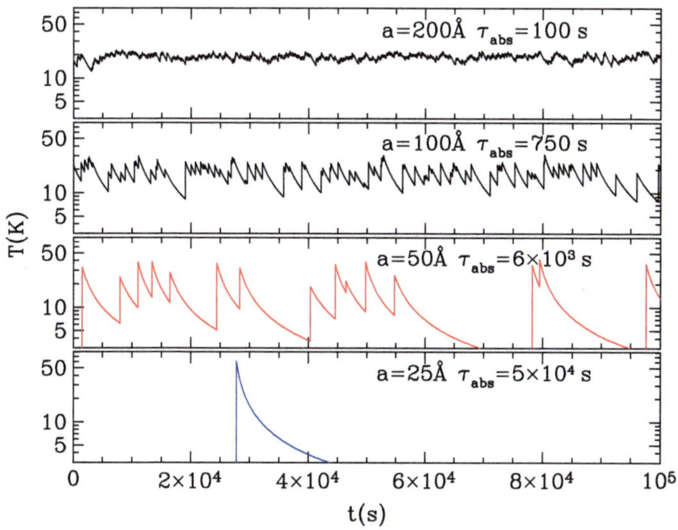

Fig. 2 The time evolution of temperature within a day ($\sim 8.6 \times 10^4$ s) for PAH/graphitic grains of different sizes, a, in the solar neighbourhood interstellar radiation field; $\tau_{\rm abs}$ denotes the mean time between photon absorptions. Grains with radii $a \gtrsim 20$ nm have large photon absorption rates $1/\tau_{\rm abs}$ and large heat capacities so that a single photon cannot significantly alter their temperature. Heat capacities and photon absorption rates are small for grains with radii $a \lesssim 5$ nm and temperature fluctuates strongly. Taken from Draine (2003)

for grain radii $a \gtrsim 20$ nm, individual photon absorptions are relatively frequent, and the grain heat capacity is large enough that the temperature excursions following individual photon absorptions are relatively small; it is reasonable to approximate the grain temperature as being constant in time. Grains with radii $a \lesssim 5$ nm, however, raise their temperature appreciably after absorption of a single photon and since photon absorption rates are small, they can cool down before absorbing another photon. As a result, the temperature raises to well above the mean value. A PAH molecule of 100 carbon atoms (corresponding to a size of $\sim$0.6 nm)[5] will even be heated to $T \approx 785$ K by a photon of $h\nu = 6$ eV, while its "equilibrium" temperature would just be $\sim$22 K. The stochastic heating is further discussed in Appendix 2. The grain equilibrium temperatures raise modestly in regions with high photon flux. This is illustrated in Fig. 3 for graphite and silicate grains. When increasing the interstellar radiation field by six times, the particles larger than 10–20 nm increase their equilibrium temperatures by $\sim (6^{1/6} - 1) \approx 35\%$.

Infrared observations from satellites have determined the diffuse ISM brightness for a broad spectral interval (Fig. 4). Initially the emission observed at 12 and 25 μm suggested the presence of nanodust, since it exceeds the emission from large

[5] The term "PAH size" refers to the radius a of a spherical grain with the same carbon density as graphite (2.24 g cm^{-3}) and containing the same number of carbon atoms $N_{\rm C}$: $a \equiv 0.13\, N_{\rm C}^{1/3}$ nm.

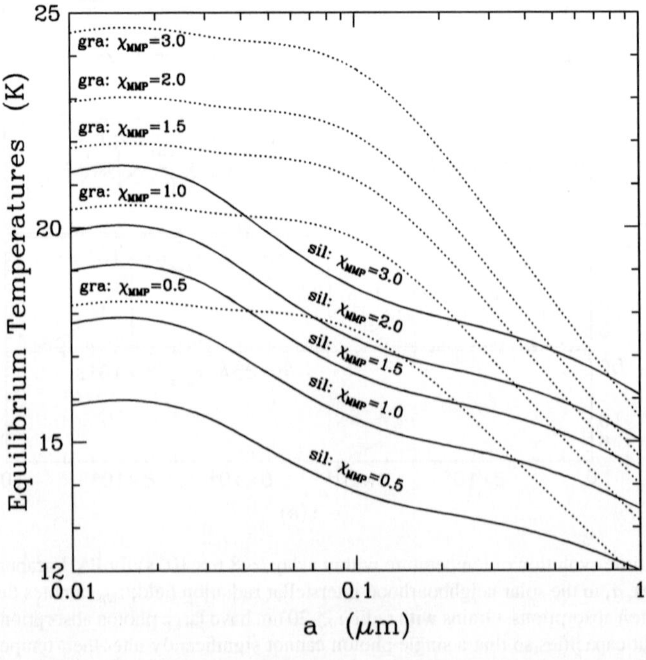

Fig. 3 Equilibrium temperatures for graphite (*dotted lines*) and silicate grains (*solid lines*) in environments with various starlight intensities, $\chi_{\rm MMP}$, in units of the Mathis et al. (1983) solar neighbourhood interstellar radiation field. Grains larger than 10–20 nm attain temperatures 12 K $< T <$ 25 K and therefore do not emit appreciably at $\lambda < 60\,\mu$m. Temperatures of grains with $a <$ 20 nm do not depend on their size. Taken from Li and Draine (2001a)

grains with ∼12–25 K thermal equilibrium temperature (Fig. 3) by several orders of magnitude (Boulanger and Perault 1988). Dust emission at 35 and 4.9 μm was also detected through broadband photometry (Arendt et al. 1998). Later observations revealed prominent emission features in the near-IR collectively referred to as the "UIR" bands. The "UIR" features were found at 3.3, 6.2, 7.7, 8.6 and 11.3 μm (Onaka et al. 1996; Tanaka et al. 1996; Mattila et al. 1996) and can be nicely distinguished in Fig. 5. These "UIR" bands are often attributed to PAH molecules (Leger and Puget 1984; Allamandola 1985). The "UIR" bands alone account already for ∼20% of the total IR emission, while the emission at $\lambda \lesssim 60\,\mu$m accounts for $\gtrsim$35% and that at $\lambda \gtrsim 60\,\mu$m accounts for $\lesssim$65%.

We now discuss a particular model to explain the different ISM dust observations. Figure 5 shows a comparison of the observed emission with the emission calculated from the silicate-graphite-PAH model (Li and Draine 2001a). In this model, ∼15% of the carbon is locked up in the PAH component and the near-IR and "UIR" spectrum are best reproduced by PAHs with a log-normal size distribution peaking at $a \sim$0.6 nm, corresponding to ∼100 carbon atoms. At $\lambda \lesssim 60\,\mu$m the emission is predominantly from PAHs and graphite grains with $a <$ 25 nm. Especially the emission at $\lambda = 12$, 25 μm and shorter wavelengths requires stochastically heated

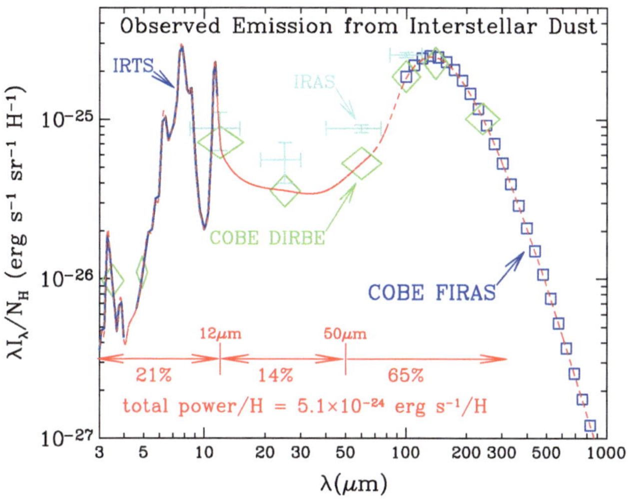

Fig. 4 Observed diffused emission of interstellar dust normalized to the hydrogen column density N_{H}, $\sim 5.1\times 10^{-24}$ erg s^{-1} H^{-1}. Bars at the *bottom* indicate the relative contributions of the spectral intervals to the total IR emission. The data are from the *Infrared Astronomical Satellite* (IRAS), the *Infrared Telescope in Space* (IRTS), the *Spitzer Space Telescope* (Spitzer) and the *Cosmic Background Explorer* (COBE) satellite with its *Far Infrared Absolute Spectrophotometer* (FIRAS) and its *Diffuse Infrared Background Experiment* (DIRBE). The crosses denote observations from IRAS (Boulanger and Perault 1988), squares from COBE-FIRAS (Finkbeiner et al. 1999), diamonds from COBE-DIRBE (Arendt et al. 1998) and the heavy curve from IRTS (Onaka et al. 1996; Tanaka et al. 1996). Taken from Draine (2003)

nanodust, but even at $\lambda = 60\,\mu$m, the grains with $a < 25$ nm contribute ~70% of the total emission. The model does not include silicate nanodust, since nondetection of the 9.7 μm silicate Si–O emission feature in the diffuse ISM (see Fig. 4) indicates its low abundance (see Li and Draine 2001b).

The large silicate and graphite grains of $a > 25$ nm together dominate the emission at $\lambda > 60\,\mu$m, accounting for ~65% of the total IR power. The far-IR emission at $\lambda > 100\,\mu$m can be closely approximated by a modified blackbody of $I_\lambda \propto \lambda^{-\beta} B_\lambda(T)$ with $\beta \approx 1.7$ and $T \approx 19.5$ K (see Draine 1999) or by model emission calculated from large grains with $a > 25$ nm.

4 Microwave Emission from Rotationally Excited Nanodust

Mappings of the microwave sky have revealed unexpected emission at 10–100 GHz frequencies, as shown in Fig. 6 which basically complements the long wavelength part of the diffuse ISM emission from vibrationally excited, submicrometre-sized dust shown in Fig. 4. The spectral variation and absolute value of this "anomalous" component of the diffuse foreground microwave emission are very different from

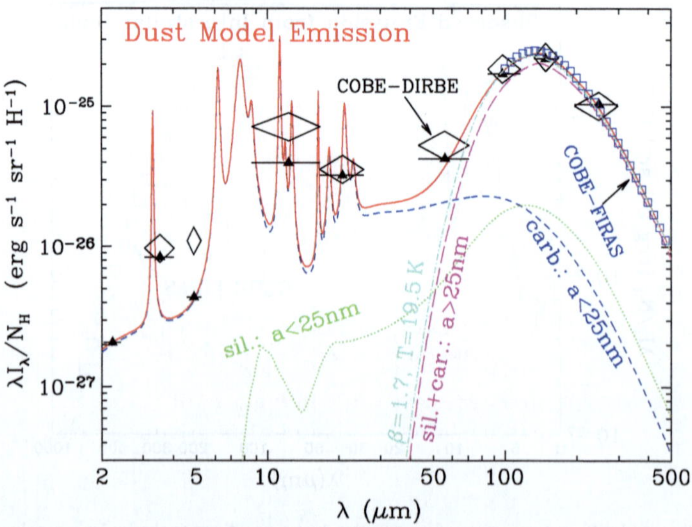

Fig. 5 Calculated dust model emission in comparison to diffuse ISM observations. *Dotted green line*: silicate grains with $a < 25$ nm. *Dashed blue line*: graphite and PAH grains with $a < 25$ nm. *Long dashed magenta line*: the sum of silicate and graphite grains with $a > 25$ nm; *Dot-dashed cyan line*: a modified blackbody of $I_\lambda \propto \lambda^{-\beta} B_\lambda(T)$ with $\beta \approx 1.7$ and $T \approx 19.5$ K approximating the far-IR emission at $\lambda > 100$ μm. The *red solid curve* shows the model spectrum obtained by adding up these different dust emissions. *Triangles* show the latter convolved with the DIRBE filters. *Diamonds* denote observational data from DIRBE (Arendt et al. 1998) and squares from FIRAS (Finkbeiner et al. 1999); see caption to Fig. 4 for the abbreviations. Taken from Li and Draine (2001a)

those of the traditional diffuse emissions at these frequency ranges (e.g., the free–free, synchrotron and thermal dust emission have power-law-like spectra at microwave frequencies) and cannot easily be explained with them.

The spatial distribution of this microwave emission is correlated with interstellar dust emission at 100 and 140 μm (see Draine (2003)), and even better correlated with the mid-IR emission (Casassus et al. 2006; Ysard et al. 2010; Vidal et al. 2011).

This suggests its origins from the dust, but extrapolating the 100–3,000- μm far-IR emission of large dust ($a > 25$ nm) to the microwave frequencies provides values far below the observed microwave emission (see Fig. 6). For these reasons, the electric dipole radiation from rapidly spinning nanograins has become the best explanation for the "anomalous" microwave emission.

A spinning grain with an electric dipole moment μ radiates power $P = 2\omega^4\mu^2 \sin^2\theta/3c^3$, where c is the speed of light and θ the angle between the angular velocity ω and μ. The angular velocity is $\omega = J/I$, where J is the grain angular moment, and I is the moment of inertia of the grain. It steeply decreases with grain size, since for spherical grains, $I \propto a^5$ and in interstellar environments only nanograins can be driven to rotate fast enough to emit at microwave frequencies. For a PAH grain of

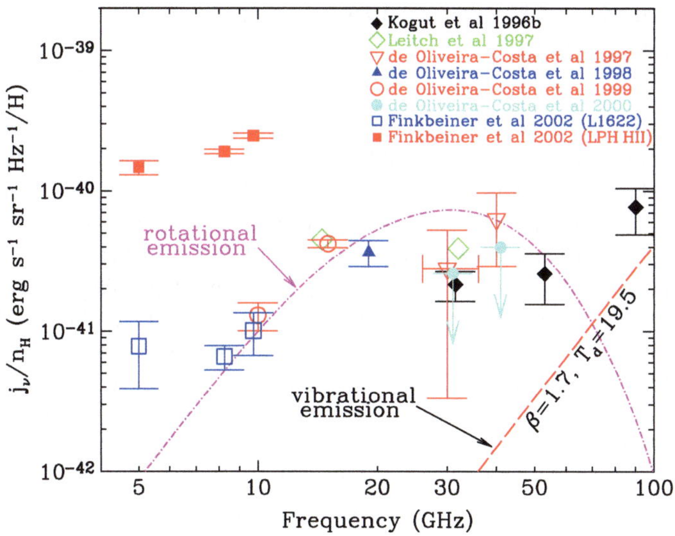

Fig. 6 Galactic foreground microwave emission in comparison to calculated rotational electric dipole emission of the nanograins that account for the "UIR" (see *dashed blue line* in Fig. 5). *Symbols*: observational determinations of "anomalous microwave emission". *Dot-dashed line*: model rotational emission spectrum of nanodust (Draine and Lazarian 1998). *Dashed line*: low-frequency tail of the emission from large grains (mostly with $a > 25$ nm). Taken from Draine (2003)

radius $a = 1$ nm in the diffuse ISM, J peaks at $\sim$2,000$\hbar$ (see Draine and Lazarian 1998).

As described by Draine and his coworkers (Draine and Lazarian 1998; Hoang et al. 2010), a number of physical processes, including collisions with neutral atoms and ions, "plasma drag" (due to interaction of the electric dipole moment of the grain with the electric field produced by passing ions), and absorption and emission of photons, can drive nanograins to rapidly rotate, with rotation rates reaching tens of GHz. The rotational electric dipole emission from these spinning nanograins, the very same grain component (i.e., PAHs) required to account for the "UIR" emission and the observed 12 and 25- μm continuum emission, was shown to be capable of accounting for the "anomalous" microwave emission (Draine and Lazarian (1998); see Fig. 6). Vidal et al. (2011) found that the microwave emission at 31 GHz of the LDN 1780 translucent cloud correlates better with the 12, 25 μm emission than with the 100 μm emission, which supports that emission at this frequency originates from nanodust.

We should note that although the electric dipole radiation from spinning nanodust provides the best explanation for the "anomalous" microwave emission, other physical mechanisms (e.g., hot free–free emission, hard synchrotron radiation or magnetic dipole emission) could still be contributing at some level (e.g., see Draine 1999; Draine and Lazarian 1999).

5 Extended Red Emission: Photoluminescence from Nanodust

Dust *emission* at optical wavelengths, not expected from its vibrational excitation, is also seen in the red part of the visible spectra of a wide variety of dusty environments. This brightness, which cannot be explained by simple light scattering of dust, is called "extended red emission" (ERE). It is characterized by a broad, featureless band between ~540 and 950 nm, with a width of 60 nm $\lesssim$ FWHM $\lesssim$ 100 nm and a peak of maximum emission at 610 nm $\lesssim \lambda_p \lesssim$ 820 nm, depending on the physical conditions of the environment where the ERE is produced (see Fig. 7). The peak wavelength λ_p varies from source to source and within a given source with distance from the illuminating star. The ERE width appears to increase as λ_p shifts to longer wavelengths (Darbon et al. 1999). The ERE has been seen in various different regions which, in terms of UV photon densities, span a range of six orders of magnitudes and, in terms of dust, represent both heavily processed interstellar dust, as well as relatively "freshly" produced dust (e.g., see Witt and Vijh 2004).

The ERE is commonly attributed to photoluminescence (PL) by some component of interstellar dust, a process in which absorptions of UV photons are followed by electronic transitions associated with the emission of optical or near-IR photons. For a given material the wavelength of photoluminescence varies with the size of the grains. The ERE is powered by UV/visible photons, as demonstrated by Smith and

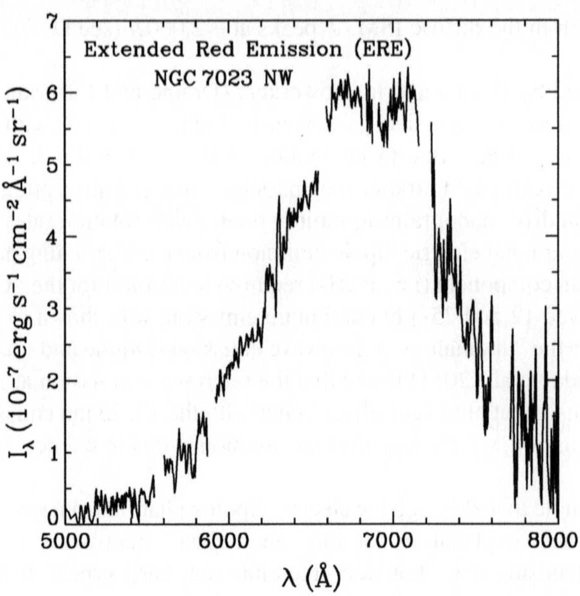

Fig. 7 Observed photoluminescence spectrum—the "extended red emission" (ERE) arising from an unidentified nanodust species in the north-west (NW) filament of NGC 7023, a reflection nebula. Data taken from Witt and Vijh (2004)

Witt (2002) who found that the maximum ERE intensity in any given environment is closely correlated with the density of UV photons.

The true nature of the ERE carriers still remains unknown, although over a dozen candidates have been proposed over the past decades. For a proposed candidate to be valid, it must luminesce in the visible with its spectrum matching that of the observed ERE. But this is not sufficient. As many candidate materials luminesce in the visible after excitation by UV photons, along with the carrier abundance, the efficiency for photoluminescence—the quantum efficiency for the conversion of UV photons absorbed by the ERE carrier to ERE photons—represents one of the strongest constraints.

Gordon et al. (1998) placed a lower limit on the photon conversion efficiency $\eta_{\rm PL}$ (measured by the number ratio of luminesced photons to exciting UV photons) to be approximately $(10 \pm 3)\%$ (also see Szomoru and Guhathakurta 1998). This lower limit was derived from the correlation of ERE intensity with HI (neutral hydrogen) column density at high Galactic latitudes (Gordon et al. 1998), with the assumption that *all UV absorption is due to the ERE carrier candidate* (in other words, assuming that *all the UV photons absorbed by dust lead to the production of ERE*). Since other absorbing interstellar dust components, not likely associated with ERE, are known, the actual luminescing efficiency must be substantially larger than 10%, perhaps in the vicinity of 50% or even higher (Smith and Witt 2002). This poses a serious challenge to materials once thought to be promising ERE candidates, as their luminescing efficiencies are $<1\%$ (see Wada et al. 2009; Godard and Dartois 2010).

All these suggest that the ERE carriers are very likely in the nanometre size range because (a), in general, nanograins are expected to luminesce efficiently through the recombination of the electron–hole pair created upon absorption of an energetic photon, since in such small systems the excited electron is spatially confined and the radiationless transitions that are facilitated by Auger- and defect-related recombination are reduced and (b) small nanograins may be photolytically more unstable and/or more readily photoionized in regions where the radiation intensity exceeds certain levels of intensity and hardness, and thus resulting in both a decrease in the ERE intensity and a redshift of the ERE peak wavelength.[6] Observationally, Darbon et al. (1999) and Smith and Witt (2002) showed that the ERE peak wavelength is indeed shifted towards longer wavelengths with increasing UV radiation density.

Several materials have been proposed as ERE carriers: carbon nanoparticles (Seahra and Duley 1999), silicon nanoparticles (Witt et al. (1998); Ledoux et al. (1998); but see Li and Draine (2002)), nanodiamonds (Chang et al. 2006) and

[6]This is because (1) photoionization would reduce the luminescence of nanograins and (2) the smaller grains would be selectively removed due to size-dependent photofragmentation (Smith and Witt 2002). Due to the quantum confinement effect, the band gap of a semiconductor-like nanograin is *smaller* (and therefore the wavelength of luminescing photons is *longer*) for a *bigger* nanograin (see Sect. 4 in Li 2004).

PAH clusters (Berné et al. 2008). But none of them satisfies all the observational requirements (see Li 2004).

6 Photoelectric Heating of the ISM Gas by Nanodust

Observations of the 21 cm line[7] show that the diffuse neutral atomic hydrogen gas (HI) in the ISM is in two distinct phases with temperatures ~100 K ("Cold Neutral Medium") and ~6,000 K ("Warm Neutral Medium"). What heats the gas to temperatures of ~100 or ~6,000 K? One possible answer to this question provides indirect evidence for the existence of an appreciable quantity of nanodust in the ISM. It has long been recognized that photoelectrons ejected from grains heat the interstellar gas: After absorbing photons with sufficiently large energy, the dust emits photoelectrons, which transfer kinetic energy through inelastic collisions to the gas.

The photoelectrons are important for heating the gas because (a) photons with energies below the ionization potential of H (~13.6 eV) do not couple directly to the gas and (b) other heating sources such as cosmic rays, magnetic fields and turbulence are not important as a global heating source for the diffuse ISM (e.g., the cosmic ray flux is too low by a factor of ~10 to account for the interstellar gas heating; Watson 1972).

In the diffuse ISM, nanodust (and, in particular, angstrom-sized PAH molecules) are much more efficient in heating the gas than large grains (see Tielens 2008) since (a) the mean free path of an electron in a solid is just ~1 nm and therefore photoelectrons created inside a large grain rarely reach the grain surface and (b) the total far-UV absorption is dominated by the nanodust component (see Sect. 3). Theoretical studies have shown that grains smaller than 10 nm are responsible for $\gtrsim$96% of the total photoelectric heating of the gas, with half of this provided by grains smaller than 1.5 nm (Bakes and Tielens 1994; Weingartner and Draine 2001b). Figure 8 shows the calculated photoelectric heating rate as a function of grain size. The size is quantified by the number, N_C of carbon atoms and by the graphite-equivalent spherical radius $a \sim 0.13\, N_C^{1/3}$ nm.

Observations confirm the dominant role of nanodust in the gas heating through the photoelectric effect. For instance, Habart et al. (2001) studied the major cooling lines, [CII] 158 μm and [OI] 63 μm, of L 1721, an isolated cloud illuminated by a B2 star in the ρ Ophiuchi molecular complex. Because of the energy balance between heating and cooling, the [CII] 158 μm and [OI] 63 μm cooling lines (which dominate the gas cooling) reflect the heating input to the gas. They found that the spatial distributions of the gas cooling lines closely correlate with that of the mid-IR emission attributed to nanodust (and PAHs).

[7] The 21-cm line originates in the hyperfine splitting of the parallel and antiparallel spin states of the electron (relative to the spin of the proton) in the electronic ground state (1*s*) of atomic H.

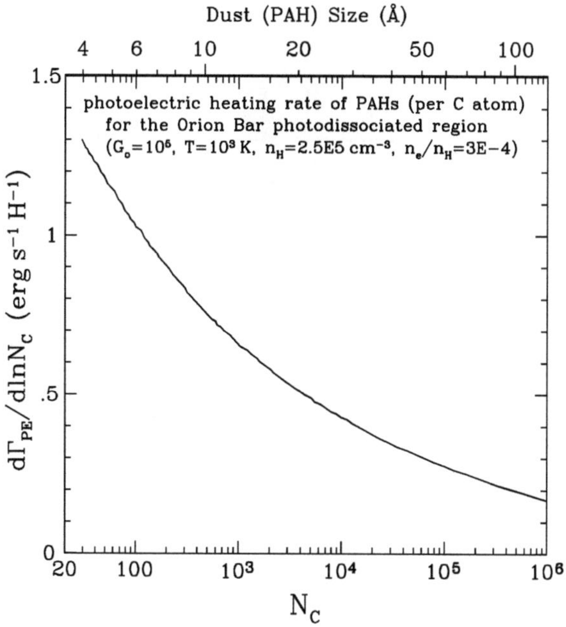

Fig. 8 The photoelectric heating rate Γ_{PE} of the interstellar gas in the Orion Bar photodissociated region as a function of PAH size. The lower axis shows the number, N_C of carbon atoms and the upper axis the graphite-equivalent spherical radius. The rates are presented in such a way that equal areas under the curve correspond to equal contributions to the heating. Typically, approximately half of the heating originates from PAHs and PAH clusters (with $N_C < 10^3$ or $a < 1.3$ nm). The other half comes from grains with sizes $1.3\,\mathrm{nm} < a < 10\,\mathrm{nm}$. Larger grains do not contribute noticeably to the heating. Data taken from Bakes and Tielens (1994)

7 Direct Evidence: Presolar Nanodust in Primitive Meteorites

Based on their isotopic anomalies, presolar grains (such as graphite, silicate, silicon carbide SiC, silicon nitride Si_3N_4, and refractory oxides including corundum Al_2O_3, spinel $MgAl_2O_4$) that predate the solar system have been identified in primitive meteorites, a class of meteorites that essentially remained chemically unaltered since their formation in the solar nebula (e.g., see Lodders 2005). Their path from formation to detection in the laboratory is sketched in Fig. 9, which also shows examples for the two observed nanometre-sized grain types.

Presolar nanodiamonds of radii $a \sim 1$ nm were found to be rich in primitive carbonaceous meteorites, with an abundance as much as $\sim 0.1\%$ of the total mass in some primitive meteorites, more abundant than any other presolar grains by

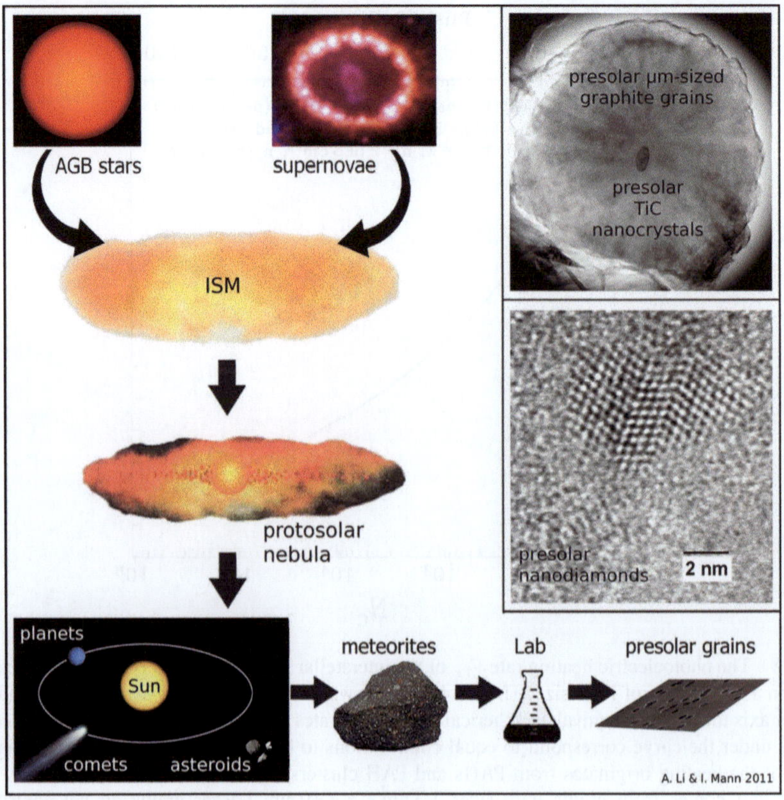

Fig. 9 A schematic illustration of the history of presolar grains from their condensation in stellar winds of *asymptotic giant branch* (AGB) stars or in supernova ejecta to our solar system. These grains survived all the violent processes occurring in the ISM and in the early stages of solar system formation and were incorporated into meteorite parent bodies, from which they are extracted in the laboratories. Inserted are the *transmission electron microscopy* (TEM) images of presolar nanodiamond grains (Banhart et al. 1998) and a presolar TiC nanocrystal within a micrometer-sized presolar graphite spherule (Bernatowicz et al. 1991)

over two orders of magnitude (Lewis et al. 1987).[8] Presolar titanium carbide (TiC) nanocrystals were also seen in primitive meteorites. With a mean radius of ~3.5 nm, they occur as nanometre-sized inclusions within micrometer-sized presolar graphitic spherules (Bernatowicz et al. 1991).

These presolar nanograins, after their condensation in stellar outflows from carbon-rich evolved stars (e.g., TiC nanocrystals) or in ejecta from supernova explosion (e.g., nanodiamonds), and prior to their incorporation into the parent

[8]The identification as presolar grains by measuring isotope ratios does not work for single nanodiamonds and there is a debate whether a fraction of them may actually have formed after the formation of the solar system (e.g., see Ott 2007).

bodies of meteorites during the early stages of solar system formation, they must have had a sojourn in the ISM out of which the solar system formed. However, neither nanodiamonds nor TiC nanocrystals could be representative of the bulk composition of nanodust in the ISM (Draine 2003).

8 Summary and Comparison to Nanodust in the Solar System

Based on the discussions in the previous sections we can summarize the major observational evidence for the presence of nanodust in the ISM: (1) the far-UV extinction at $\lambda^{-1} > 6\,\mu m^{-1}$ caused by the absorption of nanodust (Sect. 2), (2) the $\sim$2–60 μm near- and mid-IR spectral and continuum emission from stochastically heated nanograins through vibrational relaxation (Sect. 3), (3) the $\sim$10–100 GHz "anomalous" Galactic foreground microwave emission from rotationally excited nanograins through electric dipole radiation (Sect. 4) and (4) the $\sim$5,400–9,500 Å broad, featureless ERE band from electronically excited nanograins through photoluminescence (Sect. 5). The presence of an appreciable amount of nanograins in the ISM is also indirectly inferred from the photoelectric heating of interstellar gas by nanodust (i.e., PAHs; Sect. 6) and finally presolar grains are found in interplanetary dust and meteorites of the solar system. Our list is not complete. The nanodust population (particularly PAHs) may also be responsible for the lower gas-phase deuterium abundance of D/H $\approx$ 7–22 ppm in the Galactic ISM compared to the primordial value of D/H $\approx$ 26 ppm, through depleting the "missing" D onto PAHs (Draine 2006).

Based on the preceding discussion the physical processes that allow the direct or indirect observation of nanodust are UV light-scattering, stochastic heating, electric dipole radiation of rotating nanodust, photoluminescence and finally photoelectric heating of the surrounding gas.

We now discuss the possible occurrence of these processes in the solar system. It is worth noting at the beginning that the nanodust in the solar system is not identical to the nanodust in the ISM. The grains with $a < 5$ nm in the local interstellar cloud in the vicinity of the Sun do not enter the solar system, as they are deflected by the magnetic field that builds up in the outer solar system due to interaction of the local ionized ISM gas with the solar wind (Mann 2010). The nanodust in the solar system is produced locally from the interplanetary dust cloud or from solar system objects.

- *Light-scattering in far UV:* Scattering from interplanetary dust particles generated the Zodiacal light, which is the predominant diffuse brightness of the nightsky to wavelengths as short as $\lambda \sim 0.3\,\mu m$. In contrast to the dust in the ISM, the geometric cross section of the dust size distribution in the interplanetary medium has its maximum in the 1–100-μm interval and smaller dust contributes only little to the observed brightness. This is supported by the red colour of the Zodiacal light relative to the solar spectrum (Pitz et al. 1979), which does not indicate any significant contribution from light scattered by nanodust. The brightness at shorter wavelength is dominated by the emission of unresolved stars (Leinert et al. 1998). Dust emission in the X-ray was discussed after the ROSAT survey established comets as a class of X-ray sources. The discussed dust-related X-ray signals are X-ray fluorescence and scattering by nanodust and X-ray emission caused by high-velocity impacts of nanodust. The X-ray flux from nanodust in the solar system is estimated by Kharchenko and Lewkow in this book.

- *Stochastic heating:* For the nanograins in the inner solar system (Mann et al. 2007), it is more likely for them to attain an equilibrium temperature (compared to the same dust in the diffuse ISM). This is because the solar system nanodust is exposed to a far more intense radiation field: at distance $r_{\rm h}$ from the Sun, the 912 Å–1- μm solar radiation intensity is $\approx 7.6 \times 10^7$ $(r_{\rm h}/{\rm AU})^{-2}$ times that of the local interstellar radiation field, where $r_{\rm h}$ is the distance from the Sun. Figure 10 shows the temperature probability distribution functions of amorphous silicate dust using the dielectric functions of Draine and Lee (1984) for selected sizes ($a = 3.5$ Å, 5 Å, 1 nm, 2 nm, 5 nm) illuminated by the Sun at $r_{\rm h} = 0.1, 1$ AU. To facilitate comparison, we plot $\mathrm{d}P/\mathrm{d}\ln T$ in the same T and $\mathrm{d}P/\mathrm{d}\ln T$ ranges for $r_{\rm h} = 0.1$ AU and $r_{\rm h} = 1$ AU. The emission is also illustrated in the same λ and j_λ ranges (except the latter differs by a factor of $r_{\rm h}^{-2}$). We see that silicate dust with $a \gtrsim 2$ nm attains an equilibrium temperature of $T_{\rm eq} \approx 282$ K at $r_{\rm h} = 1$ AU. At $r_{\rm h} = 0.1$ AU, the $\mathrm{d}P/\mathrm{d}\ln T$ distribution functions for grains as small as $a = 1$ nm are already like a delta function, peaking at $T_{\rm eq} \approx 1{,}081$ K. At $r_{\rm h} = 1$ AU, the silicate emission spectra for $a = 2, 3, 5$ nm are almost identical. This is because in the entire UV to far-IR wavelength range, these nanograins are in the Rayleigh regime and their $Q_{\rm abs}/a$ values are independent of size, therefore they obtain an almost identical equilibrium temperature. At $r_{\rm h} = 0.1$ AU, this even applies to smaller grains (e.g., $a = 1$ nm). To summarize, in the solar system at $r_{\rm h} < 1$ AU, the stochastic heating effect is small for dust larger than ~1 nm in radius; for dust smaller than ~5 Å, it may not survive since the stochastic heating would lead to temperatures exceeding ~2,000 K. This is roughly the sublimation temperature of silicate dust, though we point out that the nanodust can have lower sublimation temperature than the bulk material (see Kimura, 2012 in this book).
- *Electric dipole radiation of rotating dust:* We are not aware of any studies of the rotational dynamics of nanodust in the solar system. We do not expect to see strong microwave emission from the dust in the solar system. Although the ions in the solar system may deliver more angular momentum to a grain than in the diffuse ISM (because of their large abundance and large kinetic energy), the dust will not be driven to rotate as fast as in the diffuse ISM, due to (1) the large grain size of the solar system nanodust population, (2) the strong rotational damping caused by photon emission in the solar system and (3) the small number densities of nanodust in the directions facing away from the Sun (see other chapters of this book). The nanodust in the solar system seems to be in the nm size range (see other chapters of this book), while the microwave emission in the ISM arises predominantly from angstrom-sized PAHs. Note that the angular velocity $\omega \propto 1/I$, while $I \propto a^5$.

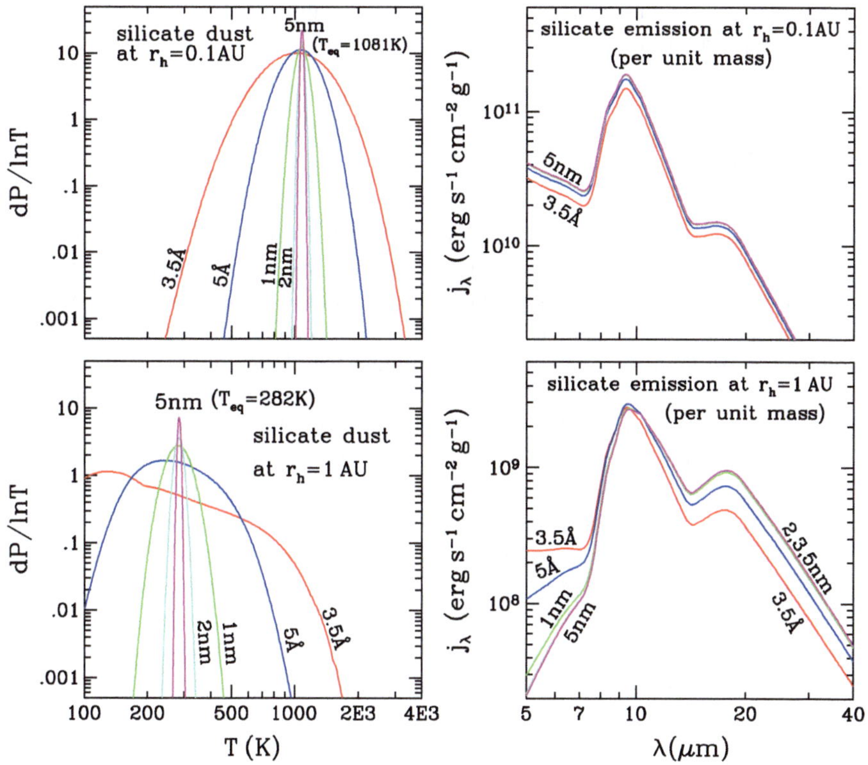

Fig. 10 Temperature probability distribution $dP/d\ln T$ and emission for silicate grains at distances $r_h = 0.1$, 1 AU from the Sun. The distributions at $r_h = 1$ AU are much broader than that for the same dust at $r_h = 0.1$ AU. The grains with $a > 1$-nm peak at the same temperature (which is their equilibrium temperature), as expected from their Rayleigh scattering nature (see Sect. 8)

- *Photoluminescence:* As far as observations in the solar system are concerned a coronal emission that appeared similar to the ERE was speculated to result from silicon nanocrystals near the Sun (Habbal et al. 2003), but this was challenged on the basis of the dust composition and emission properties (Mann and Murad 2005). As opposed to the ISM the nanodust in the solar system most likely is heterogeneous in composition and covers a broad size interval. This broadens wavelength at which the photoluminescence is observed, which makes its detection near the Sun less likely. At this point, we are not aware of an observational result in the optical or IR range confirming the existence of nanodust in the solar system and the detection of nanodust in the interplanetary medium in a similar way like the ERE is unlikely (see Mann and Czechowski, 2012 in this book).
- *Photoelectric heating:* For the solar system nanodust, photoelectric heating does not occur because the nanodust is embedded in the high temperature solar wind electrons. Instead, the presence of nanodust in the solar wind would rather lead to

cooling, because the solar wind ions charge-exchange and are decelerated when passing a nanograin.

9 Conclusion

Observational data, particularly the near- and mid-IR emission data, allow us to constrain the composition, size distribution and quantity of the nanodust population in the ISM: (1) PAHs and nanometre-sized graphitic grains are the dominant nanodust species in the ISM; they contain ~15% of the total interstellar carbon, produce the IR emission at $\lambda < 30\,\mu m$ (including the "UIR" bands) as well as the microwave emission; (2) Nano silicate grains are not important: they *at most* account for ~5% of the total interstellar silicon as indicated by the nondetection of the 9.7-μm silicate Si–O emission band in the Galactic diffuse ISM (Li and Draine 2001b). Presolar nanograins (i.e., nanodiamonds and TiC nanocrystals) that are identified in primitive meteorites and interplanetary dust are not an abundant population of the ISM nanodust, but provide the most direct evidence for interstellar nanodust. They were present in the local ISM at the time of the formation of the solar system. The physical processes involving nanodust in the ISM are less important for the nanodust in the solar system. Nonetheless, considering the stochastic heating process suggests that the size of the silicate nanodust in the inner solar system is possibly constrained by sublimation.

Appendix 1: Interstellar Extinction

The existence of small solid dust grains in interstellar space was established in 1930 when Trumpler showed that the stars in distant open clusters appear fainter than could be accounted for just by the inverse square law, and many stars in the galactic plane appear redder than expected from their spectral types; he interpreted these observations in terms of interstellar extinction and selective absorption (i.e., reddening) caused by "fine cosmic dust particles of various sizes" (Trumpler 1930).

The interstellar extinction curve is most commonly derived utilizing the "pair" method. As illustrated in Fig. 11, this technique involves photometric or spectrophotometric observations of two stars of identical spectral types, with one star located behind a dust cloud and another star (in ideal case) unaffected by interstellar dust, so that there is no obscuration between the observer and the star. Let F_λ be the observed flux from the reddened star, and $F_{o,\lambda}$ be the flux from the unreddened star. If both stars are located at an identical distance, the extinction A_λ—measured in "magnitudes"—is

$$A_\lambda \equiv 2.5 \log_{10} [F_{o,\lambda}/F_\lambda] \approx 1.086\tau_\lambda, \quad (1)$$

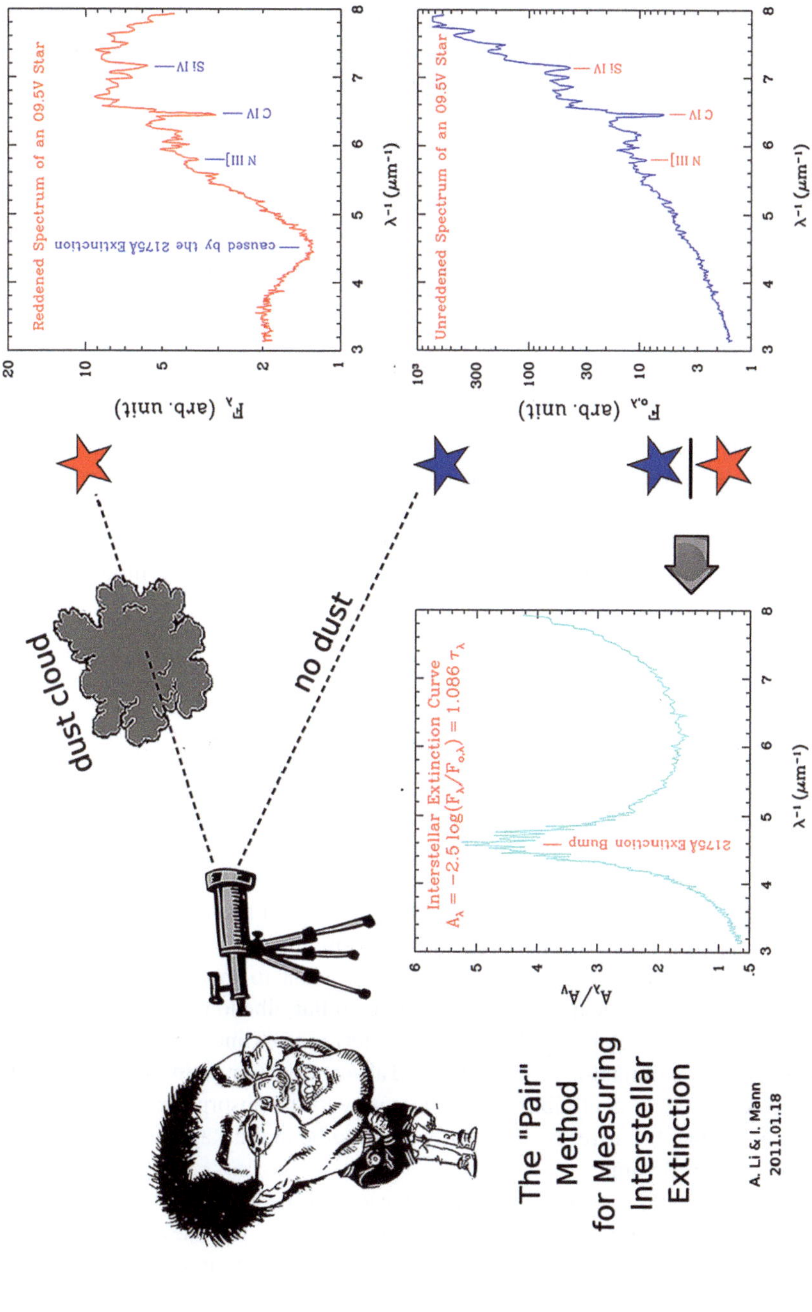

Fig. 11 A conceptual illustration of the "pair" method

where τ_λ is the optical depth. For the ISM within a few kilo pc from the Sun within the Galactic plane and within ~100 pc of the galactic plane, the mean visual extinction per unit path-length has long been determined quite accurately to $\langle A_V/L\rangle \approx 1.8\,\mathrm{mag\,kpc^{-1}}$ (e.g., see Kapteyn 1904).

As it is often not possible to find reddened/unreddened star pairs of identical spectral types which are also located at identical distances, one often measures the colour excess

$$E(\lambda - V) \equiv A_\lambda - A_V = 2.5\log_{10}\left[\frac{F_{o,\lambda}/F_{o,V}}{F_\lambda/F_V}\right] \tag{2}$$

from normalized stellar fluxes, with the V band being the usual choice for the normalization purpose. The total-to-selective extinction ratio

$$R_V \equiv A_V/E(B-V) \tag{3}$$

suggested by Cardelli et al. (1989) is frequently used to characterize the galactic extinction curves.

For particles in the Rayleigh regime (i.e., $2\pi a/\lambda \ll 1$) their extinction cross sections per unit volume, $C_{\rm ext}(a,\lambda)/V$, are independent of size. Therefore, the observational quantity $A_\lambda/N_{\rm H}$ ($\mathrm{mag\,cm^{-2}H^{-1}}$)—the extinction per unit H column—only constrains $V_{\rm tot}/n_{\rm H}$, the total dust volume per H nuclei of this grain component:

$$A_\lambda/N_{\rm H} = 1.086\int C_{\rm ext}(a,\lambda)\,n_{\rm H}^{-1}\,({\rm d}n/{\rm d}a)\,{\rm d}a = 1.086\,(V_{\rm tot}/n_{\rm H})\,(C_{\rm ext}/V)\,, \tag{4}$$

where dn is the number of grains in the size interval $[a,\ a+{\rm d}a]$, and $N_{\rm H}$ ($n_{\rm H}$) is the hydrogen column (volume) density. This explains why the MRN silicate–graphite model with a lower size cut-off of $a_{\rm min} = 5$ nm (Mathis et al. 1977), which was frequently used before the presence of nanodust was confirmed, could also closely reproduce the observed extinction curve. The MRN model fitted the extinction curve using a mixture of silicate and graphite grains with a simple power-law size distribution: ${\rm d}n/{\rm d}a \propto a^{-3.5}$ for 5 nm $\lesssim a \lesssim 0.25\,\mu$m.

The 2,175 Å bump is thought to be predominantly due to absorption, as indicated by the broad minimum near 2,175 Å of the interstellar albedo (Whittet 2003). The interstellar albedo is defined as the ratio of scattering to extinction. The interstellar extinction is the combination of scattering and absorption. For grains in the Rayleigh limit, the scattering is negligible in comparison with the absorption (see Li 2009). This suggests that its carrier is sufficiently small to be in the Rayleigh limit, with a size $a \ll \lambda/2\pi \approx 35$ nm.

Finally, we should note that a smooth extension of the MRN ${\rm d}n/{\rm d}a \propto a^{-3.5}$ size distribution down to $a = 3$ Å is not sufficient to account for the observed 12 μm and 25- μm emission (see Draine and Anderson 1985; Weiland et al. 1986) and that an extra population of nanometre-sized dust is required (e.g., see Désert et al. 1990; Dwek et al. 1997; Li and Draine 2001a). In the recent Weingartner and Draine

(2001a) model, for instance, the grain size distribution extends from a few angstroms to a few micrometers, with ~6% of the total dust mass in grains smaller than 2 nm.

Appendix 2: Stochastic Heating

Typically large dust particles in the interstellar medium reach equilibrium temperature for which the rate of radiative cooling equals the time-averaged rate of energy absorption. If the emissivity of the dust material is roughly constant with wavelength (which is more the case for a large particle) then the spectral slope of the thermal emission brightness is roughly that of a blackbody (Planck curve) with the location of maximum emission being determined by the temperature (i.e., modified blackbody or grey body). For interpreting astronomical observations a dust temperature is often assumed to be the equilibrium temperature and one denotes as colour temperature of an object the temperature of a blackbody with peak emission at the same wavelength as the observed brightness. The required dust (colour) temperature to generate peak emission in the mid-IR is, for instance, ~300 K for the 12- μm emission and ~150 K for the 25- μm emission. For temperatures of nanodust, in contrast, the Debye model more adequately considers a finite number of vibrational states of the atomic lattice to calculate emission. The Debye temperature Θ (with a dimension of kelvin) characterizes the low-temperature heat capacity U of a solid.

Nanograins are small enough that their time-averaged internal energy is smaller than or comparable to the energy of the starlight photons that heat the grains. Stochastic absorption of photons therefore results in transient "temperature spikes", during which much of the energy deposited by the starlight photon is reradiated in the near- or mid-IR.

The "temperature spike"—the maximum temperature to which a nanograin can reach upon an absorption of a photon, is sensitive to its heat capacity (which is $\propto a^3$). When illuminated by a radiation field, the observed intensity I_λ from the transiently heated nanograins is

$$I_\lambda = N_{\rm H} \int C_{\rm abs}(a,\lambda)\, n_{\rm H}^{-1}\, ({\rm d}n/{\rm d}a)\, {\rm d}a \int B_\lambda(T)\, ({\rm d}P/{\rm d}T)_a\, {\rm d}T, \qquad (5)$$

where $C_{\rm abs}(a,\lambda)$ is the absorption cross section for a grain of radius a at wavelength λ, $B_\lambda(T)$ is the Planck function at temperature T and $({\rm d}P/{\rm d}T)_a$ is the dust temperature distribution function which is sensitive to grain size a (see Li 2004). We see that although in the IR wavelength range $C_{\rm abs}(a,\lambda)/V$ is independent of grain size a, $({\rm d}P/{\rm d}T)_a$ is a sensitive function of a and therefore I_λ allows us to constrain the size distribution of the nanodust component through $({\rm d}P/{\rm d}T)_a$.

Figure 12 shows the temperature probability distributions ${\rm d}P/{\rm d}\ln T$ for PAH ions of selected sizes (at $a > 5$ nm their optical properties approach that of graphite; see Li and Draine 2001a). The distributions are shown for $\chi_{\rm MMP} = 1$

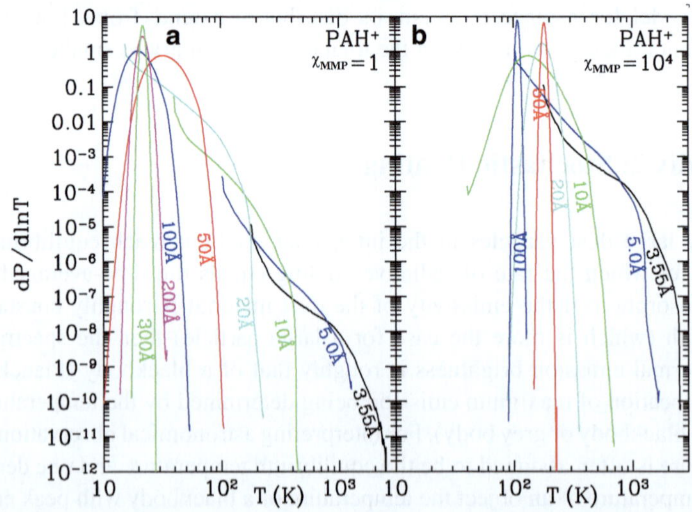

Fig. 12 Temperature probability distribution $\mathrm{d}P/\mathrm{d}\ln T$ for selected PAH ions heated by starlight with $\chi_{\mathrm{MMP}} = 1$ and $\chi_{\mathrm{MMP}} = 10^4$ (in unit of the MMP local interstellar radiation field). It is interesting to note that in the case of $\chi_{\mathrm{MMP}} = 1$, the $a = 20$ nm grain and the $a = 30$ nm grain (even the $a = 10$ nm grain as well) peak at the same temperature (which is their equilibrium temperature). This is expected from their Rayleigh scattering nature (see the caption of Fig. 3). However, in the case of $\chi_{\mathrm{MMP}} = 10^4$, the $a = 5$ nm grain and the $a = 10$ nm grain do not peak at the same temperature, although they both attain equilibrium temperatures. This is because they were "designed" to have different optical properties: PAHs with $a \gtrsim 10$ nm have graphitic properties, while those with $a \lesssim 5$ nm have PAH properties (see Li and Draine 2001a). Taken from Draine and Li (2007)

and $\chi_{\mathrm{MMP}} = 10^4$, where χ_{MMP} is the starlight intensity in units of the interstellar radiation field given by Mathis et al. (1983) (MMP). We see in Fig. 12a that small grains undergo extreme temperature excursions (e.g., the $a = 3.55$ Å PAH occasionally reaches $T > 2{,}000$ K), whereas larger grains (e.g., $a = 30$ nm) have temperature distribution functions that are very strongly peaked and like a delta-function, corresponding to only small excursions around an equilibrium temperature. It is apparent that when the rate of photon absorption increases, the "equilibrium" temperature approximation becomes valid for smaller grains; e.g., for $\chi_{\mathrm{MMP}} = 10^4$ one could approximate a $a = 5$- nm grain as having an equilibrium temperature $T_{\mathrm{eq}} \approx 150$ K whereas for $\chi_{\mathrm{MMP}} = 1$ the temperature excursions are very important for this grain (see Draine and Li 2007). This is also the reason for stochastic heating not being important in the inner solar system.

Whether a grain will undergo stochastic heating depends on (1) the grain size, (2) the optical properties of the dust, (3) the thermal properties (e.g., Debye temperature) of the dust, (4) the starlight intensity, and (5) the hardness of the starlight, which measures the relative amount of short-wavelength ("hard") photons compared to long-wavelength ("soft") photons. For a *smaller* grain with a *smaller* UV/visible absorptivity and a *larger* Debye temperature Θ, exposed to starlight of a *lower* intensity and a *harder* spectrum, it is *more* likely for this grain to be stochastically heated by single photons. This is because (1) the specific heat of a grain (at a given temperature) is proportional to a^3/Θ^3, a single photon (of a given energy) would therefore result in a *more* prominent temperature spike for a *smaller* grain with a *larger* Θ; (2) the photon absorption rate is proportional to the starlight intensity and the absorptivity of the dust in the UV/visible wavelength range, a *smaller* grain with a *smaller* UV/visible absorptivity when exposed to a more *dilute* radiation field will have a *smaller* photon absorption rate and will therefore more likely undergo stochastic heating; (3) a *more* energetic photon would cause a grain to gain a *larger* temperature raise, grains are therefore *more* likely to experience transient heating when exposed to a *harder* radiation field.

Before the stochastically heated nanodust was observed in the diffuse ISM, it was already seen in some other objects. Andriesse and de Vries (1976) presented the first IR emission evidence for nanodust in the dust cloud in M 17, a star-forming nebula. They found that the 8–20- μm emission spectrum is similar over a distance of $\sim 2'$ through the source, suggesting a constant dust temperature. Since large, submicron-sized grains would attain equilibrium temperatures that decrease with distance from the illuminating source, Andriesse (1978) interpreted this as due to stochastically heated grains of ~1 nm. A more definite piece of observational evidence was provided by Sellgren et al. (1983) who observed three visual reflection nebulae, clouds of dust that scatter the light of nearby stars, their near-IR observations (at 1.25–4.8 μm) consisting of emission features at 3.3 and 3.4 μm and a smooth continuum characterized by a colour temperature ~1,000 K. Both the 3.3- μm feature and the colour temperature of the continuum show very little variation from source to source and within a given source with distance from the central star. Sellgren (1984) argued that this emission could not be explained by thermal emission from dust in radiative equilibrium with the central star since otherwise the colour temperature of this emission should fall off rapidly with distance from the illuminating star; instead, she proposed that this emission is due to stochastically heated nanodust.

Acknowledgements We thank M. Köhler for her great help in preparing for Figs. 9 and 11. We thank B.T. Draine, J. Gao, B.W. Jiang, S. Kwok and N. Meyer-Vernet for their helpful comments and suggestions. We also thank A.N. Witt and U.P. Vijh for providing us the ERE data of NGC 7023, F. Banhart and T.J. Bernatowicz, respectively, for providing us the TEM images of presolar nanodiamonds and presolar graphite with TiC nanocrystals embedded. A.L. is supported in part by a NSF grant (AST-1109039) and a NASA Herschel Theory program.

References

Allamandola, L.J., Tielens, A.G.G.M., and Barker, J.R.: 1985, *Astrophys. J.* **290**, L25.
Andriesse, C.D.: 1978, *Astron. Astroph.* **66**, 169.
Andriesse, C.D. and de Vries, J.S.: 1976, *Astron. Astroph.* **46**, 143.

Arendt, R.G., Odegard, N., Weiland, J.L., Sodroski, T.J., Hauser, M.G., Dwek, E., Kelsall, T., Moseley, S.H., Silverberg, R.F., Leisawitz, D., Mitchell, K., Reach, W.T., and Wright, E.L.: 1998, *Astrophys. J.* **508**, 74.
Bakes, E.L.O. and Tielens, A.G.G.M.: 1994, *Astrophys. J.* **427**, 822.
Banhart, F., Lyutovich, Y., Braatz, A., Jager, C., Henning, T., Dorschner, J., and Ott, U.: 1998, *Meteoritics and Planetary Science Supplement* **33**, 12.
Berné, O., Joblin, C., Rapacioli, M., Thomas, J., Cuillandre, J.-C., and Deville, Y.: 2008, *Astron. Astroph.* **479**, L41.
Bernatowicz, T.J., Amari, S., Zinner, E.K., and Lewis, R.S.: 1991, *Astrophys. J.* **373**, L73.
Boulanger, F. and Perault, M.: 1988, *Astrophys. J.* **330**, 964.
Bradley, J., Dai, Z.R., Erni, R., Browning, N., Graham, G., Weber, P., Smith, J., Hutcheon, I., Ishii, H., Bajt, S., Floss, C., Stadermann, F., and Sandford, S.: 2005, *Science* **307**, 244.
Cardelli, J.A., Clayton, G.C., and Mathis, J.S.: 1989, *Astrophys. J.* **345**, 245.
Casassus, S., Cabrera, G.F., Förster, F., Pearson, T.J., Readhead, A.C.S., and Dickinson, C.: 2006, *Astrophys. J.* **639**, 951.
Cecchi-Pestellini, C., Malloci, G., Mulas, G., Joblin, C., and Williams, D.A.: 2008, *Astron. Astroph.* **486**, L25.
Chang, H.-C., Chen, K., and Kwok, S.: 2006, *Astrophys. J.* **639**, L63.
Darbon, S., Perrin, J.-M., and Sivan, J.-P.: 1999, *Astron. Astroph.* **348**, 990.
Desert, F.-X., Boulanger, F., and Puget, J.L.: 1990, *Astron. Astroph.* **237**, 215.
Draine, B.T.: 2006, *Astrophysics in the Far Ultraviolet: Five Years of Discovery with FUSE* **348**, 58.
Draine, B.T.: 2003, *Annual Review of Astronomy and Astrophysics* **41**, 241.
Draine, B.T.: 1999, *3K Cosmology* **476**, 283.
Draine, B.T. and Anderson, N.: 1985, *Astrophys. J.* **292**, 494.
Draine, B.T. and Lazarian, A.: 1999, *Astrophys. J.* **512**, 740.
Draine, B.T. and Lazarian, A.: 1998, *Astrophys. J.* **494**, L19.
Draine, B.T. and Lee, H.M.: 1984, *Astrophys. J.* **285**, 89.
Draine, B.T. and Li, A.: 2007, *Astrophys. J.* **657**, 810.
Draine, B.T. and Li, A.: 2001, *Astrophys. J.* **551**, 807.
Duley, W.W. and Seahra, S.: 1998, *Astrophys. J.* **507**, 874.
Dwek, E., Arendt, R.G., Fixsen, D.J., Sodroski, T.J., Odegard, N., Weiland, J.L., Reach, W.T., Hauser, M.G., Kelsall, T., Moseley, S.H., Silverberg, R.F., Shafer, R.A., Ballester, J., Bazell, D., and Isaacman, R.: 1997, *Astrophys. J.* **475**, 565.
Finkbeiner, D.P., Davis, M., and Schlegel, D.J.: 1999, *Astrophys. J.* **524**, 867.
Godard, M. and Dartois, E.: 2010, *Astron. Astroph.* **519**, A39.
Gordon, K.D., Witt, A.N., and Friedmann, B.C.: 1998, *Astrophys. J.* **498**, 522.
Greenberg, J.M.: 1968, *Nebulae and Interstellar Matter*, 221.
Habart, E., Verstraete, L., Boulanger, F., Pineau des Forêts, G., Le Peintre, F., and Bernard, J.P.: 2001, *Astron. Astroph.* **373**, 702.
Habbal, S.R., Arndt, M.B., Nayfeh, M.H., Arnaud, J., Johnson, J., Hegwer, S., Woo, R., Ene, A., and Habbal, F.: 2003, *Astrophys. J.* **592**, L87.
Hoang, T., Draine, B.T., and Lazarian, A.: 2010, *Astrophys. J.* **715**, 1462.
Joblin, C., Leger, A., and Martin, P.: 1992, *Astrophys. J.* **393**, L79.
Kapteyn, J.C.: 1904, *The Astronomical Journal* **24**, 115.
Kharchenko, V and Lewkow, N.: 2012, Charge-Exchange and X-ray Processes with Nanodust Particles, In: Mann, I., Meyer-Vernet, N., Czechowski, A., (eds.) *Nanodust in the Solar System: Discoveries and Interpretations*, 194.
Kimura, Y.: 2012, Phenomena of Nanoparticles in Relation to the Solar System. In: Mann, I., Meyer-Vernet, N., Czechowski, A., (eds.) *Nanodust in the Solar System: Discoveries and Interpretations*, 46.
Ledoux, G., Ehbrecht, M., Guillois, O., Huisken, F., Kohn, B., Laguna, M.A., Nenner, I., Paillard, V., Papoular, R., Porterat, D., and Reynaud, C.: 1998, *Astron. Astroph.* **333**, L39.

Leger, A. and Puget, J.L.: 1984, *Astron. Astroph.* **137**, L5.
Leinert, C., Bowyer, S., Haikala, L.K., Hanner, M.S., Hauser, M.G., Levasseur-Regourd, A.-C., Mann, I., Mattila, K., Reach, W.T., Schlosser, W., Staude, H.J., Toller, G.N., Weiland, J.L., Weinberg, J.L., and Witt, A.N.: 1998, *Astronomy and Astrophysics Supplement Series* **127**, 1.
Lewis, R.S., Ming, T., Wacker, J.F., Anders, E., and Steel, E.: 1987, *Nature* **326**, 160.
Li, A.: 2004, *Astrophysics of Dust* **309**, 417.
Li, A. In: Mann, I., Nakamura, A.M., and Mukai, T.: 2009, *Lecture Notes in Physics, Berlin Springer Verlag* **758**.
Li, A. and Draine, B.T.: 2002, *Astrophys. J.* **564**, 803.
Li, A. and Draine, B.T.: 2001, *Astrophys. J.* **554**, 778.
Li, A. and Draine, B.T.: 2001, *Astrophys. J.* **550**, L213.
Lodders, K.: 2005, *Chemie der Erde / Geochemistry* **65**, 93.
Mann, I.: 2010, *Annual Review of Astronomy and Astrophysics* **48**, 173.
Mann, I. and Czechowski, A.: 2012, Causes and Consequences of the Existence of Nanodust in Interplanetary Space, In: Mann, I., Meyer-Vernet, N., Czechowski, A., (eds.) *Nanodust in the Solar System: Discoveries and Interpretations*, 219.
Mann, I. and Murad, E.: 2005, *Astrophys. J.* **624**, L125.
Mann, I., Murad, E., and Czechowski, A.: 2007, *Planetary and Space Science* **55**, 1000.
Mathis, J.S., Mezger, P.G., and Panagia, N.: 1983, *Astron. Astroph.* **128**, 212.
Mathis, J.S., Rumpl, W., and Nordsieck, K.H.: 1977, *Astrophys. J.* **217**, 425.
Mattila, K., Lemke, D., Haikala, L.K., Laureijs, R.J., Leger, A., Lehtinen, K., Leinert, C., and Mezger, P.G.: 1996, *Astron. Astroph.* **315**, L353.
Onaka, T., Yamamura, I., Tanabe, T., Roellig, T.L., and Yuen, L.: 1996, *Publications of the Astronomical Society of Japan* **48**, L59.
Ott, U.: 2007, *Space Science Reviews* **130**, 87.
Pitz, E., Leinert, C., Schulz, A., and Link, H.: 1979, *Astron. Astroph.* **74**, 15.
Platt, J.R.: 1956, *Astrophys. J.* **123**, 486.
Purcell, E.M.: 1969, *Astrophys. J.* **158**, 433.
Seahra, S.S. and Duley, W.W.: 1999, *Astrophys. J.* **520**, 719.
Sellgren, K.: 1984, *Astrophys. J.* **277**, 623.
Sellgren, K., Werner, M.W., and Dinerstein, H.L.: 1983, *Astrophys. J.* **271**, L13.
Smith, T.L. and Witt, A.N.: 2002, *Astrophys. J.* **565**, 304.
Stecher, T.P.: 1965, *Astrophys. J.* **142**, 1683.
Steglich, M., Jäger, C., Rouillé, G., Huisken, F., Mutschke, H., and Henning, T.: 2010, *Astrophys. J.* **712**, L16.
Szomoru, A. and Guhathakurta, P.: 1998, *Astrophys. J.* **494**, L93.
Tanaka, M., Matsumoto, T., Murakami, H., Kawada, M., Noda, M., and Matsuura, S.: 1996, *Publications of the Astronomical Society of Japan* **48**, L53.
Tielens, A.G.G.M.: 2008, *Annual Review of Astronomy and Astrophysics* **46**, 289.
Trumpler, R.J.: 1930, *Publications of the Astronomical Society of the Pacific* **42**, 214.
Vidal, M., Casassus, S., Dickinson, C., Witt, A.N., Castellanos, P., Davies, R.D., Davis, R.J., Cabrera, G., Cleary, K., Allison, J.R., Bond, J.R., Bronfman, L., Bustos, R., Jones, M.E., Paladini, R., Pearson, T.J., Readhead, A.C.S., Reeves, R., Sievers, J.L., and Taylor, A.C.: 2011, *Monthly Notices of the Royal Astronomical Society* **414**, 2424.
Wada, S., Mizutani, Y., Narisawa, T., and Tokunaga, A.T.: 2009, *Astrophys. J.* **690**, 111.
Watson, W.D.: 1972, *Astrophys. J.* **176**, 103.
Weiland, J.L., Blitz, L., Dwek, E., Hauser, M.G., Magnani, L., and Rickard, L.J.: 1986, *Astrophys. J.* **306**, L101.
Weingartner, J.C. and Draine, B.T.: 2001, *Astrophys. J.* **548**, 296.
Weingartner, J.C. and Draine, B.T.: 2001, *The Astrophysical Journal Supplement Series* **134**, 263.
Whittet, D.C.B.: 2003, *Dust in the galactic environment, 2nd ed. by D.C.B. Whittet. Bristol: Institute of Physics (IOP) Publishing, 2003 Series in Astronomy and Astrophysics, ISBN 0750306246.*.

Witt, A.N. and Vijh, U.P.: 2004, *Astrophysics of Dust* **309**, 115.
Witt, A.N., Gordon, K.D., and Furton, D.G.: 1998, *Astrophys. J.* **501**, L111.
Xiang, F.Y., Li, A., and Zhong, J.X.: 2011, *Astrophys. J.* **733**, 91.
Ysard, N., Miville-Deschênes, M.A., and Verstraete, L.: 2010, *Astron. Astroph.* **509**, L1.

Phenomena of Nanoparticles in Relation to the Solar System

Yuki Kimura

Abstract The physical properties of nanometer-sized particles are strongly affected by their surface due to larger surface to volume ratio and very different from bulk materials. For example, the melting point decreases as small as the half value and the diffusion coefficients become as large as nine orders. Surface free energy considerations can be as important—or even more important—than bulk thermodynamic quantities. Because the diameters of newly formed dust particles are of the order of nanometers, their physical properties, which induce a fusion growth as nanoparticles never occur in bulk, must be considered when considering the formation of cosmic dust. In this chapter, we exhibit the properties of nanoparticles that have a great potential for understanding phenomena in the current solar system and its evolution.

1 What is a Nanoparticle?

It would appear obvious that a nanoparticle has a size of the order of nanometers; however, the actual size of the nanoparticle range depends on the study discipline. When studies on tiny particles actively began in the fields of solid-state physics and quantum physics, nanoparticles were originally known as ultrafine particles. In solid-state physics, many studies on nanoparticles were initiated following a theoretical study by Kubo (1962), who showed that extremely small metallic particles have different physical properties from those of the bulk material because of their discrete electronic states; this is known as the Kubo effect.

Y. Kimura (✉)
Department of Earth and Planetary Materials Science, Graduate School of Science, Tohoku University, Sendai 980–8578, Japan
e-mail: ykimura@m.tohoku.ac.jp

I. Mann et al. (eds.), *Nanodust in the Solar System: Discoveries and Interpretations*, Astrophysics and Space Science Library 385, DOI 10.1007/978-3-642-27543-2_3,

Initially, a nanoparticle was defined as a particle that consists of a countable number of atoms, although this definition is a conceptual one. In 2008, the International Council of Chemical Associations (ICCA) defined a nano-object as *a material with one, two, or three external dimensions in the nanoscale, where nanoscale is defined as the size range from approximately 1–100 nm* (ISO TS 27687 CCA Core Elements of a Regulatory Definition of Manufactured Nanomaterials). In accordance with this definition, a nanoparticle is "a material with three external dimensions in the size range from ∼1–100 nm in diameter." Recently, several institutes have attempted to define a nanoparticle to avoid any confusion. The definition that appears to have achieved a consensus is *nanoparticles are particles that have characteristic properties different from those of the bulk material.* In accordance with this definition, nanofibers and nanoplates are one-dimensional and two-dimensional nanomaterials, respectively.

Nanoparticles within a radius range of ∼1–50 nm can be formed directly by condensation from a gas phase. This is observed in laboratory experiments and also occurs in the solar system. Model calculations suggest, for instance, that dust particles condense from a vapor phase behind a planetesimal bow shock in the solar nebula or during high-speed collision events on asteroids (Miura et al. 2010). Figure 1 shows transmission electron microscope (TEM) images of three different kinds of nanoparticles with various sizes produced by smoke experiments (Kimura et al. 2011a; Kimura and Tsukamoto 2011b). Figure 1a shows nanoparticles of metallic manganese with three different crystal habits: tristetrahedron, rhombic dodecahedron, and rod-shaped. It has been recognized that even nano-sized particles can have crystalline habits that are similar to those of bulk crystals, as can be clearly seen in TEM images. Lattice images are observable, demonstrating that nanoparticles can be crystalline materials; an example of this can be seen in Fig. 1b, which shows a particle of silicon carbide. When nanoparticles are produced from a vapor phase, they condense exclusively from the supercooled gas, because the initial condensate has no sites available for heterogeneous nucleation. When the degree of supercooling and the cooling rate are sufficiently high, condensates cannot crystallize because of their rapid growth and amorphous particles are formed exclusively, as shown for silica in Fig. 1c.

Figure 2 shows how the total number of atoms depends on the particle size. The total number of atoms in a particle several nanometers in radius is 10^2–10^4, which can be a countable number. The ratio of surface atoms to the total number of atoms is greater than 0.2 when a nanoparticle is less than ∼5 nm in radius. Surface atoms have dangling bonds that are energetically unstable and cause surface reconstruction. At this size scale, the surface energy can be relatively and significantly higher than the corresponding value for the bulk material . Therefore, nanoparticles have characteristic properties that differ from those of the bulk material, as will be described in the next section.

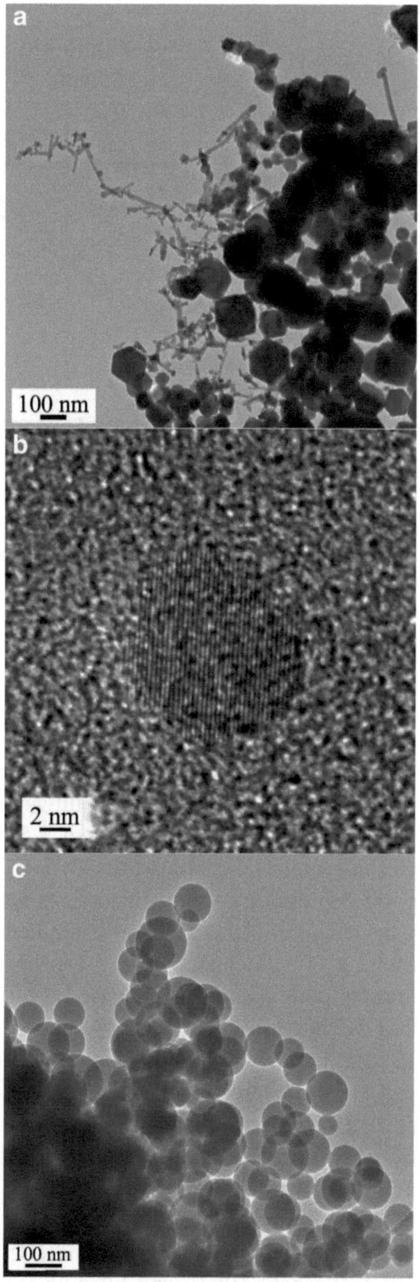

Fig. 1 Typical TEM images of laboratory-synthesized nanoparticles of (**a**) manganese (Mn), (**b**) silicon carbide (SiC), and (**c**) silica (SiO_x), respectively, recorded by using a JEOL 3200FS TEM with a 300-kV electron beam at the Institute for Molecular Science in Japan

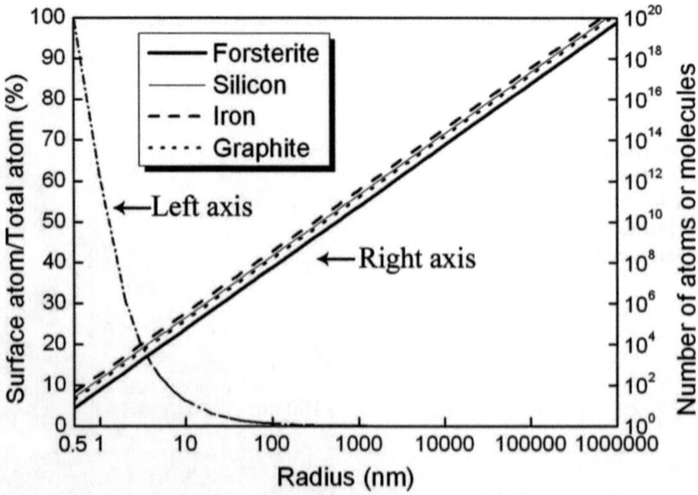

Fig. 2 Relationship between the particle radius and the total numbers of atoms or molecules for forsterite (Mg_2SiO_4; *right axis*) and the relationship between the particle diameter and the ratio of the number of atoms located on the surface to the total number of atoms in a single particle (*left axis*). The particles are assumed to be spherical

2 Physical Properties of Nanoparticles

In the 1950s, it was discovered by means of diffractometry that the melting points of thin films of lead, tin, or bismuth are lower than those of the bulk metals (Takagi 1954). Thus, the melting point of a 5-nm-thick film is 559 K for lead (600 K for the bulk metal), 475 K for tin (505 K), and 521 K for bismuth (544 K). Melting-point depression of nanoparticles is also a well-known phenomenon, and that of gold nanoparticles is the most studied (Sambles 1971; Buffat and Borel 1976). Melting of gold nanoparticles has been examined directly under TEM, and the melting point was determined to be 1,100 K for 2.5-nm-radius particles (Lee et al. 2009). This decreases to $\sim$700 K for 1-nm-radius gold particles (Buffat and Borel 1976). Similarly, the melting point of silver decreases from 1,233 to 667 K as the particle size decreases from the macroscopic scale to a radius of 7.5 nm (Kashiwase et al. 1977). Depression of the melting point has also been reported on the basis of experiments and molecular-dynamic simulations for various other materials such as indium (In) (Coombes 1972; Boiko et al. 1969; Skripov et al. 1981), copper (Cu), aluminum (Al), and germanium (Ge) (Gladkich et al. 1966), as well as the metals mentioned above: lead (Ben et al. 1995), tin (Blackman and Curxon 1959; Wronski 1967), and bismuth (Olson et al. 2005). Attempts have also been made using molecular-dynamic simulations in order to clarify the melting points of self-standing (substrate-free) small particles of gold (Ercolessi et al. 1991), nickel (Ni) (Yue et al. 2001), copper (Delogu 2005), and aluminum (Alavi and Thompson 2006). Some theoretical studies have been performed that attempted to explain the depression of the melting point by using phenomenological models (Buffat and Borel 1976; Couchman and Jesser 1977), the Lindemann melting rule (Couchman 1979; Couchman and Ryan 1978; Hoshino and Shimamura 1979; Zhang et al. 2000), or the self-consistent statistical method (Lubashenko 2010) among others. Even after these extensive studies on the depression of the melting point, no comprehensive explanation has been developed. However, the melting-point depression can be described simply by means of the Lindemann criterion (1910). The melting point of a nanoparticle with a radius r can be described as follows (Zhang et al. 2000): $T_m(r)/T_m(\infty) = \exp[-(\alpha - 1)/(r/r_0 - 1)]$, where $T_m(r)$ and $T_m(\infty)$ are the melting point of a nanoparticle with radius r and that of the corresponding bulk materials, respectively; r_0 is the critical radius at which all atoms of the particle are located on its surface; and α is the ratio of the mean-square displacement of atoms on the surface to that in the interior of crystal (Jiang et al. 1999). In the case of nanoparticles with free surfaces, the mean-square displacement of the surface atoms is larger than that of the interior atoms of the nanoparticles and α is greater than unity (Kashiwase et al. 1977; Harada et al. 1980; Ohshima et al. 1980; Zhang et al. 2000). Therefore, $T_m(r)$ decreases as r decreases.

The diffusion coefficients of metals also change with particle size. For example, that of copper atoms in gold at 300 K changes from 2.4×10^{-28} in the bulk material to 8.3×10^{-19} ($m^2\,s^{-1}$) in nanoparticles (Mori et al. 1991). A large increase in the diffusion coefficients was predicted by a microcanonical molecular-dynamic study using the two-dimensional binary Morse model. This model shows a result that diffusion of interior atoms can be caused by the presence of a melting surface, even if they are solid like (Shimizu et al. 2001). The inward diffusion from the surface is accelerated by the frequent onset of gliding motions of the surface atoms. Accordingly, it is strongly dependent on the surface Debye temperature. Shimizu and co-workers also showed that diffusion is dependent on the particle size. The activation energy increases and the diffusion coefficient decreases with increasing particle size. These results agree closely with experimental observations. Unfortunately, such data are only available for metallic materials at this time. The physical properties of materials of astronomical interest, such as oxides, silicates, or carbonaceous materials, have received little attention so far. It can, however, be assumed that the melting points of nanoparticles of these materials are also likely to be lower than the corresponding bulk values. In particular, when the size is of the order of 1 nm, this phenomenon is significant and can result in the unexpected growth of nanoparticles, as described in the next section.

It is also known that nanoparticles with metastable crystalline structures can be produced; for instance, iron and chromium nanoparticles with a size of less than 50 nm in radius have an A-15 type crystal structure whereas the body-centered cubic (bcc) structure is the stable crystal form of these metals at room temperature and ambient pressure (Kimoto and Nishida 1967; Kido et al. 2005) (Fig. 3).

Nanoparticles also display significant changes in their optical properties. Gold nanomaterials, for example, lose the metallic luster and thin films appear blue and nanoparticles have a black (dark brown) color. The difference in color is caused by a surface plasmon resonance. This surface resonance is the result of collective

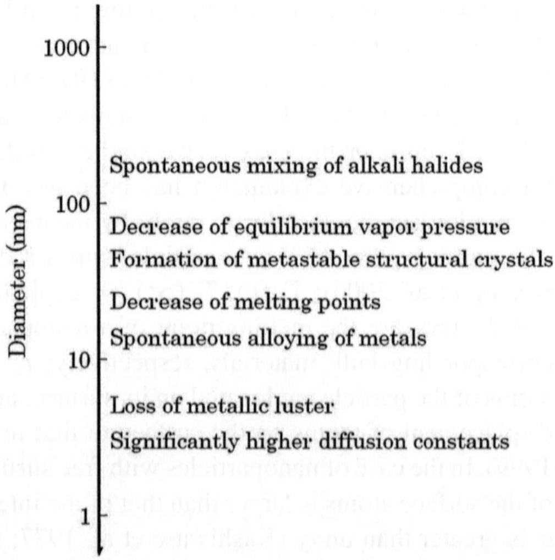

Fig. 3 Summary of characteristic properties of nanoparticles that appear at sizes smaller than that shown on the left-hand side of the arrow

oscillations of free electrons in the surface of the material. The scattering of free electrons on the particle surface affects their response to optical excitation if the particles are smaller than the electronic mean-free path in the bulk metal; for example, the threshold radius for gold is approximately 10 nm (Alvarez et al. 1997). A surface plasmon resonates with the electromagnetic wave and absorbs light. The frequency of the collective oscillation of surface atoms is lower than that of the bulk material (Bohren and Huffman 1983) and, as a result, the color of nanoparticles differs from that of the bulk material.

It has also been reported that the refractive index of nanoparticles shows a geometry-dependent shift as a result of surface plasmon effects (Chen et al. 2008; Karakouz et al. 2009). These shifts or changes in optical properties suggest that the efficiency of the Poynting–Robertson effect (Burns et al. 1979), which causes deceleration of dust in bound orbit around the Sun as a result of the nonradial component of the radiation pressure force, should be different from the value calculated from the bulk properties. Optical constants of thin films of metals have been measured at visible wavelength, and they have indices of refraction, n values of larger than unity, meaning that they are transparent (Johnson and Christy 1974). In the case of insulating particles, surface phonons affect the optical properties, particularly at infrared wavelengths. If the optical properties of alumina, for instance, differ from those of the bulk material, the initial acceleration efficiency of newborn nanodust around a star will be changed.

In some cases, the infrared spectrum is also changed. The optical properties of nanoparticles are markedly dependent on their geometry because of the presence of surface modes (Bohren and Huffman 1983). In addition, differences in the surface structure of nanoparticles from that of the bulk material also affect their infrared features. Because surface atoms have dangling bonds, they can bond with each other to form new structures. This phenomenon is called surface reconstruction

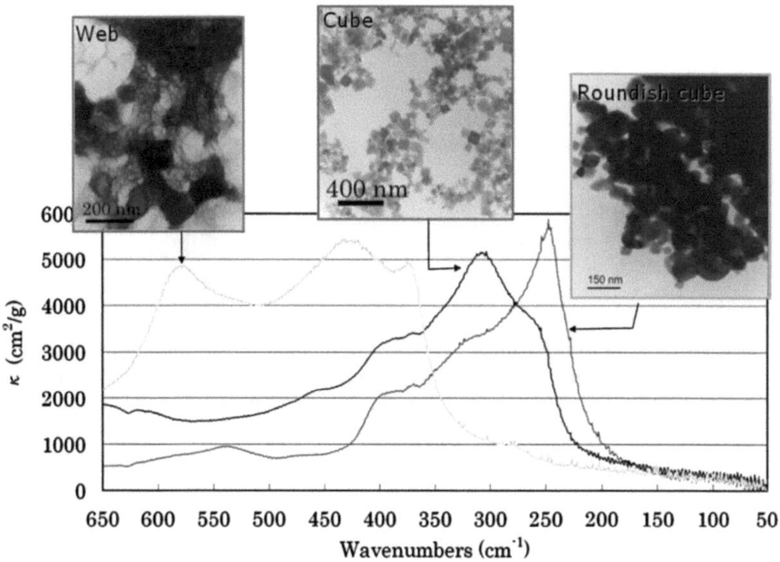

Fig. 4 Infrared spectra in the range 650–50 cm^{-1} for grains of magnesium sulfide (MgS) with three different geometrical shapes. All three samples were produced in a laboratory from the same commercial MgS powder. Spider-web MgS has three characteristic peaks at 376, 427, and 580 cm^{-1} (26.6, 23.4, and 17.2 μm). Cubic MgS shows four characteristic peaks at 460, 400, 311, and 262 cm^{-1} (21.7, 25.0, 32.1, and 38.2 μm). Rounded cubic MgS has a main peak at 247 cm^{-1} (40.5 μm) with two shoulders at 318 and 400 cm^{-1} (31.4 and 25.0 μm). (Figures adapted from Kimura et al. 2005a)

(Brommer et al. 1992). Because the reconstructed surface has a different structure from that of the interior of the particle, its infrared features differ from those of the bulk material. In particular, in the case of nanoparticles, their significantly greater surface-to-mass ratio confers significant new features. Figure 4 shows examples of far-infrared spectra and the corresponding TEM images of magnesium sulfide nanoparticles produced under various condensation conditions (Kimura et al. 2005a). The typical crystal habit of magnesium sulfide is cubic with a NaCl-type structure, as shown in the central TEM image. However, crystals can adopt various shapes depending on the environmental conditions under which they were formed, such as temperature, concentration, presence of impurities, and degree of supersaturation (e.g., see Kobatake et al. 2008). As clearly shown in Fig. 4, the spectral features of nanometer-sized particles markedly depend on their shapes and, hence, the environment in which they are formed. To date, the 6.8-μm features of young stellar objects (Kimura and Nuth 2005b), the 21-μm features of asymptotic giant branch stars (Kimura et al. 2005c), and an absorption bump at 217 nm in the interstellar extinction curve (Kimura et al. 2004) have been explained on the basis of the properties of nanoparticles.

The macroscopic behavior of nanoparticles also differs from that of a typical macro-scale powder; in other words, the behavior of a powder composed of

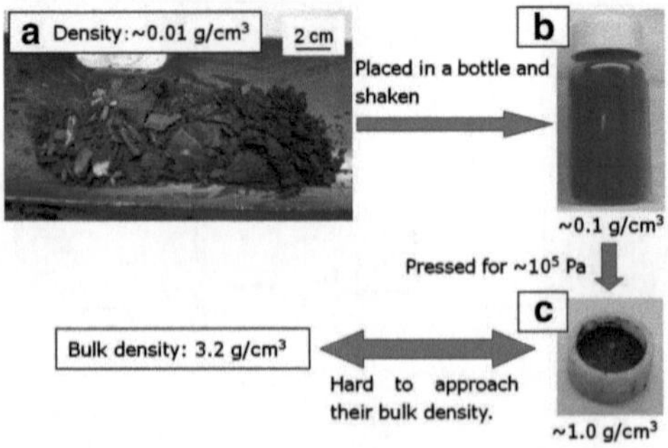

Fig. 5 Magnesiosilicate smoke particles produced from a gaseous mixture of Mg, SiO, H_2, and O_2 by using the condensation flow apparatus managed by J. A. Nuth III at the Astrochemistry Laboratory of the NASA Goddard Space Flight Center. The as-prepared sample is very fluffy. (**a**) After collection. (**b**) Charged in a bottle and shaken. (**c**) Charged into a 5-mm-diameter alumina pan and compressed

nanoparticles is different from that of the corresponding granular material. For example, aggregates of as-prepared silicate nanoparticles produced by smoke experiments are very fluffy and have a density of only $\sim$0.01 g cm^{-3}, which is similar to that of an aerogel, whereas the density of the bulk material is 3.2 g cm^{-3} (Fig. 5). When the as-prepared nanoparticle sample is collected, placed in a bottle, and shaken, its density increases immediately to about 0.1 g cm^{-3}. The agglomerated smoke particles can have an appearance similar to that of a extraterrestrial regolith. When the sample is compressed at $\sim$10^5 kg m^{-2}, its density approaches $\sim$1.0 g cm^{-3}. In the case of magnesiosilicate, the density of the sample remains unchanged even if the sample is heated to 1,500 K. Since the bulk density of the material is 3.2 g cm^{-3}, the porosity of the compressed nanoparticulate material is roughly 70%, which is greater than that of most asteroids.

3 Significant Phenomena of Nanoparticles

Physical properties that appear exclusively in nanoparticles induce significant phenomena that depend on the particle size. To obtain a clear picture of nanodust in the Solar System, we must understand how nanoparticles behave. Spontaneous alloying is probably the most widely known singular phenomenon of nanoparticles. When copper atoms are deposited onto a gold nanoparticle with 2 nm in radius, the copper atoms diffuse into the gold nanoparticle within 20 s and form an alloy without the application of external energy. The diffusion constant is more than nine orders of magnitude greater than that in the bulk phase. If copper atoms diffuse into the gold nanoparticle with a diffusion constant of bulk, it takes 2–3 years to be alloy. In the case of 5-nm-radius particles, the alloy phase is formed only at the surface of the initial gold particle, whereas 2-nm-radius particles are alloyed throughout. For gold particles with radii larger than 15 nm, spontaneous alloying does not occur. It has been confirmed that the spontaneous alloying takes place by a purely solid-state process rather than by a quasi-melting process (Mori et al. 1994). Spontaneous alloying readily occurs when the ratio of the atomic radii of the two elements is less than 14%. It is enhanced for materials with a negative heat of solution. It has been observed in several binary clusters, such as Au–Ni, In–Sb, Au–Zn, Au–Al, and Sb–Au (Yasuda et al. 1992, 1993b; Yasuda and Mori 1992, 1993a, 1994a,b; Mori et al. 1994; Mori and Yasuda 1999). To elucidate these phenomena, some theoretical simulations have been performed (Okada et al. 1995, 1996a,b; Shimizu et al. 1998, 2001). Spontaneous mixing phenomena have also been shown to occur in ionic crystals in an alkali halide system, for example when KCl is deposited onto previously formed nanoparticles of RbCl (Kimura et al. 2000, 2002). As shown in Fig. 6a, the electron-diffraction rings from each plane show that the resulting particles have only one lattice distance and therefore exist as a single phase. Accordingly, it can be concluded that mixtures of crystals of KCl and RbCl are produced spontaneously by diffusion of ions. Successive evaporation of RbCl and KCl results in intermixing at room temperature and produces a solid–solution crystal with radius of 100 nm, which is roughly 100 times larger than that of a metallic alloying system, as shown in Fig. 6a. In the case of the KBr–KI system, however, electron-diffraction patterns clearly show that the crystal has two different lattice constants with the same crystallographic relationship, as shown in Fig. 6b. In this case, KI grows epitaxially on the KBr particles, and no solid–solution crystal is produced. As a result of systematic experiments, conditions for spontaneous mixing were identified as a difference in ionic radii of less than 32% (compared with 14% for metals) and a size smaller than 100 nm in radius. The latter value is considerably larger than the value of 5 nm for metal diffusion. These phenomena can induce unexpected growth processes of nanoparticles, as described in the next section.

Unfortunately, there is no report of any study concerning crystals of a predominantly covalent character, such as an oxide or a silicate. However, similar kinds of reaction with large amounts of atomic diffusion at room temperature can be seen on some nanoparticles of astronomical interest. One example is iron sulfide. Pyrrhotite, one of the polymorphs of iron sulfide, has been found in greater quantities than the troilite polymorph in chondritic interplanetary dust particles, although thermodynamic modeling clearly shows that troilite should have formed in the solar nebula (Kerridge 1976). Furthermore, troilite is the stable phase when the metal is present. Zolensky and Thomas (1995) therefore proposed an explanation for the presence of pyrrhotite, in which it is formed by further sulfidation of exfoliated troilite from preexisting iron grains. If we take into account the larger diffusion constant of sulfur atoms in iron, a different route for the formation of pyrrhotite grains in an iron-rich environment can be considered on the nanoscale, based on the solid–solid reaction between particles of iron and sulfur, as shown in Fig. 7 (Kimura et al. 2005d). When sulfur particles are deposited onto iron particles, as shown in Fig. 7a, the iron atoms diffuse into the surface of the sulfur and formed pyrrhotite. A void then forms at the center of the particle as a result of diffusion as shown in Fig. 7b. FeS crystallites with single pyrrhotite phase are thus produced

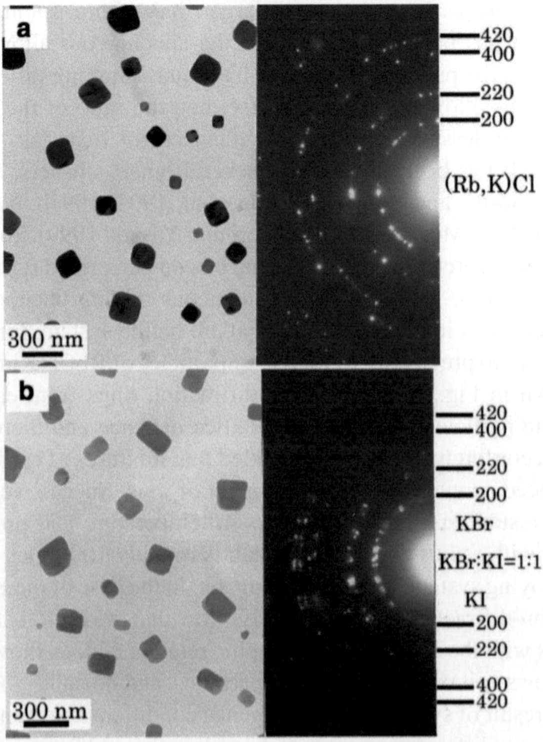

Fig. 6 TEM images and electron-diffraction (ED) patterns of alkali halide island crystals on an amorphous carbon substrate. (**a**) KCl and (**b**) KI were successively evaporated onto RbCl and KBr, respectively, in vacuum. The mean thickness of all four materials was 5 nm. Island crystals were formed rather than a continuous film. In case (**a**), each plane produces a single ED ring, showing that solid–solution crystals are formed even at room temperature in the solid phase. In case (**b**), ED spots with two different lattice constants can be seen with the same crystallographic orientation, showing that KI grows epitaxially on the KBr crystals (Kimura 2008a)

by solid–solid reaction, in contrast to the predicted formation of a troilite phase in such an iron-rich environment. We do not claim that all meteoritic pyrrhotite has been produced by this reaction, but it is one possible mechanism to make nuclei of pyrrhotite, particularly in photodissociated hydrogen sulfide gas.

Figure 8 shows a TEM image of a particle with similar void-shell structure to that seen in Fig. 7b. However, the process for its formation is completely different (Saito and Kimura 2009). The hollow interior could be formed by coagulation of vacancies formed by electronic excitation and/or by knock-out of carbon atoms following irradiation by plasma particles, such as protons or He^+ ions, during condensation of polycyclic aromatic hydrocarbons in a plasma field. The bonds in the globules are broken by inner-shell, valence, and plasmon excitations induced by electron irradiation in a plasma field, and the excess energy accumulated during relaxation is converted directly into kinetic energy (Yasuda et al. 2008). Indeed, the

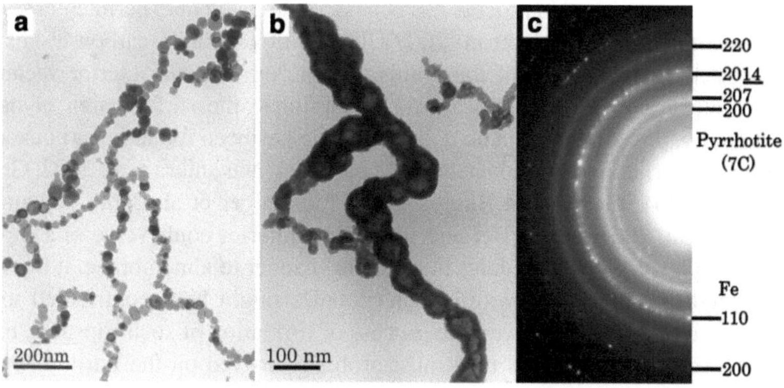

Fig. 7 (**a**) Typical TEM image of metallic iron particles produced by the gas evaporation method. (**b**) TEM image of iron sulfide particles after a solid–solid reaction. (**c**) The electron-diffraction pattern corresponding to (**b**) shows the formation of pyrrhotite, which is a stable phase in a sulfur-rich environment, although metallic iron is still present (Kimura et al. 2005d)

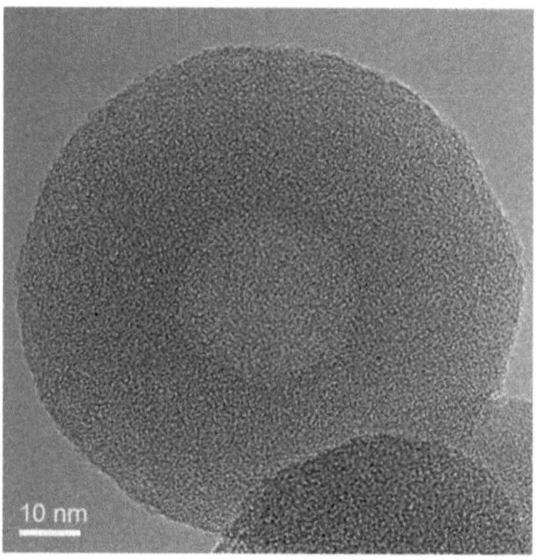

Fig. 8 Typical high-resolution TEM image of a hollow globule produced in the laboratory from He and a benzene (C_6H_6) plasma. The central region of the grain has a brighter contrast, suggesting that it has a hollow center (Saito and Kimura 2009). Reproduced by permission of the AAS

energy of electrons ($\sim 10^4$ K = $\sim 10^2$ kJ mol^{-1}) in the experimental plasma is of a similar magnitude to the binding energy of hydrocarbons (C–H: 339–434 kJ mol^{-1}, C–C: 344–468 kJ mol^{-1}). As a result, excited carbon atoms are displaced from their equilibrium sites to form vacancies. This kind of effect occurs most efficiently with nanometer-sized particles, because the shallow interatomic potentials that result from phonon softening of lattice vibrations (lattice softening) reduce the activation energy for atomic diffusion (Marks and Ajayan 1986; Ahayan and Marks 1998). Although vacancies can be formed anywhere in the globules, the ease with which vacancies at or near the surface should vanish by escaping from the surface should increase as a result of lattice softening and recombination with displaced carbon atoms from the interior; this process has been demonstrated in experimental studies on a semiconductor (Yasuda et al. 2008). In addition, mobile carbon atoms can be segregated at the surface of the grains. On the other hand, interior vacancies can form clusters and grow into a void to minimize the surface area of defect sites. Actually, this kind of organic hollow globule has been found in carbonaceous chondrite meteorites and interplanetary dust particles (Nakamura et al. 2002; Gravie and Buseck 2004; Garvie and Buseck 2006; Messenger et al. 2008; Nakamura-Messenger et al. 2006, 2008). Because the hollow interior could serve as a capsule to store complex organic molecules in a similar manner to a membrane, it has been hypothesized that inclusions within such globules might have contributed to the origins of life on Earth. However, the process of formation of such capsules, based on a nanoscale view, suggests they most probably arrived on the Earth as empty vessels.

There are other cases in which nanoscale-specific properties give rise to unexpected phenomena related to the formation of nanodust; these include, for example, the diffusion of organic molecules into amorphous magnesiosilicate nanoparticles to assist the crystallization of forsterite (Kimura et al. 2008a). Nanodust sometimes behaves in a manner similar to a liquid, even in solid materials, as described in the next section. We therefore need to take account of this property in relation to the formation of nanodust.

4 Growth Mechanism of Nanoparticles

In the current Solar System, nanodust has been formed by release from comets, from vapor-phase condensation in volcanic plumes, and by fragmentation of asteroids or smaller bodies through collisions. Nanodust originating from comets retains information about the process of its in the solar nebula. In contrast, nanodust formed in volcanic plumes and by fragmentation has its origins in the existing Solar System. Although there are some reports of experimental studies on fragmentation by impact and on measurements of the size distribution of the resulting fragments (Nakamura and Fujiwara 1991), the size range of such particles is of the order of microns. In materials science, 10-nm-sized nanoparticles have been successfully produced by fragmentation in a bead mill. In the case of surface comminution by frictional forces, although the size distribution has a double peak, the smaller peak can include $\sim$10-nm-sized particles. In this case, where the collision frequency is limited, most of the mass is present in larger fragments and, therefore, the efficiency of formation of nanoparticles may be low, as it takes a lot of energy to produce that much surface from a bulk material. When intense collisions between bodies in the Solar System cause evaporation of minerals, nanoparticles may be formed with a relatively greater efficiency in the ejected gases. Because the energy barrier for heterogeneous nucleation is much lower than that for homogeneous nucleation, the vapor condenses readily onto the surfaces of the surviving fragments before nanoparticles can be formed through homogeneous nucleation. Therefore,

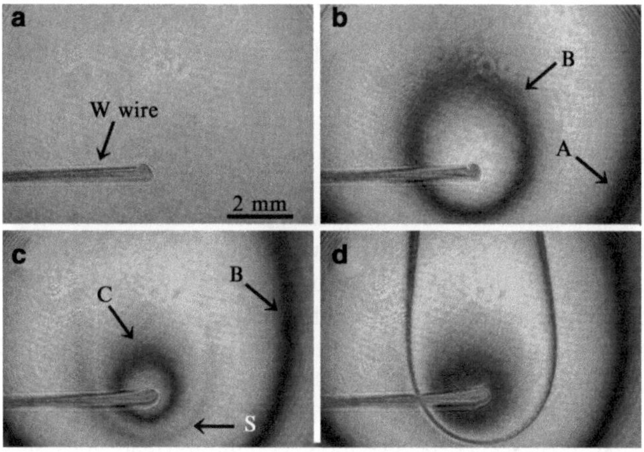

Fig. 9 Typical infinite interferograms recorded along the axis of a tungsten wire with a source temperature of (**a**) room temperature, (**b**) 698 K, (**c**) 1,373 K, and (**d**) 1,461 K in a gaseous mixture of Ar 9.0×10^3 Pa and O_2 1.0×10^3 Pa. The tungsten wire (diameter 0.3 mm) was 70 mm long and was aligned parallel to the optical path. *Arrows* A, B, and C indicate temperatures of 320, 500, and 1,150 K, respectively. These temperatures were determined by another experiment using identical experimental conditions and the same tungsten wire, but in an oxygen-free environment. Therefore, the actual temperature of profile C is higher than 1,150 K as a result of the relatively higher refractive index of the evaporated tungsten oxide vapor. The temperature distribution around the evaporation source is visualized in (**b**). The interferogram in (**c**) suggests that nucleation occurred homogeneously below the evaporation source indicated by *arrow* S. Smoke can be clearly seen between the two fringes B and C in (**d**). Nucleated smoke particles rise as a result of convection generated by the ambient gas (Kimura et al. 2011a)

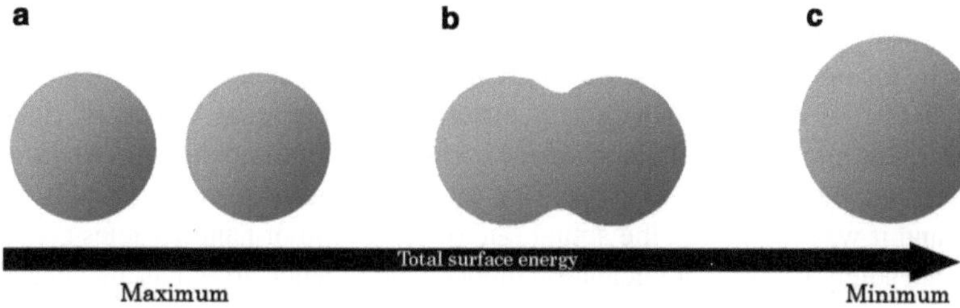

Fig. 10 Schematics of a fusion growth of nanoparticles. When two nanoparticles (**a**) contact, they fuse together (**b**) and form a single particle (**c**). Their total surface area decreases 20% from (**a**) to (**c**), whereas total mass is same. They behave like a droplet to minimize their total surface free energy due to high diffusion constant

if nanodust is to be formed by fragmentation resulting from a collision, any solid fragments must be quickly removed from the evaporated gas atmosphere and/or the ejected gas must be at a relatively high pressure to prevent scavenging of the refractory vapor by such fragments.

The significantly higher diffusion constants and the lower melting points of nanoparticles induce a unique process for the formation of nanoparticles that is fusion growth (or liquid-like coalescence; Fig. 10). When two nanoparticles come into contact, they fuse together like droplets to decrease their total surface energy and form a larger particle. If two different kinds of particles collide, a compound single particle is formed. For example, particles of iron and nickel can collide to form particles of tetrataenite (Nagata et al. 1991). In other words, spontaneous alloying, as described above, occurs not only by vapor deposition onto a nanoparticle, but also through contact between nanoparticles. Surface-melting coalescence is another mechanism for the growth of nanoparticles. This growth occurs when the surface temperature of the particle is above the surface-melting temperature. The surface-melting temperature depends on the surface Debye temperature. When two different kinds of particle coalesce, a chemical compound can be produced at the interface by diffusion. Whether the particles grow via by the fusion growth (liquid-like coalescence) or the by surface-melting coalescence depends on the particle size; liquid-like coalescence occurs more commonly than does surface-melting coalescence in the smaller size range.

To understand the process of formation of cosmic dust particles, some experimental studies have been performed in relation to the nucleation process (Nuth and Donn 1982; Kobatake et al. 2005; Kimura and Nuth 2009; Kimura et al. 2008a,b, 2011a; Kimura and Tsukamoto 2011b; Kimura et al. 2006). Nuth and Donn (1982) obtained the nucleation rate and surface energy of silica particles and discussed these measurements in detail. When Kobatake et al. (2005) measured the gas temperature by using a thermocouple, the concentration was as expected from the amount of evaporated material captured on a substrate. This experiment demonstrated a qualitatively important phenomenon: the two-step nucleation of $MgSiO_3$. It has recently been shown that nanoparticles in very high-supersaturation environments condense in an exclusively homogeneous manner; this phenomenon was studied by in situ observation of the nucleation process of nanoparticles in a smoke by means of a Mach–Zehnder type interferometer as shown in Fig. 9 (Kimura et al. 2011a). By using such in situ observation techniques, refractive indexes, growth rates, sticking coefficients of the growth unit onto nuclei, and surface free energies can be deduced. The resulting smoke particles were also analyzed by TEM, and it was found that the actual rate of formation of nanoparticles cannot be explained solely in terms of nucleation theory. Coalescence growth is also necessary (Fig. 10).

To understand the growth processes of nanodust, we have to recognize that very different phenomena occur on nanometer scales as compared with the bulk. It can be said that the behavior of nanometer-sized particles is governed by their properties of their surfaces. A very different regime occurs in which previously negligible physical properties, in this case the surface energy, begin to dominate the behavior of the system. In other words, understanding how nanoparticles behave is important when considering the formation, growth, and composition of cosmic dust particles.

Acknowledgements We thank Prof. Joseph A. Nuth III of NASA's Goddard Space Flight Center for useful comments.

References

Ajayan, P. M., Marks, L. D.: 1988, *Phys. Rev. Lett.* **60**, 585.
Alavi, S., Thompson, D. L.: 2006, *J Phys. Chem. A* **110**, 1518.
Alvarez, M. M., Khoury, J. T., Schaaff, T. G., Shafigullin, M. N., Vezmar, I., Whetten, R. L.: 1997, *J. Phys. Chem. B* **101**, 3706.
Ben D. T., Lereah, Y., Deutscher, G., Kofman, R., Cheyssac, P.: 1995, *Phil. Mag. A* **71**, 1135.
Blackman, M., Curxon, A. E.: 1959, In: *Structure and Properties of Thin Films, Wiley, New York*, p. 217.
Bohren, C. F., Huffman, D. R.: 1983, In: *Absorption and Scattering of Right by Small Particles, Wiley Interscience*, p. 357, 367.
Boiko, B. T., Pugachev, A. T., Bratsykhin, V. M.: 1969, Sov. Phys.-Solid State 10, 2832.
Brommer, K. D., Needels, M., Larson, B. E., Joannopoulos, D.: 1992, *Physical Review Letters* **68**, 1355.
Buffat, Ph. Borel, J.: 1976, *Phys. Rev. A* **13**, 2287.
Burns, J. A., Lamy, P. L. Soter, S.: 1979, *Icarus* **40**, 1.
Chen, H., Kou, X., Yang, Z., Ni, W., Wang, J.: 2008, *Langmuir* **24**, 5233.
Coombes, C. J.: 1972, *J. Phys.* **2**, 441.
Couchman, P. R.: 1979, *Phil. Mag. A* **40**, 637.
Couchman, P. R., Jesser, W. A.: 1977, *Nature* **269**, 481.
Couchman, P. R., Ryan, C. L.: 1978, *Phil. Mag. A* **37**, 369.
Delogu, F.: 2005, *Phys. Rev. B* **72**, 205418.
Ercolessi, F., Andreoni, W., Tosatti E.: 1991, *Phys. Rev. Lett.* **66**, 911.
Garvie, L. A. J., Buseck, P. R.: 2004, *Earth Planet. Sci. Lett.* **224**, 431.
Garvie, L. A. J., Buseck, P. R.: 2006, *Meteorit. Planet. Sci.* **41**, 633.
Gladkich, N. T., Niedermayer, R., Spiegel, K.: 1966, *Phys. Status Solidi.* **15**, 181.
Harada, J., Yao, S., Ichimiya, A.: 1980, *J. Phys. Sot. Japan* **48**, 1625.
Hoshino, K., Shimamura, S.: 1979, *Phil. Mag. A* **40**, 137.
Jiang, Q., Shi, H. X., Zhao, M.: 1999, *J. Chem. Phys.* **111**, 2176.
Johnson, P. B., Christy, R. W.: 1974, *Physical Review B* **9**, 5056.
Karakouz, T., Holder, D., Goomanovsky, M., Vaskevich, A., Rubinstein, I.: 2009, *Chem. Mater.* **21**, 5875.
Kashiwase, Y., Nishida, I., Kainuma, Y., Kimoto, K.: 1977, *J. Phys.* **38, Suppl. C-2**, 157.
Kido, O., Kamitsuji, K., Kurumada, M., Sato, T., Kimura, Y., Suzuki, H., Saito, Y., Kaito, C.: 2005, *J. Crystal Growth* **275**, e1745.
Kimoto, K., Nishida, I.: 1967, *J. Phys. Soc. Jap.* **22**, 744.
Kerridge, J. F.: 1976, *Nature* **259**, 189.
Kimura, Y., Saito, Y., Nakada, T., Kaito, C.: 2000, *Phys. Low-Dim. Struct.* **1/2**, L1.
Kimura, Y., Saito, Y., Nakada, T., Kaito, C.: 2002, *Physica E: Low-dimensional Systems and Nanostructures* **13**, 11–23.
Kimura, Y., Sato, T., Kaito C.: 2004, *Carbon* **42**, 33.
Kimura, Y., Kurumada, M., Tamura, K., Koike, C., Chihara, H., Kaito, C.: 2005a, *Astronomy & Astrophysics* **442**, 507.
Kimura, Y., Miura, H., Tsukamoto, K., Li, C. Maki, T. 2011a, *Journal of Crystal Growth* **316**, 196.
Kimura, Y., Nuth III, J. A.: 2005b, *The Astrophysical Journal* **630**, 637.
Kimura, Y., Nuth III, J. A.: 2009, *The Astrophysical Journal Letters* **697**, L10.
Kimura, Y., Nuth III, J. A., Ferguson, F. T.: 2005c, *The Astrophysical Journal Letters* **632**, L159.
Kimura, Y., Nuth III, J. A., Ferguson, F. T.: 2006, *Meteoritics & Planetary Science* **41**, 673.
Kimura, Y.: 2008, *Yuseijin (in Japanese)* **17**, 62.
Kimura, Y., Miyazaki, Y., Kumamoto, A., Saito, M., Kaito, C.: 2008a, *The Astrophysical Journal Letters* **680**, L89.
Kimura, Y., Sasaki, S., Suzuki, H., Kumamoto, A., Saito, M., Kaito, C.: 2008b, *The Astrophysical Journal* **684**, 1496.

Kimura, Y., Tamura, K., Koike, C., Chihara, H., Kaito, C.: 2005d, *Icarus* **177**, 280.
Kimura, Y., Tsukamoto, K.: 2011b, *J. Jpn. Soc. Microgravity Appl.*, **28**, S9.
Kobatake, H. Tsukamoto, K., Satoh, H.: 2005, *Journal of Crystal Growth* **279**, 186.
Kobatake, H., Tsukamoto, K., Nozawa, J., Nagashima, K., Satoh, H., Dold, P.: 2008, *Icarus* **198**, 208.
Kubo, R.: 1962, *J. Phys. Soc. Jpn.* **17**, 975.
Lee, J., Lee, J., Tanaka, T., Mori, H.: 2009, *Nanotechnology* **20**, 475706.
Lindemann, F. A.: 1910, *Z Phys.* **11**, 609.
Lubashenko, V. V.: 2010, *J. Nanopart. Res.* **12**, 1837.
Marks, L. D., Ajayan, P. M.: 1986, *Ultramicroscopy* **20**, 77.
Messenger, S., Nakamura-Messenger, K., Keller, L. P.: 2008, *Lunar Planet. Sci.* **XXXIX**, # 2391.
Miura, H., Tanaka, K. K., Yamamoto, T., Nakamoto, T., Yamada, J., Tsukamoto, K., Nozawa, J.: 2010, *The Astrophysical Journal* **719**, 642.
Mori, H., Komatsu, M., Takeda; K., Fujita, H.: 1991, *Philosophical Magazine Letters* **63**, 173.
Mori, H., Yasuda, H., Kamino, T.: 1994, *Philos. Mag. Lett.* **69**, 279.
Mori, H., Yasuda, H.: 1999, *Bull. Mater. Sci.* **22**, 181.
Nagata, T., Kaito, C., Saito, Y., Funaki, M.: 1991, *Proceedings of the 15th National Institute of Polar Research Symposium on Antarctic Meteorites* **4**, 404.
Nakamura, A., Fujiwara, A.: 1991, *Icarus* **92**, 132.
Nakamura, K., Zolensky, M. E., Tomita, S., Nakashima, S., Tomeoka, K.: 2002, *Int. J. Astrobiology* **1**, 179.
Nakamura-Messenger, K., Messenger, S., Keller, L. P., Clemett, S. J., Zolensky, M. E.: 2006, *Science* **314**, 1439.
Nakamura-Messenger, K., Messenger, S., Keller, L. P., Clemett, S. J., Zolensky, M. E.: 2008, *Geochim. Cosmochim. Acta* **72**, A671.
Nuth, J. A., Donn, B.: 1982, *J. Chem. Phys.* **77**, 2639.
Ohshima, K., Hayashi, A., Harada, L.: 1980, *J. Phys. Sot. Japan* **48**, 1631.
Okada, M., Yoshimoto, H., Nishioka, K.: 1995, *Molec. Simul.* **14**, 67.
Okada, M., Yoshimoto, H., Nishioka, K., Mori, A.: 1996a, *Molec. Simul.* **16**, 179.
Okada, M., Yoshimoto, H., Nishioka, K., Mori, A.: 1996b, *Mater. Sci. Eng.* **A217–218**, 235.
Olson, E. A., Efremov, M. Yu, Zhang, M., Allen, L. H.: 2005, *J. Appl. Phys.* **97**, 034304.
Qi, Y., Tahir, C., Johnson, W. L., Goddard III, W. A.: 2001, *J. Chem. Phys.* **115**, 385.
Saito, M., Kimura, Y.:2009, *The Astrophysical Journal Letters* **703**, L147.
Sambles, J. R.: 1971, *Proc. R. Soc. A* **324**, 339.
Shimizu, Y., Sawada, S., Ikeda, K.S.: 1998, *Eur. Phys. J. D* **4**, 365.
Shimizu, Y., Ikeda, K. S., Sawada, S.: 2001, *Physical Review B* **64**, 075412.
Skripov, V. P., Koverda, V. P., Skokov, V. N.:1981 *physica status solidi (a)* **66**, 109.
Takagi, M.: 1954, *Journal of the Physical Society of Japan* **9**, 359.
Yasuda, H., Mori, H., Komatsu, M., Takeda, K., Fujita, H.: 1992, *J. Electron Microsc.* **41**, 267.
Yasuda, H. Mori, H.: 1992, *Phys. Rev. Lett.* **69**, 3747.
Yasuda, H., Mori, H.: 1993, *Intermetallics* **1**, 35.
Yasuda, H., Mori, H., Komatsu, M., Takeda, K.: 1993, *J. Appl. Phys.* **73**, 1100.
Yasuda, H., Mori, H.: 1994a, *Z. Phys. D: At., Mol. Clusters* **31**, 131.
Yasuda, H., Mori, H.: 1994b, *Z. Phys. D: At., Mol. Clusters* **31**, 209.
Yasuda, H., Tanaka, A., Matsumoto, K., Nitta, N., Mori, H.: 2008, *Phys. Rev. Lett.* **100**, 105506.
Wronski, C. R. M.: 1967, *Br. J. Appl. Pjys.* **18**, 1731.
Zhang, Z., Li, J. C., Jiang, Q. J.: 2000, *Phys. D* **33**, 2653.
Zolensky, M. E., Thomas, K. L.: 1995, *Geochim. Cosmochim. Acta* **59**, 4707.

Nanodust Dynamics in Interplanetary Space

Andrzej Czechowski and Ingrid Mann

Abstract In distinction to the case of larger grains, the dynamics of nanodust in the interplanetary medium is strongly affected by the electromagnetic forces. In this chapter we use a simple model of the solar wind, the solar magnetic field and the dust equilibrium surface charge as a basis for a study of the motion of the nanodust. The results are as follows. The nanodust produced close to the Sun can be trapped in bound non-Keplerian orbits. The grains created outside of the trapped region are accelerated to the velocities of the order of the solar wind velocity and escape to large distances. A qualitative understanding of the nanodust dynamics is possible with use of the ideas derived from the guiding centre approximation.

1 Introduction

A single unbalanced charge on a dust grain of 1-nm radius produces a charge-to-mass ratio Q/m of the order of $10^{-3}\,e/m_\mathrm{p}$ (e = elementary charge, m_p = the proton mass). The equilibrium charge on such small grains in the interplanetary space is not exactly known. However, if the approximate scaling $Q/m \sim 1/a^2$ (a = grain radius) would hold down to nanometre sizes, then from the models developed for larger grains (Mukai 1981; Kimura and Mann 1998) it follows that for a grain

A. Czechowski (✉)
Space Research Centre, Polish Academy of Sciences, Bartycka 18A, 00–716 Warsaw, Poland
e-mail: ace@cbk.waw.pl

I. Mann
Belgium Institute for Space Aeronomie, 3 Avenue Circulaire, 1180 Brussels, Belgium

Laboratoire d'Etudes Spatiales et d'Instrumentation en Astrophysique, Observatoire de Paris, Universite Pierre et Marie Curie, Universite de Paris Diderot, CNRS, Meudon, France

EISCAT Scientific Association, P. O. Box 812, SE-981 28 Kiruna, Sweden
e-mail: ingrid.mann@eiscat.se

I. Mann et al. (eds.), *Nanodust in the Solar System: Discoveries and Interpretations*, Astrophysics and Space Science Library 385, DOI 10.1007/978-3-642-27543-2_4,

of 3 nm radius $Q/m \approx 10^{-4} e/m_{\mathrm{p}}$ (10^4 C/kg) and for 10 nm $Q/m \approx 10^{-5} e/m_{\mathrm{p}}$ (10^3 C/kg) (Czechowski and Mann 2010).

In this chapter we present the results of a study of dynamics of charged nanodust particles in the interplanetary space. They follow on and extend those given in Czechowski and Mann (2010). We show that for the grains with $Q/m \sim 10^{-5}$–$10^{-4} e/m_{\mathrm{p}}$ and higher the electromagnetic forces become strong enough to dominate the dynamics. In result, charged nanodust grains behave in some respects similarly to ions if the effects of pitch angle scattering are neglected (Luhmann 2003): they can be "picked up" and accelerated by the solar wind and exhibit drifts along the heliospheric current sheet. They can also become trapped in non-Keplerian orbits that result from an interplay between the electromagnetic and the gravity forces.

The existence of the trapping region for nanodust was already proposed by Mann et al. (2007). Our study of the nanodust dynamics (Czechowski and Mann 2010, 2011) is to our knowledge the first in which the trapping mechanism was identified. The effect of electromagnetic forces on the orbits of larger grains was studied by Morfill and Grün (1979) and after them by several other researchers (Consolmagno 1980; Wallis and Hassan 1985; Krivov et al. 1998; Mann et al. 2000). Trajectories for the 10–100-nm grains with initial orbits between 0.25 and 4 AU were calculated numerically by Hamilton et al. (1996).

2 The Model

In our study we use a simple model of the solar wind and of the solar magnetic field (Figs. 1 and 2). Although simplified, it incorporates the two-component structure of the solar wind ("slow" wind at low and "fast" wind at high heliolatitudes) and the solar magnetic field with the tilted heliospheric current sheet. We solve numerically full equations of motion for the grains. We also consider a simplified description, based on the guiding centre approximation: we find that it helps to obtain a qualitative understanding of our results.

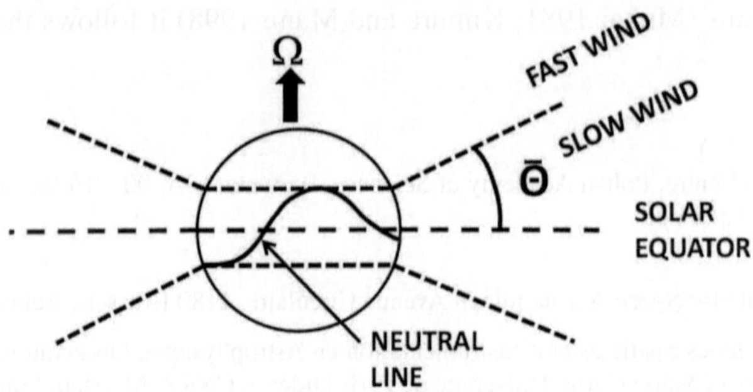

Fig. 1 Boundary conditions for solar wind. Slow wind emerges from the belt of heliolatitudes that include the tilted neutral line. The fast wind flows from northern and southern caps

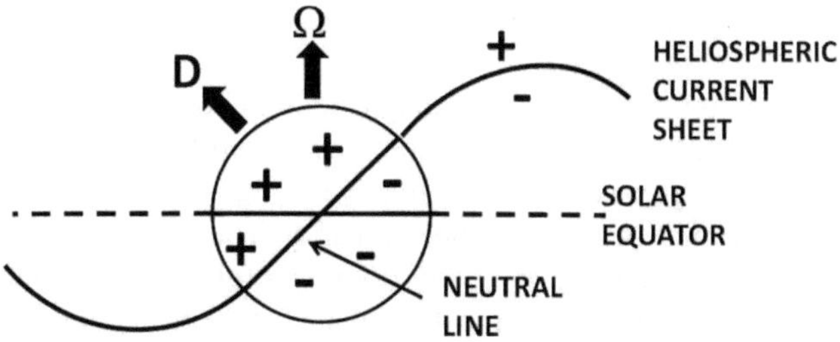

Fig. 2 Boundary conditions for solar magnetic field. The solar dipole **D** is tilted relative to the solar rotation axis. The neutral line separates the hemispheres of opposite field orientation. The heliospheric current sheet extends from the neutral line into the solar wind

We use fixed inertial system of coordinates $(\hat{\mathbf{e}}_r, \hat{\mathbf{e}}_\theta, \hat{\mathbf{e}}_\phi)$ with the origin at the centre of the Sun. The angle θ is the solar co-latitude, counted from the North pole of the solar rotation axis.

The solar wind velocity is radially directed and independent of distance: $\mathbf{V} = V\hat{\mathbf{e}}_r$. V depends only on heliolatitude: $V = 400$ km/s in the slow solar wind region ($|\text{latitude}| < \bar{\Theta}$) and $V = 800$ km/s in the fast wind region ($|\text{latitude}| > \bar{\Theta}$). In most calculations $\bar{\Theta}$ is also equal to the tilt of the heliospheric current sheet (for the case of zero tilt $\bar{\Theta}$ is set to 20°).

Our model for magnetic field is obtained under the assumption that the field is frozen in the solar wind plasma. In this case a magnetic field line follows the positions of the plasma packets emitted (at different moments of time) from the same point on the rotating surface of the Sun (the footpoint of the field line). The shape of the magnetic field line is known as the Parker spiral (Parker 1958). The unit vector tangent to the magnetic field is

$$\hat{\mathbf{b}} \equiv \frac{\mathbf{B}}{B} = \pm \frac{\hat{\mathbf{e}}_r - kr\hat{\mathbf{e}}_\phi}{(1 + k^2r^2)^{1/2}}, \tag{1}$$

where $k = (\Omega/V)\sin\theta$ with Ω the angular speed of solar rotation and θ the co-latitude. The plus sign is for outgoing and minus for incoming field.

The Maxwell equations for the magnetic field reduce to

$$\nabla \cdot \mathbf{B} = 0 \tag{2}$$

$$\partial_t \mathbf{B} = \nabla \times (\mathbf{V} \times \mathbf{B}). \tag{3}$$

The solution can be written as

$$\mathbf{B} = \hat{\mathbf{b}}|\tilde{B}_r|(1 + k^2r^2)^{1/2}/(r/\bar{r})^2, \tag{4}$$

where $\tilde{B}_r$ is the radial component of $\mathbf{B}$ at $r = \bar{r} = 1$ AU. The sign and value of $\tilde{B}_r$ can vary between different magnetic field lines, so that the field $\mathbf{B}$ is in general nonstationary. However, in our calculations we choose $|\tilde{B}_r|$ to be constant over each of the solar wind regions ($|\tilde{B}_r|$ ≈35 μG in the slow and ≈45 μG in the fast solar wind).

To model the sector structure of the magnetic field polarity we assume that the solar surface is divided into two hemispheres, separated by a "neutral line" which we approximate by a great circle tilted relative to the solar equator (Fig. 2). The hemispheres correspond to opposite field polarity. The sign of $\tilde{B}_r$ is positive or negative depending on the position of the footpoint of the field line with respect to the neutral line. In result, the only time dependence of the field consists of direction reversals when the boundary between the sectors of opposite polarity is crossed. This generates a discontinuity surface in the magnetic field, which is a model of the heliospheric current sheet.

We distinguish between two polarity orientations of the magnetic field. The case of incoming (negative) field northward of the neutral line (also known as $qA < 0$) corresponds to the electric field $-(1/c)\mathbf{V} \times \mathbf{B}$ pointing towards the heliospheric current sheet: we refer to it as the "focusing" field orientation (Grün et al. 1994; Gustafson and Lederer 1996; Grogan et al. 1996; Landgraf 2000). The drift of the positively charged particles is then directed towards the current sheet. The opposite orientation we call "non-focusing" or "defocusing". The field reversals occurring every 11 years exchange the "focusing" and "non-focusing" orientations. The period after the year 2000 solar maximum was "focusing".

In our calculations we assume fixed values of the charge-to-mass ratio Q/m of the nanodust particles. Since the most important forces are the Lorentz force and the gravity force, the value of Q/m is the main parameter specifying the dynamics. Most of our calculations were done for constant values $Q/m = 10^{-4}$–$10^{-5} e/m_p$.

Our assumption of radial, constant V plasma flow restricts the model applicability to r larger than the outer limit of the "intermediate zone" ~10–20 R_{Sun} (Schatten et al. 1969). For discussion of charged grain dynamics in the vicinity of the Sun see Krivov et al. (1998) but only the grains larger than 0.1 μm are considered there.

In reality the solar wind velocity and the magnetic field are highly variable on different timescales. The effect of these variations on nanodust dynamics requires a separate investigation. The present model can be considered as an approximation to time-averaged, smoothed solar wind flow and the magnetic field. It is not applicable in particular to the periods of high solar activity.

3 Equations of Motion

The full equation of motion, which we solve numerically, is

$$\frac{d\mathbf{v}}{dt} = \frac{Q}{mc}(\mathbf{v} - \mathbf{V}) \times \mathbf{B} - \frac{GM_S}{r^2}\hat{\mathbf{e}}_r + \mathbf{F}_\gamma, \tag{5}$$

where $\mathbf{v}$, Q and m are the velocity, the charge and the mass of the dust grain, $\mathbf{V}$ is the plasma velocity, $\mathbf{B}$ the magnetic field, G the gravity constant, M_S the solar mass.

The radiation force

$$\mathbf{F}_\gamma = \frac{GM_\mathrm{S}}{r^2}\beta\left(\left(1-\frac{v_\mathrm{r}}{c}\right)\hat{\mathbf{e}}_\mathrm{r} - \frac{\mathbf{v}}{c}\right) \tag{6}$$

is proportional to the radiation pressure to gravity ratio β for the dust grain. v_r is the radial component of the grain velocity. The nonradial part of the radiation force is known to cause the Poynting–Robertson contraction of the gravity-controlled orbits of larger dust grains. It can also affect the orbits of nanodust (see Sect. 6). The radiation pressure correction to the gravity force $-\beta GM_\mathrm{S}/r^2$ for the nanodust is not negligible but small ($\beta \sim 0.1$, see discussion in the last section of this chapter).

Although our calculations use the full equation of motion, we shall refer also to the guiding centre approximation. The equations below are adapted from Northrop (1958) but restricted to the case of time-stationary field ($\partial_t \mathbf{B} = 0$). The parallel velocity of the guiding centre is given by

$$\frac{\mathrm{d}v_\parallel^G}{\mathrm{d}t} = g_\parallel - \mu\partial_\mathrm{S} B + v_\parallel^G \mathbf{V}_\perp \cdot \partial_\mathrm{S}\hat{\mathbf{b}} + \mathbf{V}_\perp \cdot (\mathbf{V}_\perp \cdot \nabla)\hat{\mathbf{b}}, \tag{7}$$

where $v_\parallel^G$ is the guiding centre velocity component parallel to the magnetic field, $g_\parallel$ the parallel component of the gravity force per unit mass acting on the grain, $\mathbf{V}_\perp = \mathbf{V} - \hat{\mathbf{b}}(\hat{\mathbf{b}} \cdot \mathbf{V})$ the plasma velocity perpendicular to $\mathbf{B}$, $\mu \equiv v_\perp'^2/2B$ (where $\mathbf{v}_\perp' = \mathbf{v}_\perp - \mathbf{V}_\perp$) the adiabatic invariant and $\partial_\mathrm{S} \equiv \hat{\mathbf{b}} \cdot \nabla$.

The perpendicular velocity of the guiding centre can be written as a sum of two terms:

$$\mathbf{v}_\perp^G = \mathbf{V}_\perp + \mathbf{V}_\mathrm{D}, \tag{8}$$

where the first term is just the perpendicular component of the plasma velocity. Note that $\mathbf{V}_\perp = c\mathbf{E} \times \mathbf{B}/B^2 \equiv u_\mathrm{E}$ of Northrop (1958) since the electric field $\mathbf{E} = -(1/c)\mathbf{V} \times \mathbf{B}$. The second term depends on the charge-to-mass ratio of the particle through the Larmor frequency $\Omega_\mathrm{L} \equiv QB/mc$:

$$\mathbf{V}_\mathrm{D} = \frac{1}{2}\frac{v_\perp'^2}{\Omega_\mathrm{L}}\frac{\hat{\mathbf{b}} \times \nabla B}{B} + \frac{1}{\Omega_\mathrm{L}}[-\hat{\mathbf{b}} \times \mathbf{g} + v_\parallel^2 \hat{\mathbf{b}} \times \partial_\mathrm{S}\hat{\mathbf{b}} + v_\parallel \hat{\mathbf{b}} \times (\mathbf{V}_\perp \cdot \nabla)\hat{\mathbf{b}} \tag{9}$$
$$+ v_\parallel \hat{\mathbf{b}} \times \partial_\mathrm{S}\mathbf{V}_\perp + \hat{\mathbf{b}} \times (\mathbf{V}_\perp \cdot \nabla)\mathbf{V}_\perp].$$

In our calculations the term $\mathbf{V}_\mathrm{D}$ in (8) is small compared to $\mathbf{V}_\perp$, so that the guiding centre stays close to the magnetic field line convected at $\mathbf{V}_\perp$.

Note that the last term in (7) can be understood as a centrifugal force. From (8) it follows that the magnetic lines of force to which the guiding centre is tied rotate around the Sun rotation axis. Imagine the guiding centre as a "bead" constrained to move along this rotating line. The rotation at the angular velocity ω should cause the centrifugal force per unit mass $-\omega \times (\omega \times \mathbf{r})$ where $\mathbf{r}$ is the position of the guiding centre. By explicit calculation one finds that the last term in (7) is equal to the parallel component of the centrifugal force: $\mathbf{V}_\perp \cdot (\mathbf{V}_\perp \cdot \nabla)\hat{\mathbf{b}} = -(\omega \times (\omega \times \mathbf{r})) \cdot \hat{\mathbf{b}}$. The angular velocity vector ω is parallel to the solar rotation axis and has the magnitude

$|\omega| = \mathbf{V}_\perp \cdot \hat{\mathbf{e}}_\phi/(r \sin\theta)$. Note that ω is not identical to the angular velocity of solar rotation. Instead, it is equal to the angular velocity (in the inertial system that we use) of the radius vector of a point moving at the velocity $\mathbf{V}_\perp$.

4 Initial Conditions

An important source of nanodust in the interplanetary space is thought to be the collisional fragmentation of larger dust grains which form the circumsolar disk of the interplanetary dust cloud (Grün et al. 1985). The initial conditions for the nanodust equations of motion used in our calculations are suggested by this production mechanism.

We take the initial velocity of the nanodust particle to be equal to that of the parent body (a larger grain) moving in circular Keplerian orbit. The velocity difference between the fragment and the parent body is neglected: since most of the nanodust produced by this mechanism originates close to the Sun, where the Keplerian velocities are large, this is a reasonable approximation. The orbits of the parent bodies we divide into two classes: the "disk" (orbits restricted to inclinations within ±12° from the ecliptic) and the "halo" (the inclinations within ±80°).

Our results show that the phase space of initial conditions consists of two regions: the trapped nanodust region around the Sun and the remaining region corresponding to nanodust particles escaping to large distances. Sample orbits of trapped and escaping particles with $Q/m = 10^{-5}$ and $10^{-6} e/m_\mathrm{p}$ are presented in Fig. 3. Note

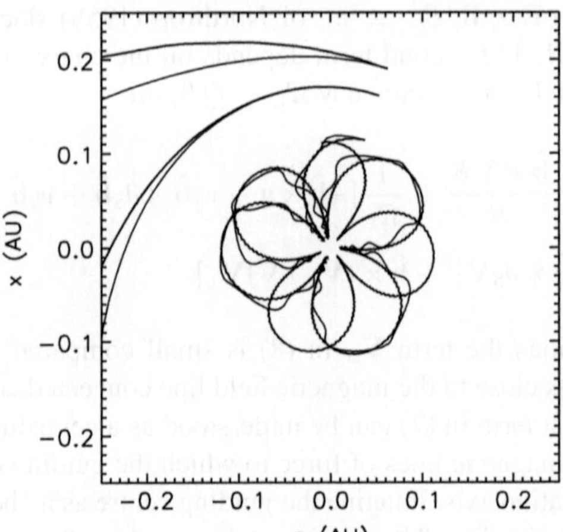

Fig. 3 Orbits of trapped and escaping nanodust particles projected onto the solar equator plane. *Thick lines* are for $Q/m = 10^{-5} e/m_\mathrm{p}$ and *thin lines* for $Q/m = 10^{-6} e/m_\mathrm{p}$. The particles are released from parent bodies in circular Keplerian orbits near the ecliptic at the distances 0.12, 0.17 and 0.2 AU from the Sun. The grains released at 0.12 AU are trapped. Note the effect of Larmor rotation for the trapped orbits

that even for a low value of $Q/m = 10^{-6} e/m_\mathrm{p}$ the orbit of the trapped grain is qualitatively similar to that for the $Q/m = 10^{-5} e/m_\mathrm{p}$ case, although the deviations due to Larmor rotation are much larger for the bigger grain.

5 Pickup and Acceleration

The interaction between the charged nanodust particle and the solar wind plasma resembles the pickup process of freshly created ions. The pickup process is illustrated in Fig. 4: the particle commences rotating around a magnetic field line carried with the plasma flow, leading to velocity oscillations with the initial amplitude determined by the difference $|\mathbf{v}_\perp - \mathbf{V}_\perp|$. The rotation is characterized by the Larmor frequency $\Omega_\mathrm{L} = QB/mc$ and the Larmor radius $R_\mathrm{L} = |\mathbf{v}_\perp - \mathbf{V}_\perp|/\Omega_\mathrm{L}$. Approximately, $\Omega_\mathrm{L} = (Q/m)10^4 B$ where Q/m is in e/m_p, B in Gauss and Ω_L in rad/s.

The pickup process works differently in the region near the Sun (where most of the nanodust is created) and beyond 1 AU.

At the distance above few AU from the Sun the solar wind flow is almost perpendicular to the magnetic field, so that $|\mathbf{V}_\perp|$ is large. The small nanodust particle (with initial Larmor radius R_L much less than the initial distance from the Sun) released in this region would then behave as shown in Fig. 5 for the $Q/m = 10^{-4} e/m_\mathrm{p}$ particle picked up at 2.5 AU (initial R_L ~0.2 AU). Its average velocity becomes equal to the solar wind velocity already within a single Larmor rotation time. The velocity evolution can be understood using the concepts following from the guiding centre approximation. In particular, the accelerated particle stays close to the magnetic field line convected with the solar wind flow. Also, the decrease in the amplitude of the velocity oscillations (approximately as $1/r^{1/2}$) can be derived from conservation of the adiabatic invariant $|\mathbf{v}_\perp - \mathbf{V}_\perp|^2/2B$.

For $Q/m = 10^{-5} e/m_\mathrm{p}$ the initial Larmor radius at 2.5 AU is ~2 AU and the guiding centre approximation does not work. Nevertheless, it is possible to

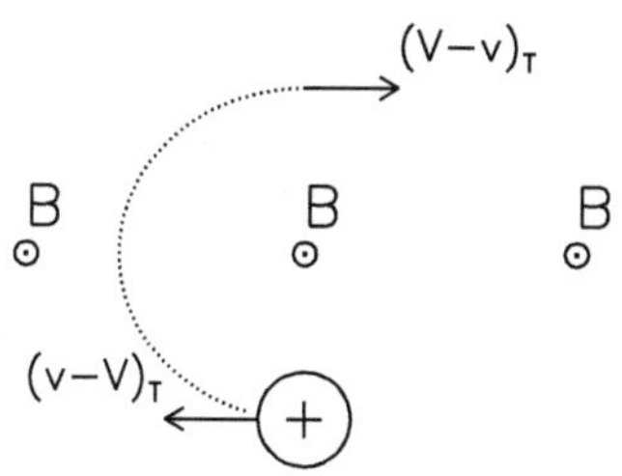

Fig. 4 The pickup process. A freshly created charged particle with the transverse component of initial velocity $\mathbf{v}_\perp$ commences rotating around the magnetic field line carried at the transverse velocity $\mathbf{V}_\perp$ of the plasma flow

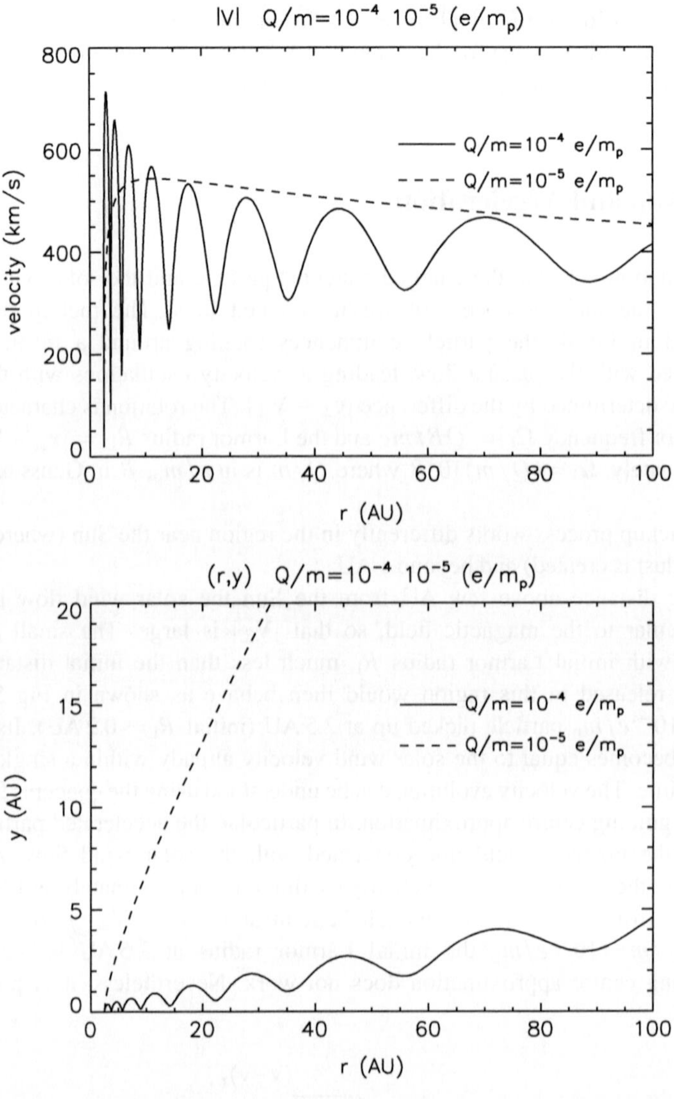

Fig. 5 The speed vs. distance profiles (*upper figure*) and the trajectories (*lower figure*; y is the coordinate perpendicular to the solar equator plane) for particles with $Q/m = 10^{-4}e/m_{\rm p}$ (*solid line*) and $10^{-5}e/m_{\rm p}$ (*dashed line*) released at 2.5 AU for non-focusing field orientation. The case $Q/m = 10^{-4}e/m_{\rm p}$ illustrates the similarity with the (scatter-free) ion pickup process. To simplify the comparison with the $Q/m = 10^{-5}e/m_{\rm p}$ case, the slow solar wind region was assumed to extend to all latitudes

understand its velocity evolution as shown in Fig. 5 using a simple model that permits analytical solution (see the Appendix).

Close to the Sun the magnetic field becomes almost radial and $\mathbf{V}_\perp$ is small. The average velocity acquired by a freshly picked up particle within single Larmor

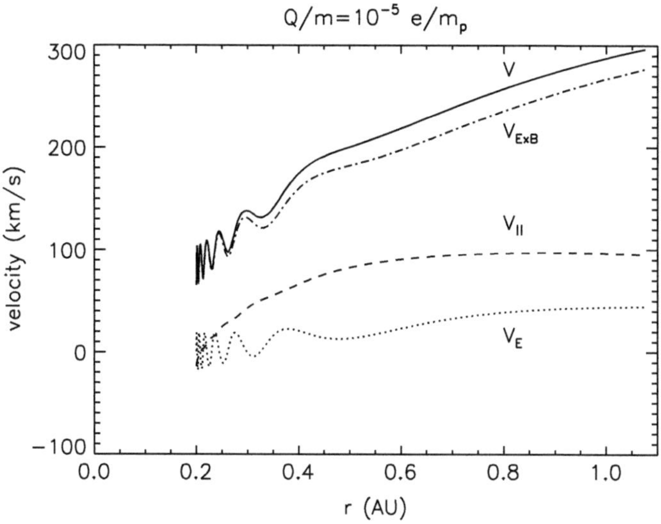

Fig. 6 The velocity evolution for the case of $Q/m = 10^{-5}e/m_\mathrm{p}$ grain released at 0.2 AU from the Sun. The total velocity v and the velocity components along the magnetic field $v_{||}$, along the electric field v_E and along the $\mathbf{E} \times \mathbf{B}$ direction $v_{\mathbf{E}\times\mathbf{B}}$, are shown. The velocity is dominated by the component $v_{\mathbf{E}\times\mathbf{B}}$ (which is also the direction of the perpendicular component $\mathbf{V}_\perp$ of the solar wind velocity) with a smaller contribution from $v_{||}$. The component v_E is the one relevant for the energy increase

rotation period is of the order $V_\perp$. Acceleration to total solar wind velocity occurs well after the pickup and involves acceleration of the particle also in the direction parallel to magnetic field (Fig. 6). The electric field $\mathbf{E} = -(1/c)\mathbf{V} \times \mathbf{B}$ is perpendicular to $\mathbf{B}$. In the absence of non-electromagnetic forces this implies $\hat{\mathbf{b}} \cdot \mathrm{d}\mathbf{v}/\mathrm{d}t = 0$. Nevertheless, $\mathrm{d}v_{||}/\mathrm{d}t \neq 0$ because the field direction is non-uniform, so that $\mathrm{d}\hat{\mathbf{b}}/\mathrm{d}t \neq 0$ along the particle trajectory.

Figure 6 shows also the evolution of the velocity component v_E parallel to the electric field. Although small compared to other components, it is important for energy conservation, since the time derivative of the kinetic + gravitational energy per mass of a charged particle is given by $(Q/m)\mathbf{v} \cdot \mathbf{E}$. The detailed discussion of energy conservation in the model is included in Czechowski and Mann (2010).

Within our model assumptions, the difference $|\mathbf{v}_\mathrm{orb} - \mathbf{V}|_\perp$ ($\mathbf{v}_\mathrm{orb}$ is the orbital velocity for circular Keplerian orbit) passes through zero at about 0.17 AU from the Sun (in the ecliptic plane). At the distance larger than 0.17 AU from the Sun the solar wind motion in the perpendicular direction is faster than the initial motion of the grain, similarly to the situation at large distances. On the other hand, inwards from 0.17 AU, the dust particle in Keplerian orbit moves faster than the solar wind in the direction perpendicular to the magnetic field. This leads to some interesting effects (not discussed here).

The speed vs. distance profiles inwards from 1 AU for the escaping particles of different Q/m created near the Sun (initial distances 0.17 and 0.3 AU) are shown in Fig. 7. For the smallest grains ($Q/m \geq 10^{-5} e/m_{\mathrm{p}}$) the velocity value reached at 1 AU is about 300 km/s. Beyond 1 AU (not shown) the velocity continues to increase and approaches the solar wind velocity (400 km/s). The velocity reached at 1 AU is lower for the larger grains: the decrease occurs over the range of Q/m between 10^{-5} and $10^{-6} e/m_{\mathrm{p}}$. For $Q/m = 10^{-7} e/m_{\mathrm{p}}$ the velocity does not increase with distance: nevertheless, even for this value of Q/m the particle escapes under the action of the Lorentz force from the initial bound orbit and moves to large distance from the Sun. Note, however, that a particle of this Q/m would have the radius of the order 100 nm = 0.1 μm and the radiation pressure to gravity ratio $\beta \sim 1$ rather than $\beta = 0.1$ assumed in our calculation. Acceleration by the radiation pressure force would be important for this case.

Another effect seen in Fig. 7 is that the acceleration becomes less efficient for the focusing field orientation. This is caused by encounters with the heliospheric current sheet and will be discussed in a later section.

6 The Trapping Mechanism

Before proceeding to further discussions of the ejected nanodust we discuss now the trapping mechanism. A simple but useful model of the trapping mechanism follows by neglecting the term $\mathbf{V}_{\mathrm{D}}$ in (8). This implies that the co-latitude θ (or, equivalently, the parameter k) becomes a constant of motion. The motion of the guiding centre is then restricted to the two-dimensional phase plane. As the phase plane coordinates one can use the distance s along the field line and the parallel velocity of the guiding centre $v_{\parallel}^{G} \equiv \mathrm{d}s/\mathrm{d}t$. Another choice (to be followed here) is the radial distance of the guiding centre from the Sun r and the radial velocity of the guiding centre $v_{\mathrm{r}}^{G} \equiv \mathrm{d}r/\mathrm{d}t$. In this section we denote v_{r}^{G} by v to simplify the formulae. The relation between v and $v_{\parallel}^{G}$ is

$$v = \frac{v_{\parallel}^{G}}{(1+k^2r^2)^{1/2}} + \frac{Vk^2r^2}{1+k^2r^2}. \tag{10}$$

In Czechowski and Mann (2010) the last term in this relation was omitted by mistake. See also the Erratum (Czechowski and Mann 2011).

After evaluating explicitly all terms in (7) one obtains

$$\frac{\mathrm{d}v_{\parallel}^{G}}{\mathrm{d}t} = -\frac{GM_{\mathrm{S}}(1-\beta)}{r^2}\frac{1}{h^{1/2}} + \frac{2\mu\tilde{B}_{\mathrm{r}}\tilde{r}^2}{r^3}\left(1-\frac{k^2r^2}{2h}\right) - v_{\parallel}^{G}V\frac{k^2r(1+h)}{h^2} + \frac{V^2k^2r}{h^{5/2}}, \tag{11}$$

where $h = 1 + k^2r^2$. Using (10) leads to the expression for $\mathrm{d}v/\mathrm{d}t$ as a function of v and r:

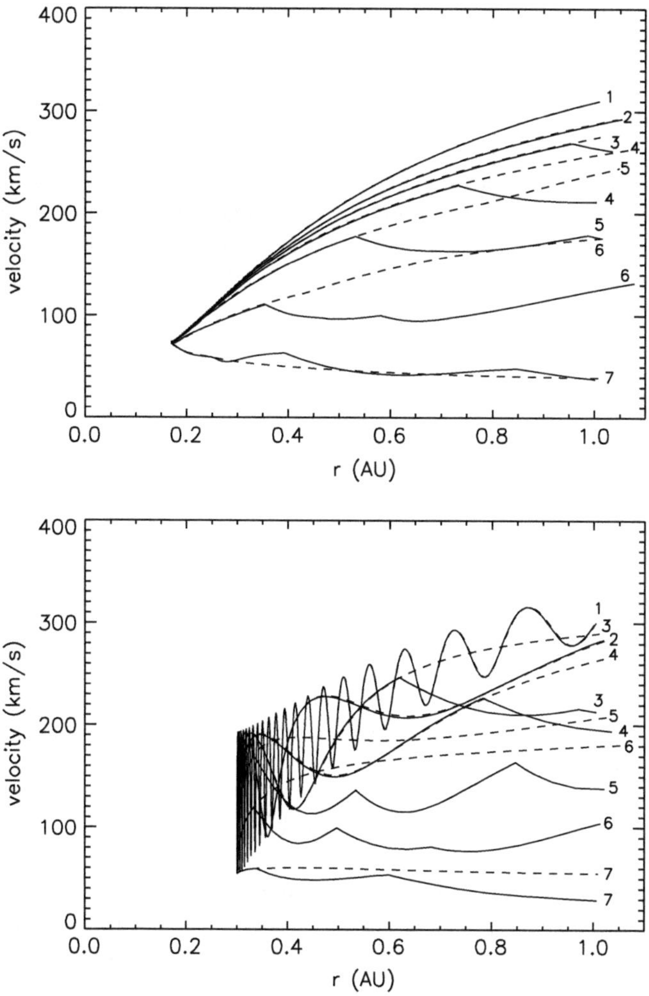

Fig. 7 The effect of Lorentz force acceleration on the nanodust of different size within 1 AU. The velocity of the grain is plotted as a function of distance from the Sun. The grains with $Q/m = 10^{-7}$–$10^{-4} e/m_p$ are released from a circular orbit of radius 0.17 AU (*upper panel*) and 0.3 AU (*lower panel*) close to the solar equator plane. The lines are numbered in the order of decreasing Q/m: (1) 10^{-4}, (2) 10^{-5}, (3) 7×10^{-6}, (4) 5×10^{-6}, (5) 3×10^{-6}, (6) 10^{-6}, (7) 10^{-7}. *Solid line* corresponds to focusing and *dashed line* to non-focusing field. Sharp "kinks" in the velocity profiles mark the crossings of the current sheet. Velocity oscillations seen for $Q/m = 10^{-4}$ and $10^{-5} e/m_p$ are caused by Larmor rotation. For particles starting at 0.3 AU the amplitude of these oscillations is large because of large initial difference in transverse velocity between the nanodust particle and the solar wind. For 0.17 AU case this difference is very small and the oscillations are not visible

$$\frac{\mathrm{d}v}{\mathrm{d}t} = -\frac{GM_S(1-\beta)}{r^2 h} + \frac{2\mu \tilde{B}_r \bar{r}^2}{r^3 h^{1/2}}\left(1 - \frac{k^2 r^2}{2h^{3/2}}\right) + \frac{V^2 k^2 r}{h} - \frac{k^2 r}{h} v^2. \tag{12}$$

Together with the equation $\mathrm{d}r/\mathrm{d}t = v$ this defines the dynamical system in the (r, v) phase plane.

To understand the behaviour of the nanodust orbits qualitatively it is enough to use an even simpler model. Formally, it is obtained by assuming that $kr \ll 1$. Although this approximation is not necessary we use it here because it makes all expressions simpler. The equations for $\mathrm{d}v/\mathrm{d}t$ and $\mathrm{d}r/\mathrm{d}t$ simplify to

$$\dot{r} = v \tag{13}$$

$$\dot{v} = W(r) - k^2 r v^2, \tag{14}$$

where the dots denote time differentiation and

$$W(r) = -\frac{GM_S(1-\beta)}{r^2} + \frac{2\mu \tilde{B}_r \bar{r}^2}{r^3} + V^2 k^2 r. \tag{15}$$

The three terms in $W(r)$ correspond to the gravity force, the magnetic mirror force and the centrifugal force, respectively. The discussion in this section is based on the system defined by (13)–(15). We refer to it as the "2-D trapping model".

In the (r,v) phase plane the null isoclines (the lines along which one of the derivatives $\dot{r}$ or $\dot{v}$ is zero) can be found explicitly. The $\dot{r} = 0$ isocline is the r axis. The $\dot{v} = 0$ isocline is given by

$$v = \pm (W(r)/k^2 r)^{1/2}. \tag{16}$$

The $\dot{r} = 0$ and $\dot{v} = 0$ null isoclines cross each other at the fixed points: the points at which both $\dot{r} = 0$ and $\dot{v} = 0$ simultaneously hold. Since one of the null isoclines is the $v = 0$ line, the fixed points lie on the r axis. Also, from (14) it follows that the fixed points are given by $W(r) = 0$. $W(r)$ can be rewritten as

$$W(r) = \frac{GM_S(1-\beta)}{r^2}\left[-1 + \frac{r_2}{r} + \left(\frac{r}{r_1}\right)^3\right], \tag{17}$$

where

$$r_1^3 = \frac{GM_S(1-\beta)}{V^2 k^2} \tag{18}$$

$$r_2 = \frac{2\mu \tilde{B}_r \bar{r}^2}{GM_S(1-\beta)} \tag{19}$$

The fixed point positions on the r axis are given by the roots of $W(r) = 0$. The real roots exist provided $r_1 > (4^{4/3}/3) r_2$. Assuming that $r_1 \gg r_2$, the approximate positions of the fixed points are $r = r_1$ and $r = r_2$ (see Fig. 8). Note that $r = r_1$ is

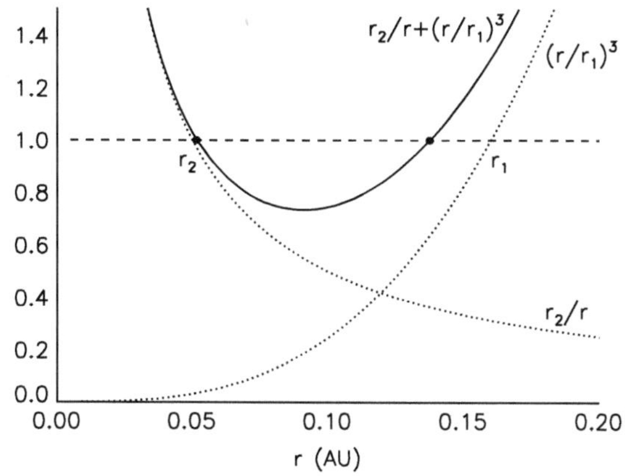

Fig. 8 Solution of the equation $W = 0$ for the fixed points (the values of r at which the velocity $\dot{r}$ and radial acceleration $\dot{v}$ vanish simultaneously). True fixed points (marked by *dots*) are the solutions of $r_2/r + (r/r_1)^3 = 1$ where r_1 and r_2 are given by (18) and (19). The approximate fixed points r_1 and r_2 are the solutions of $(r/r_1)^3 = 1$ and $r_2/r = 1$, respectively

the point at which the gravity force balances the centrifugal force, and $r = r_2$ the point at which the magnetic mirror force is in balance with the gravity force.

The behaviour of the phase plane trajectories near the fixed points can be found by linearizing (13) and (14). The outer fixed point is a saddle point (one eigenvalue positive, one negative) and the inner fixed point is a node with both eigenvalues purely imaginary. That is, the trajectories go around and do not enter the inner fixed point. In Czechowski and Mann (2010) the inner point was wrongly identified as a spiral sink because the equation corresponding to (14) contained a term linear in v. When the correct relation between $v_{\|}$ and the radial velocity is used, this linear term is cancelled (Czechowski and Mann 2011).

Figures 9 and 10 show parts of the phase plane of the 2-D trapping model. The null isoclines are shown (the dashed lines): note that the $\dot{v} = 0$ isocline has two disjoint parts. Also shown are the trapped trajectories following from the 2-D trapping model (solid lines) and calculated numerically using the full equation of motion (dotted lines). The latter include also the effect of the Poynting–Robertson force, which is absent from the 2-D trapping model.

The 2-D trapping model provides a simple way to estimate the extent of the trapping region. The results are consistent with the behaviour of the nanodust trajectories calculated using the complete equations of motion.

First thing to note is that the trapping region boundary in space depends on the value of the adiabatic invariant μ. That is, the point of release does not suffice to determine whether the particle is trapped: the answer also depends on the initial velocity.

However, as long as $r_2 \ll r_1$, the outer fixed point position (and the outer boundary of the trapping region) is well approximated by r_1. Taking in (18) $\beta = 0.1$ and $V = 400\,\text{km/s}$ gives $r_1 \approx 0.16\,\text{AU}$ in the ecliptic plane (co-latitude $\theta = \pi/2$)

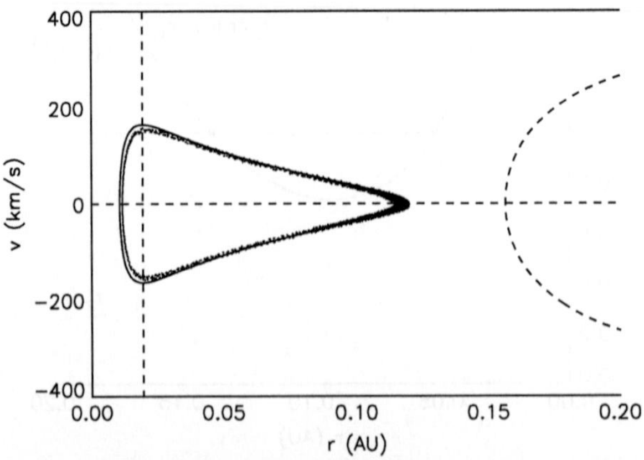

Fig. 9 Phase space (r, v) of the two-dimensional trapping model. The $\dot{r} = 0$ isocline is identical with the r axis ($v = 0$ line). The $\dot{v} = 0$ line has two branches. The fixed points occur where the $\dot{r} = 0$ and $\dot{v} = 0$ isoclines (the *dashed lines*) cross each other. Also shown is the trajectory of the nanodust particle released at 0.12 AU from the Sun from a circular Keplerian orbit near the ecliptic. The *solid line* is the orbit calculated in the 2-D trapping model. The *dotted line* is the numerical solution of the full equation of motion for $Q/m = 10^{-4} e/m_p$

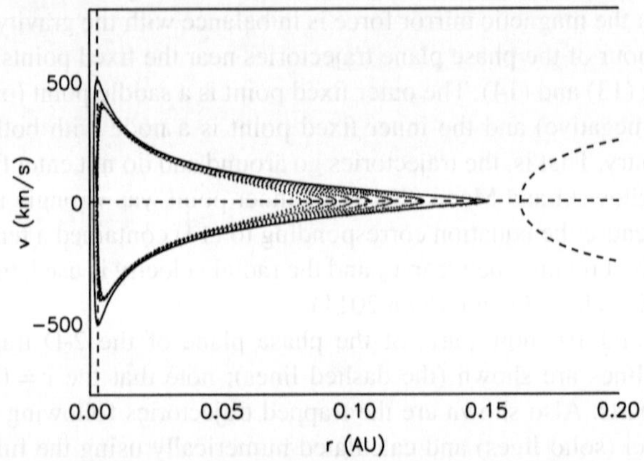

Fig. 10 As Fig. 9 but for the release at 0.15 AU from the Sun. The solution of the full equation of motion differs from that of the 2-D trapping model due to effect of the Poynting-Robertson force. This effect is much larger than in Fig. 9 because the perihelion of the orbit in the present case is very close to the Sun

and the co-latitude dependence $r_1 \sim \sin\theta^{-2/3}$, so that the trapping region is wider for the particle released at high latitudes from inclined orbits. This agrees with the results from calculations based on full equations of motion for the nanodust particles with $Q/m = 10^{-4} e/m_p$ released from inclined circular orbits at the point of maximum latitude (Fig. 11). For the grains with lower Q/m the boundary of the trapping region near the ecliptic is still well approximated by r_1, although at

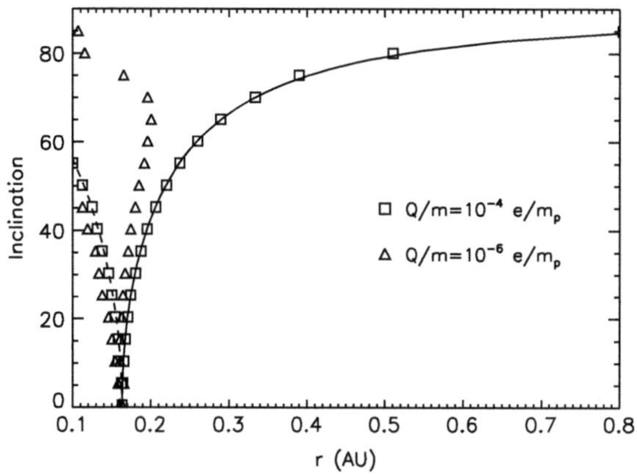

Fig. 11 Outer limit of the trapped region for nanodust as a function of the parent body orbital inclination. The curves show the results from the trapping model. The *solid curve* is the limit for particles released at the point of the orbit with maximum ecliptic latitude. *Dashed curve* is for particles released at the point where the orbit crosses the ecliptic. The limits derived from the nanodust trajectories calculated using the full equations of motion are shown for comparison (*squares*: $Q/m = 10^{-4}e/m_{\mathrm{p}}$, *triangles*: $Q/m = 10^{-6}e/m_{\mathrm{p}}$)

higher latitudes the trapped region is less wide than follows from the $r_1 \sim \sin\theta^{-2/3}$ behaviour.

If a particle is released from the orbit with the inclination α at the point where the orbit is passing through the ecliptic plane, the limits of the trapped region depend significantly on the value of μ. The effect can be deduced with help of Fig. 8. The initial $|v_\perp|'$ and therefore the adiabatic invariant μ increase with the inclination of the initial orbit, so that r_2 (but not r_1) increases and so does the sum $r_2/r + (r/r_1)^3$ (the solid line in Fig. 8). The true outer fixed point moves inwards, so that the trapped region contracts with increasing α. This is also confirmed by the behaviour of the orbits following from the full equations of motion (Fig. 11).

At $r_2 = (3/4^{4/3})r_1$ the two fixed points coincide. For $r_2 > (3/4^{4/3})r_1$ the trapped region is absent.

The position of the inner fixed point can be used to estimate the upper limit on the perihelion distance of the trapped orbit. Let the nanodust particle be created inside the trapping region, at the initial distance $r_0 < r_1$ from the Sun, with the initial velocity corresponding to Keplerian orbit with low inclination. From the behaviour of the adiabatic invariant it then follows that the distance of the inner fixed point from the Sun behaves as in Fig. 12: as long as r_0 is larger than ~0.06 AU, the inner fixed point is moving outwards when r_0 decreases. The calculations using the full equations of motion agree with this result (Fig. 12): the lowest value of the perihelion distance occurs for the particles created near the outer boundary of the trapping region.

Because the perihelion distance of the trapped nanodust is very small (few solar radii) most of these particles will be destroyed by sublimation during the first orbit

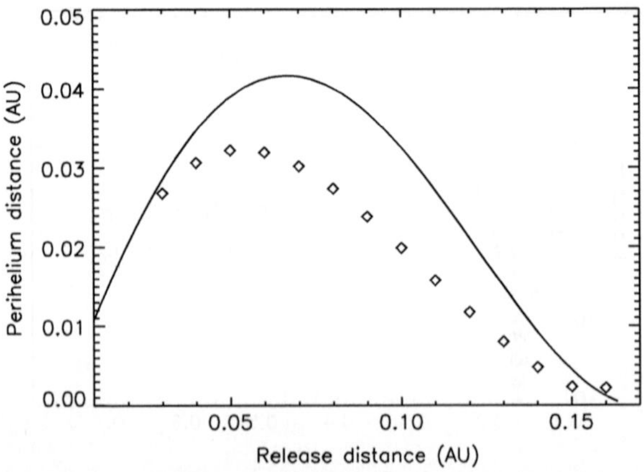

Fig. 12 The approximate position r_2 (*solid line*) of the inner fixed point as a function of the initial heliocentric distance for a trapped particle near the ecliptic. The *diamonds* show the calculated perihelion distances during the first orbit calculated from the full equations of motion for $Q/m = 10^{-4}e/m_{\rm p}$ nanodust. According to the phase space model (Fig. 9), r_2 should provide an upper limit for the perihelion distance

(Czechowski and Mann 2010). This conclusion was based on the sublimation rates considered by Krivov et al. (1998) and Mann et al. (2004).

7 Nanodust at 1 AU

The nanodust originating outside the trapping region becomes accelerated and escapes away from the Sun. This population may possibly explain the STEREO plasma wave measurements near Earth's orbit (Meyer-Vernet and Zaslavsky 2012).

We use our calculations of the nanodust trajectories to estimate the velocity distributions of particles in this dust population arriving at 1 AU. This requires calculating the trajectories for an appropriate set of initial conditions (positions and velocities). Also, it is necessary to know the probability that a nanodust particle is produced with these initial conditions.

We assume that the nanodust originates from collisional fragmentation of larger dust grains that form the circumsolar dust cloud. The production rate of nanodust depends therefore on the dust collision rate and the distribution of fragments created in a collision.

The main population of the grains of the circumsolar cloud with sizes larger than several micrometer forms a disk confined to the vicinity of the ecliptic plane. There may also be a halo extending to large ecliptic latitudes. The number density of the dust $n_{\rm d}(r)$ increases towards the Sun and this increase is probably faster than $n_{\rm d} \sim 1/r$ (which would follow from the equilibrium with respect to the Poynting–Robertson inward drift). The collision rate is proportional to $n_{\rm d}^2$ times the average

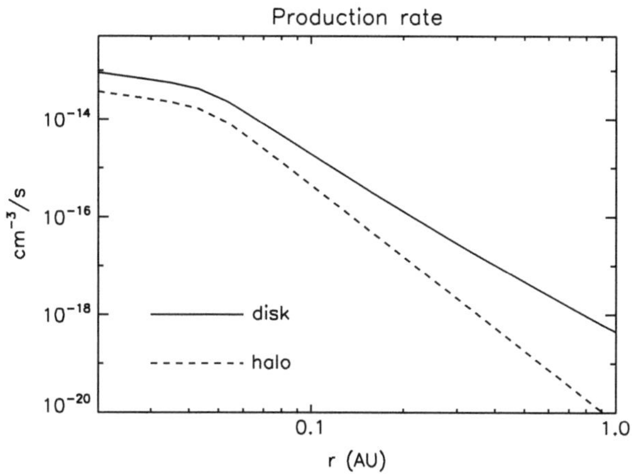

Fig. 13 Assumed production rates ρ_d (*disk*) and ρ_h (*halo*) of nanodust in the 1–5-nm size range as a function of distance from the Sun. This figure is adapted from Czechowski and Mann (2010)

relative orbital velocity of the grains v_{rel}. Assuming that $v_{rel} \sim r^{-1/2}$ the collision rate would then increase faster than $r^{-5/2}$. The nanodust production rate would increase even faster, because the mass of fragments following from a collision, for the same values of target and projectile mass, grows with the collision velocity. In result the nanodust production would be concentrated in the vicinity of the Sun.

Figure 13 shows the spatial distribution of the nanodust production rate assumed in our calculations (Czechowski and Mann 2010). The model is based on the fragmentation rates used in Mann and Czechowski (2005). The "disk" production rate (solid line) applies only in the region within ±13.5° from the ecliptic. The "halo" production rate (dashed line) is applicable outside the disk.

The initial velocities of the nanodust particles correspond to Keplerian circular orbits of their parent bodies. The starting point on the orbit is parametrized by the angle ϕ, and the orbit by its inclination ϵ and the longitude of the ascending node ψ. A grid of initial values of ϕ, ϵ, ψ and r is selected and the trajectories calculated. The production rates $\rho_d(r)$ and $\rho_h(r)$ define the weights for calculating the averages over the ensemble. The formulae used for calculating the averages can be found in Czechowski and Mann (2010). A sample of trajectories originating in the region with high production rate is shown in Fig. 14.

The results are presented in Table 1. Each row corresponds to an ensemble of 9016 particles. Listed are the velocity components: v_r, v_ϕ, v_θ (v_θ is defined to be positive when directed towards the solar equator) and the total velocity v averaged over the ensemble. The values of the corresponding dispersion $((< v^2 > - < v >^2)^{1/2}$ etc.) are given after the ± signs. The fraction of the "escaping" nanodust particles (those that reach the distance 0.9 AU from the Sun) is given by f_{esc} and f_{obs} is the fraction of the "escaping" particles that arrives at 0.9 AU close to the ecliptic plane (within ±13° from the solar equator).

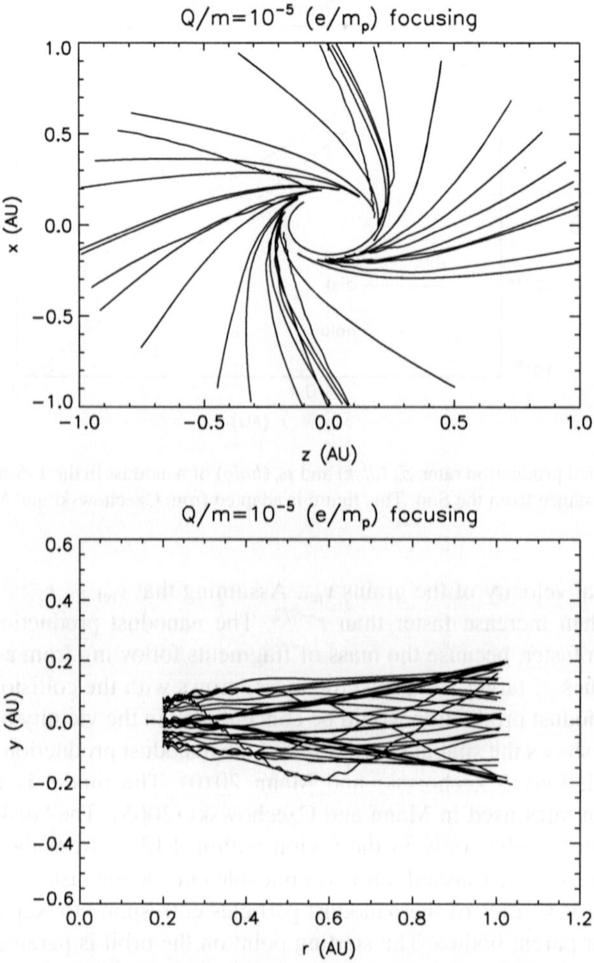

Fig. 14 Selected trajectories from the ensemble used to derive the velocity distributions at 1 AU projected onto the ecliptic plane (z, x) and onto the (r, y) plane (y is perpendicular to the ecliptic). The plots include only the trajectories starting within 0.25 AU from the Sun

For nanodust with $Q/m = 10^{-4}$–$10^{-5} e/m_p$) the results are as follows:

- The escaping grains fraction f_{esc} is about 0.16: that is, most of the produced nanodust is trapped.
- Most of the escaping grains in this size range (45–88%) arrives at 1 AU within 13° from the ecliptic plane.
- Average radial velocity component at 1 AU is 238–273 km/s, about twice the azimuthal component (104–135 km/s). The deviation from radial motion is in the prograde direction. The average latitudinal component is much smaller, but with a large spread.

Table 1 Velocity averages and dispersions at 1 AU

Q/m (e/m_p)	a (nm)	f_{esc}	f_{obs}	$< v_r >$ (km/s)	$< v_\phi >$ (km/s)	$< v_\theta >$ (km/s)	$< v >$ (km/s)
	Tilt = 20°	Focusing					
5×10^{-4}	1.5	0.42	0.99	276 ± 40	140 ± 52	0.24 ± 56	316 ± 61
10^{-4}	3	0.16	0.88	272 ± 39	134 ± 46	2.4 ± 57	308 ± 58
10^{-5}	10	0.16	0.6	238 ± 46	104 ± 28	-22 ± 64	269 ± 48
2×10^{-6}	20	0.16	0.33	140 ± 33	53 ± 11	1.7 ± 88	174 ± 29
4×10^{-7}	50	0.2	0.42	66 ± 21	26 ± 4	-0.6 ± 50	87 ± 22
	Tilt = 20°	Antifocusing					
10^{-4}	3	0.16	0.87	273 ± 37	135 ± 46	7 ± 57	310 ± 57
10^{-5}	10	0.16	0.45	257 ± 20	114 ± 18	-36 ± 43	287 ± 24
2×10^{-6}	20	0.17	0.003	86 ± 42	60 ± 8	-106 ± 75	170 ± 27

- The effect of magnetic field orientation (focusing vs. nonfocusing) on the distributions at 1 AU is small in this size range. For larger grains ($Q/m = 2\ 10^{-6} e/m_p$) the effect becomes strong: in particular, for the non-focusing field case only 0.3% of the grains arrives at 1 AU close to the ecliptic plane.

The small value of f_{esc} suggests that the flux of the escaping nanodust may be strongly variable: a small change in the outer boundary of the trapping region (caused possibly by a transient change in magnetic field) may lead to a large change in the flux of escaping grains.

The assumed nanodust production rates together with the calculated fractions of escaping and observed grains can be used for an estimation of the flux of nanodust arriving at 1 AU. This was attempted in Czechowski and Mann (2010) but the result was too high by about two orders of magnitude due to an error in calculation. The uncertainty in the flux estimation from the model is high (orders of magnitude) since the value of the nanodust production rate is not well known. A further discussion is given in Mann and Czechowski (2012).

8 Interaction with the Heliospheric Current Sheet

The heliospheric current sheet is the boundary between the sectors with different orientation of the solar magnetic field. Observations by the spacecraft show that its thickness (near 1 AU from the Sun) is of the order few 10^3–10^4 km (Winterhalter et al. 1994; Zhou et al. 2005), that is much less than typical Larmor radii for the nanodust.

The main effect of the heliospheric current sheet on the dynamics of nanodust is that it causes the charged particles that encounter the sheet to drift along its surface in the direction perpendicular to the magnetic field. The mechanism of the drift is illustrated by Fig. 15: it follows from the fact that the direction of Larmor rotation is reversed on crossing the sheet. A similar drift is known to be important for transport of the cosmic ray ions in the heliosphere (Jokipii et al. 1977; Kota 1979; Jokipii and Thomas 1981).

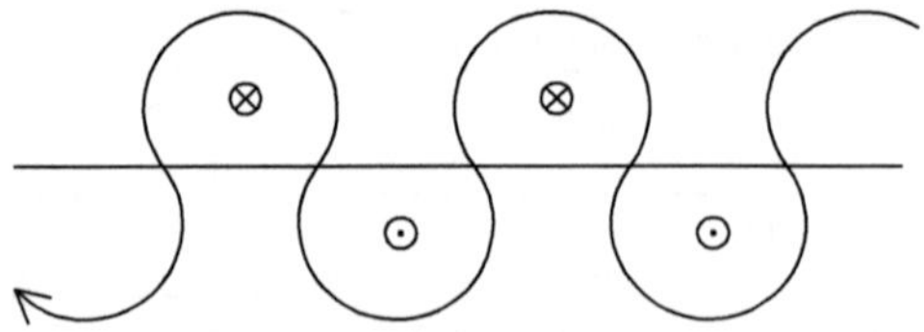

Fig. 15 Drift along the current sheet. On crossing the sheet the magnetic field changes direction causing the corresponding change in the direction of Larmor rotation

To interact with the current sheet the nanodust particles must encounter it. Our calculations of the nanodust trajectories inward from 1 AU (see the previous section) show that, in the focusing field case, of about 5,000 trajectories corresponding to the escaping particles, 2,494 trajectories for the $Q/m = 10^{-5}e/m_{\mathrm{p}}$ grains and only 296 trajectories for for the $Q/m = 10^{-4}e/m_{\mathrm{p}}$ grains encounter the current sheet.

If a particle would be strictly tied to a single magnetic field line carried by the plasma flow, it could encounter the current sheet only if the magnetic field line would pass within the Larmor radius distance from the sheet surface. In fact, the particle motion includes also a drift (the $\mathbf{V}_{\mathrm{D}}$ term in the guiding centre approximation, see (9)) that takes it away from the magnetic field line. The effect of the heliospheric current sheet on the nanodust motion depends on this drift magnitude.

We find that a convenient tool to study both the drift along the heliospheric current sheet and the drift away from the magnetic field line is provided by the "footpoint projection". This consists of projecting any point ($\mathbf{r}$,t) on the particle trajectory to the footpoint (on the solar surface) of the magnetic field line that crosses the trajectory at ($\mathbf{r}$,t). In this projection only the motion away from the magnetic field line is visualized: the motion along the field line corresponds to the same footpoint. Since in our model the shape of the neutral line that generates the current sheet does not change with time (fixed tilt) the projection of the heliospheric current sheet at any time is given by the same neutral line.

Figure 16 shows examples of the $Q/m = 10^{-4}e/m_{\mathrm{p}}$ and $10^{-5}e/m_{\mathrm{p}}$ nanodust trajectories in the footpoint projection. One thing to note is that the drift away from the magnetic field line takes the particles towards the current sheet if the field is in focusing configuration (trajectories 3 and 4), and away from it if the field is defocusing (5 and 6). This explains why the effect of the heliospheric current sheet is absent for the defocusing field. The signs of the drift can be explicitly checked to agree with the formula for $\mathbf{V}_{\mathrm{D}}$ (9) following from the guiding centre approximation.

The drift along the heliospheric current sheet particles reduces the effect of acceleration for nanodust particles within 1 AU. As discussed in Sect. 5, the accelerated particle velocity has the largest component along $\mathbf{V}_{\perp}$, the perpendicular component of the solar wind velocity at which the magnetic field lines are convected. The drift along the heliospheric current sheet has the opposite direction (Fig. 16), so that the total particle velocity is reduced by the drift.

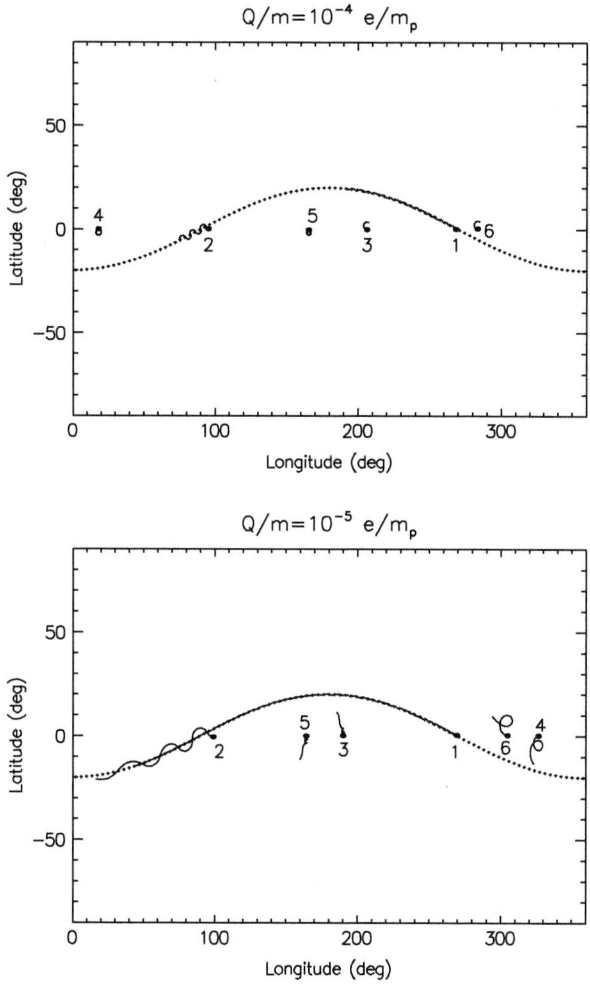

Fig. 16 Selected trajectories of the $Q/m = 10^{-4}e/m_{\rm p}$ and $10^{-5}e/m_{\rm p}$ grains projected to the footpoints of the magnetic field lines. The initial points are marked by the *filled circles*. The *dotted line* shows the projection of the heliospheric current sheet. The Sun rotation direction is towards the right. Illustrated are: drift along the heliospheric current sheet (1, 2); drift in the focusing field case (3, 4); drift for the defocusing field (5, 6). The drift is towards the current sheet for the focusing and away from it for the defocusing field case. The selected $Q/m = 10^{-4}e/m_{\rm p}$ trajectories start at larger initial distances from the Sun than the $Q/m = 10^{-5}e/m_{\rm p}$ ones to make the Larmor rotation more visible

9 Nanodust from Planets

In addition to nanodust production from collision between dust grains in the circumsolar cloud, the nanodust can be also released from the planets. The magnetospheres of giant planets can accelerate the nanodust to large velocities, as observed in the vicinity of Jupiter (Hamilton and Burns 1993; Horányi et al. 1993; Grün et al. 1994; Krüger et al. 2006; Meyer-Vernet et al. 2009).

Our model is not applicable inside or near the magnetospheres, but it can be used to investigate the motion of the nanodust after it leaves the magnetosphere and its vicinity.

Consider the case of the nanodust emitted from the Jupiter magnetosphere. The results of calculation of the trajectories for a sample of nanodust particles with the initial velocity assumed to be 400 km/s and directed radially away from the planet are shown in Figs. 17 and 18. In the defocusing configuration the nanodust is swept by the solar wind in the direction away from the Sun. On the other hand, for the focusing field configuration, the nanodust of $Q/m = 10^{-5}e/m_p$ emitted from Jupiter can reach the orbit of the Earth. This is possible because of the drift along the heliospheric current sheet.

The flux of the nanodust from Jupiter at the Earth orbit would probably be much less than the flux observed by STEREO. From the Ulysses measurements Krüger et al. (2006) find that at the distance 4 AU from Jupiter the flux of stream particles is well below 0.01 $m^{-2}\,s^{-1}$. Assuming their suggestion of the $1/r^2$ behaviour of the flux it then follows that the flux of Jupiter stream particles at the Earth orbit would be about two orders of magnitude below the STEREO measurements.

A question of interest is what would happen to the nanodust emitted from the Earth magnetosphere. Since the orbits of STEREO spacecraft pass close to the Earth orbit, it is important to find out if the nanodust from the Earth can contribute to STEREO observations. Figure 19 shows the results for the case of two samples of 400 nanodust particles with $Q/m = 10^{-4}e/m_p$ and $10^{-5}e/m_p$, respectively, starting at the initial speed 20 km/s relative to Earth, for the focusing and the non-focusing magnetic field orientations. The initial velocities are uniformly distributed over the sphere of radial directions away from the Earth. The plot shows the positions of the grains relative to Earth at the moment of time when the particles reach the distance 1.1 AU from the Sun. We use the R,T,N system of coordinates with the axes ($\hat{e}_R$, $\hat{e}_T$, $\hat{e}_N$) where $\hat{e}_R$ is directed along the Sun–Earth line, $\hat{e}_T$ in the direction of the Earth orbital motion and $\hat{e}_N = \hat{e}_R \times \hat{e}_T$. In Fig. 19 R_T and R_N denote the components of the particle positions along the $\hat{e}_T$, $\hat{e}_N$ axes. It can be seen that after being picked up by the solar wind the particles form narrow beams which bypass the ecliptic plane. The conclusion is that they can easily be missed by the spacecraft near the Earth orbit.

For the focusing configuration the beams move into the northern hemisphere (R_N >0) and for the non-focusing case into the southern hemisphere (R_N <0). None of the trajectories in the samples encounter the current sheet.

10 Summary and Discussion

In this chapter a model of the solar wind and the solar magnetic field was used as a basis for a study of the dynamics of the nanodust particles.

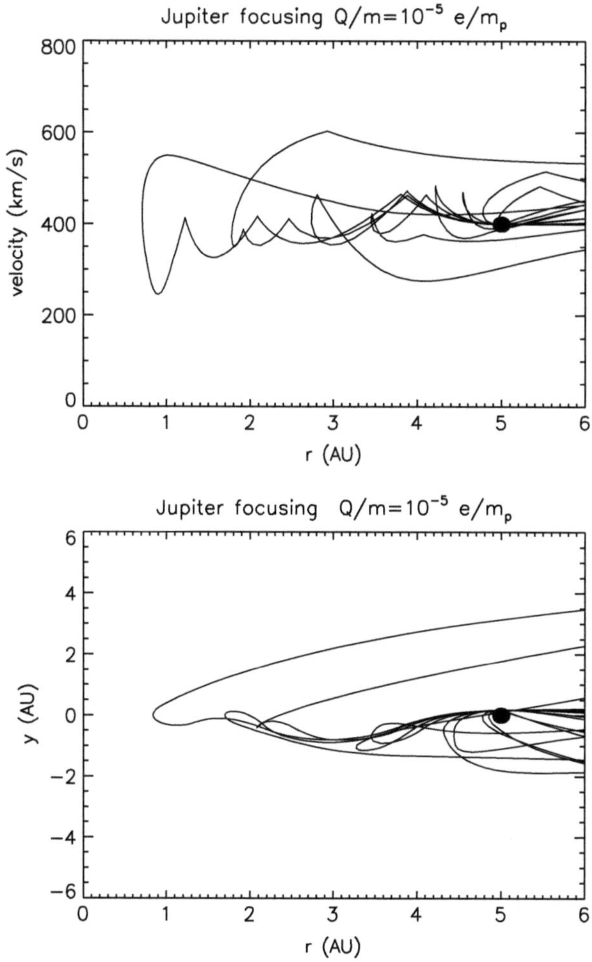

Fig. 17 The trajectories of a sample of $Q/m = 10^{-5}$ e/m_{p} nanodust emitted from Jupiter magnetosphere at the radially directed initial velocity of 400 km/s relative to planet. The plots show the velocity and the vertical coordinate y (perpendicular to ecliptic) as functions of the distance from the Sun. Solar magnetic field is in focusing configuration. Jupiter position is shown by the *filled circle*. Some of the trajectories can be seen to reach 1 AU with the corresponding velocity profiles showing the effect of current sheet encounters

The main results are:

- There is a trapping region in the vicinity of the Sun: the nanodust particles created inside this region are trapped in bound, non-Keplerian orbits.
- The nanodust particles created outside this region become accelerated to the velocities comparable to that of the solar wind and escape away from the Sun.

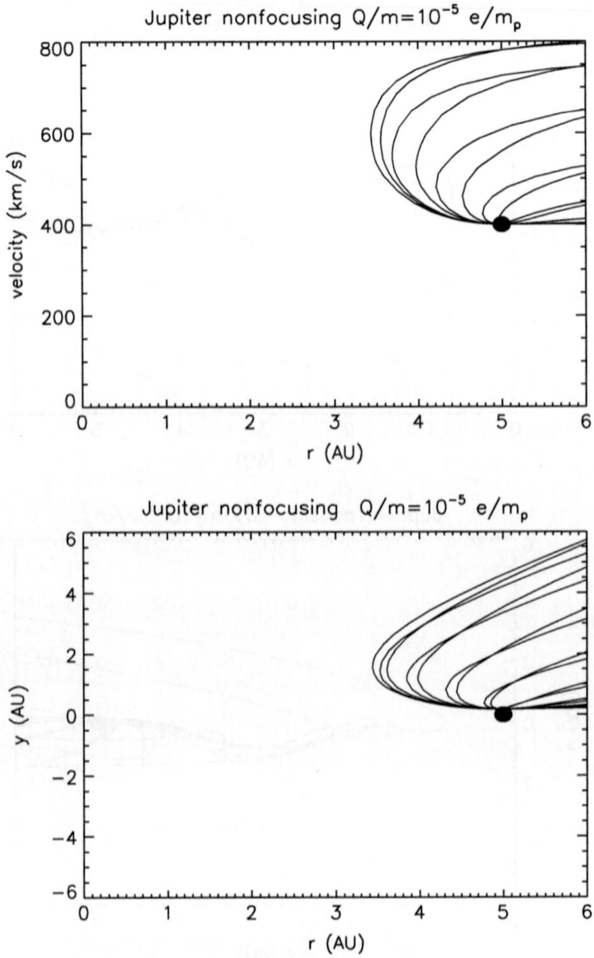

Fig. 18 As Fig. 17 but for defocusing field. All particles move away from the Sun

The phenomenon of trapping can be qualitatively (and quantitatively as well) understood in terms of a two-dimensional phase space model derived from the guiding centre approximation. The guiding centre approximation is also useful to understand the dynamics of the escaping particles, even though it is not strictly applicable.

The model implies that the average velocity of the nanodust arriving at 1 AU is of the order of 300 km/s, with the radial component of ∼270 km/s and the azimuthal ∼130 km/s. The deviation from the radial motion is in the prograde direction. The average latitudinal component of the nanodust velocity at 1 AU is small (few km/s).

The heliospheric current sheet has a significant effect on the nanodust motion if the magnetic field orientation is "focusing" (incoming field in the northern

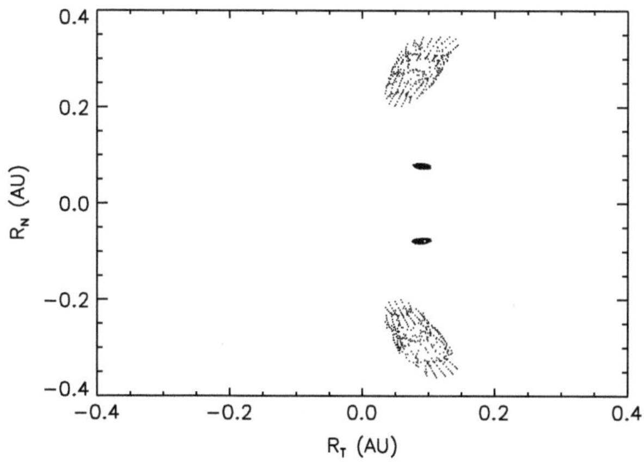

Fig. 19 Calculated positions in the (T,N) plane of four samples ($Q/m = 10^{-4}e/m_p$ and $10^{-5}e/m_p$, focusing and non-focusing field) of 400 nanodust particles emitted from the Earth in different radial directions at the speed 20 km/s. The positions are shown at the moment of time when the particles reach 1.1 AU from the Sun. The trajectories form narrow beams represented by the groups of dots (compact for $Q/m = 10^{-4}e/m_p$, loose for $Q/m = 10^{-5}e/m_p$). In the focusing configuration the beams move north ($R_N > 0$) and in the non-focusing case south ($R_N < 0$)

hemisphere). The drift along the current sheet reduces the effect of acceleration. It can also facilitate penetration of the nanodust grains emitted from the Jovian magnetosphere to the vicinity of the Earth orbit. A related effect applies to the interstellar grains penetrating into the heliosphere through the region of wavy heliospheric current sheet (Czechowski and Mann 2003; Landgraf 2000; Landgraf et al. 2003; Krüger et al. 2007).

A simplification used in our calculation is taking the charge-to-mass ratio Q/m to be constant during the motion. This means that we have disregarded two effects: the change in the equilibrium charge during the motion and the charge fluctuations. For the region within 1 AU these effects should not change significantly our results (see Czechowski and Mann 2010): the equilibrium charge does not vary much with distance, while the frequency of charge oscillations is much less than the Larmor rotation frequency even for the very small (1 nm) particles so that the effect of charge oscillations approximately averages out. Note also that the values of Q/m that we use correspond to $Q \sim 17e$ for 3 nm and $Q \sim 63e$ for 10 nm, so that random charge fluctuations are unlikely to change Q/m by a large factor. The effect on particle dynamics may be still non-negligible. Stochastic charging of dust in planetary rings was considered by Schaffer and Burns (1995).

The radiation pressure to gravity ratio β was set to 0.1 throughout this chapter. For nanometre size particles the value of β is not known. In this chapter we use the same value also for larger particles ($Q/m = 10^{-7}e/m_p$) in order to compare the effects of the Lorentz force for different Q/m.

The actual value of β depends on the light scattering properties of the grains. From theoretical calculations assuming typical compositions of interplanetary dust it follows that, for the grain masses $>10^{-11}$ kg, the value of β is <0.1 and behaves as $1/a$ ($a =$ radius of the grain). For smaller masses the β passes through a maximum between roughly 10^{-17} and 10^{-15} kg ($a \sim 0.1$–$0.5\,\mu$m). A model calculation shows that for compact particles of typical interplanetary dust composition the maximum values of β are between 0.8 and 1 (Wilck and Mann 1996). For further decreasing dust masses the value of β steeply drops down.

For particles with charge-to-mass ratio $Q/m = 10^{-7}e/m_{\mathrm{p}}$ the size is $\sim$0.1 μm and $\beta \sim 1$. These particles are therefore ejected outward by the radiation pressure force (β-meteoroids or β particles). The final velocities reached by the β-meteoroids are far below those reached by the nanodust particles with sizes 10 nm or less.

As a caveat to these considerations we point out that our knowledge of the composition and light scattering properties of the nanodust in the interplanetary dust cloud is incomplete. We also did not discuss the processes that lead to destruction of the nanodust that is trapped near the Sun.

Appendix: Pickup and Acceleration Model

The model is defined by the following assumptions:

1. The magnetic field is purely azimuthal: $\mathbf{B} = B(r)\hat{\mathbf{e}}_\phi$ with $B(r) \sim 1/r$.
2. The solar wind velocity is constant and radially directed: $\mathbf{V} = V\hat{\mathbf{e}}_{\mathrm{r}}$.
3. The gravity and radiation forces in (5) are neglected.
4. Only the solutions with $v_\phi = 0$ are considered.

The $\nabla \cdot \mathbf{B} = 0$ and $\nabla \times (\mathbf{V} \times \mathbf{B})$ (freezing-in) equations are satisfied in the model. Since there is no change in polarity between the northern and the southern hemispheres, the model should be only applied to the motion that can be restricted to one hemisphere.

The equations of motion following from (5) are

$$r\frac{\mathrm{d}v_{\mathrm{r}}}{\mathrm{d}t} = v_\theta^2 + (\bar{\Omega}_{\mathrm{L}}\bar{r})v_\theta \tag{20}$$
$$r\frac{\mathrm{d}v_\theta}{\mathrm{d}t} = -v_{\mathrm{r}}v_\theta - (\bar{\Omega}_{\mathrm{L}}\bar{r})(v_{\mathrm{r}} - V),$$

where $\bar{\Omega}_{\mathrm{L}}$ is the value of the Larmor frequency of the particle at the reference distance $\bar{r}$, so that $\bar{\Omega}_{\mathrm{L}}\bar{r}$ is a constant with the dimension of velocity. Two additional equations $\mathrm{d}r/\mathrm{d}t = v_{\mathrm{r}}$ and $\mathrm{d}\theta/\mathrm{d}t = v_\theta/r$ complete the system in the $(r, \theta, v_{\mathrm{r}}, v_\theta)$ phase space.

If we replace formally the time variable by a new variable τ such that $\mathrm{d}\tau = \mathrm{d}t/r(t)$, the system (20) can be regarded as a two-dimensional dynamical system in the (v_r,v_θ) phase plane:

$$\frac{dv_r}{d\tau} = v_\theta^2 + (\bar{\Omega}_L\bar{r})v_\theta \tag{21}$$
$$\frac{dv_\theta}{d\tau} = -v_r v_\theta - (\bar{\Omega}_L\bar{r})(v_r - V).$$

The trajectories in the (v_r,v_θ) plane following from (21) are also projections of the trajectories of the full system from the 4-D (r, θ, v_r, v_θ) phase space. We shall see that the analysis of the asymptotic $\tau \to \infty$ behaviour of the trajectories for the 2-D system in the (v_r,v_θ) plane provides also information about the $t \to \infty$ behaviour of the trajectories of the full 4-D system.

One trajectory in the (v_r,v_θ) plane is a projection of the set of 4-D trajectories with the same initial conditions for v_r, v_θ but different for r, θ. Consider the 2-D trajectory passing at initial $\tau = 0$ through the point (v_r^0,v_θ^0). Choose the initial values for r and θ to be r^0 and θ^0. The equation $dr/dt = v_r$ is equivalent to $d\log r/d\tau = v_r$, so that $r(\tau)$ can be found along the ($v_r(\tau)$,$v_\theta(\tau)$) trajectory of the 2-D system. The time t can then be expressed through τ and the full 4-D solution $r(t)$, $\theta(t)$, $v_r(t)$, $v_\theta(t)$ found.

Equation (21) have a single fixed point at $v_r = V$, $v_\theta = 0$. Expanding around this point and keeping only the first order terms in $\delta v_r = v_r - V$ and in $\delta v_\theta \equiv v_\theta$ lead to

$$\frac{d\delta v_r}{d\tau} = (\bar{\Omega}_L\bar{r})\delta v_\theta \tag{22}$$
$$\frac{d\delta v_\theta}{d\tau} = -(\bar{\Omega}_L\bar{r})\delta v_r - V\delta v_\theta.$$

We look for the solution in the form $(\delta v_r, \delta v_\theta) = (\delta v_r^0, \delta v_\theta^0)\exp\lambda\tau$. The eigenvalue λ is obtained from the consistency condition and is given by

$$\lambda_\pm = [-V \pm (V^2 - 4(\bar{\Omega}_L\bar{r})^2)^{1/2}]/2. \tag{23}$$

Since the real part of λ is negative, the fixed point is always attractive.

For $V < 2(\bar{\Omega}_L\bar{r})$ the solution near the fixed point is oscillatory (the square root in λ is imaginary), with the amplitude of the velocity oscillations decreasing as $\exp -V\tau/2$. Since at the fixed point $v_r = V$, r behaves as Vt and τ as $(1/V)\log t$. It follows that the amplitude of the velocity oscillations around $v_r = V$, $v_\theta = 0$ goes down as $t^{-1/2}$ or equivalently as $r^{-1/2}$. This result does not require that the guiding centre approximation holds: in fact, $V < 2(\bar{\Omega}_L\bar{r})$ is true also for $Q/m = 10^{-5}e/m_p$.

For $V \geq 2(\bar{\Omega}_L\bar{r})$ there is no velocity oscillation. The approach to the fixed point is described by a composition of two power-law terms in r.

Acknowledgements A.C. acknowledges support from Polish Ministry of Science grant NN 203 513 038. I.M.acknowledges support from the Belgian Solar Terrestrial Center of Excellence.

References

Consolmagno, G.J.: 1980, *Icarus* **43**, 203.
Czechowski, A., Mann, I.: 2003, *J. Geophys. Res.* **108**, A10, 13–1.
Czechowski, A., Mann, I.: 2010, *Astrophys. J.* **714**, 89.
Czechowski, A., Mann, I.: 2011, *Astrophys. J.* **732**, 127.
Grogan, K., Dermott, S.F., Gustafson, B.A.S.: 1996, *Astrophys. J.* **472**, 812.
Grün, E., Zook, H. A., Fechtig, H., Giese, R. H.: 1985, *Icarus* **62**, 244.
Grün, E., Zook, H. A., Baguhl, M., Balogh, A., Bame, S. J., Fechtig, H., Forsyth, R., Hanner, M. S., Horányi, M., Kissel, J., Lindblad, B.-A., Linkert, D., Linkert, G., Mann, I., McDonnell, J. A. M., Morfill, G. E., Phillips, J. L., Polanskey, C., Schwehm, G., Siddique, N., Staubach, P., Svestka, J., Taylor, A.: 1994, *Nature* **362**, 428.
Grün, E., Gustafson, B., Mann, I., Baguhl, M., Morfill, G. E., Staubach, P., Taylor, A., Zook, H.A.: 1993, *Astr. & Astrophys.* **286**, 915.
Gustafson, B.A.S., Lederer, S.M.: 1996, In: Gustafson B. A. S., Hanner M. S. (eds.) *Physics, Chemistry, and Dynamics of Interplanetary Dust, IAU Colloq. 150* Astron. Soc. Pacific Conf. Ser. **104** 35.
Hamilton, D.P., Burns, J.A.: 1993, *Nature* **364**, 695.
Hamilton, D. P., Grün, E., Baguhl, M.: 1996, In: Gustafson B. A. S., Hanner M. S. (eds.) *Physics, Chemistry, and Dynamics of Interplanetary Dust, IAU Colloq. 150* Astron. Soc. Pacific Conf. Ser. **104** 31.
Horányi, M., Morfill, G., Grün, E.: 1993, *J. Geophys. Res.* **98**, 21245.
Jokipii, J.R., Levy, E.H., Hubbard, W.B.: 1977, *Astrophys. J.* **213**, 861.
Jokipii, J.R., Thomas, B.: 1981, *Astrophys. J.* **243**, 1115.
Kimura, H., Mann, I.: 1998, *Astrophys. J.* **499**, 454.
Kota, J.: 1979, In: Miyake, S. (ed.) *Proc. 16th Int. Cosmic Ray Conference*, Kyoto, **3**, 13.
Krivov, A., Kimura, H., Mann, I.: 1998, *Icarus* **134**, 311.
Krüger, H., Graps, A. L., Hamilton, D. P., Flandes, A., Forsyth, R. J., Horányi, M., Grün, E.: 2006, *Planet. Space Sci.* **54**, 919.
Krüger, H., Landgraf, M., Altobelli, N., Grün, E.: 2007, *Space Sci. Revs.* **130**, 401.
Landgraf, M., Krüger, H., Altobelli, N., Grün, E.: 2003, *J. Geophys. Res.* **108**, LIS 1.
Landgraf, M.: 2000, *J. Geophys. Res.* **105**, 10303.
Luhmann, J. G.: 2003, *Astrophys. J.* **592**, 1241.
Mann, I., Czechowski, A.: 2005, *Astrophys. J. Lett.* **621**, L73.
Mann, I., Czechowski, A.: 2012, *Causes and Consequences of the Existence of Nanodust in Interplanetary Space*, Nanodust in the Solar System: Discoveries and Interpretations, **385**, 195.
Mann, I., Krivov, A., Kimura, H.: 2000, *Icarus* **146**, 568.
Mann, I., Kimura, H., Biesecker, D.A., Tsurutani, B.T., Grün, E., McKibben, R.B., Liou, J.C., MacQueen, R.M., Mukai, T., Guhathakurta, M., Lamy, P.: 2004, *Space Sci. Revs.* **110**, 269.
Mann, I., Murad, E., Czechowski, A.: 2007, *Planet. Space Sci.* **55**, 1000.
Meyer-Vernet, N. and Zaslavsky, A.: 2012, In Situ Detection of Interplanetary and Jovian Nanodust with Radio and Plasma Wave Instruments, *Nanodust in the Solar System: Discoveries and Interpretations*, **385**, 160.
Meyer-Vernet, N., Lecacheux, A., Kaiser, M. L., Gurnett, D. A.: 2009, *Geophys. Res. Lett.* **36**, 3103.
Morfill, G. E., Grün, E.: 1979, *Planet. Space Sci.* **27**, 1269.
Mukai, T.: 1981, *Astron. Astrophys.* **99**, 1.
Northrop, T.G.: 1958, *Adiabatic Motion of Charged Particles*, J. Wiley and Sons, New York.

Parker, E.N.: 1958, *Astrophys. J.* **128**, 664.
Schaffer, L., Burns, J.A.,: 1995, *J. Geophys. Res.* **100**, 213.
Schatten, K.H., Wilcox, J.M., Ness, N.F.: 1969, *Solar Phys.* **6**, 442.
Wallis, M.K, Hassan, M.H.A.: 1985, *Astron. Astrophys.* **151**, 435.
Wilck, M., Mann, I.: 1996, *Planet. Space Sci.* **44**, 493.
Winterhalter, D., Smith, E.J., Burton, M.E., Murphy, N., McComas, D.J.: 1994, *J. Geophys. Res.* **99**, A4, 6667.
Zhou, X.-Y., Smith, E.J., Winterhalter, D., McComas, D.J., Skoug, R.M., Goldstein, B.E., Smith, C.W.: 2005, In: Fleck, B., Zurbuchen, T.H., Lacoste, H. (eds.) *Proc. Solar Wind 11-SOHO 16 Conf.* **SP-592**, ESA Noordwijk, 659.

Dynamics, Composition, and Origin of Jovian and Saturnian Dust-Stream Particles

Hsiang-Wen Hsu, Harald Krüger, and Frank Postberg

Abstract Stream particles are nanometer-sized dust particles ejected from the Jovian and Saturnian systems with speeds higher than $100\,\mathrm{km\,s^{-1}}$. Due to the large charge-to-mass ratio, their dynamics is dominated by electromagnetic forces, which lead to the high-speed ejection from the planetary magnetospheres and the dust impact bursts reported from the in situ dust measurements (i.e., the "dust stream phenomena"). Following measurements with the in situ dust detectors on board the Ulysses, Galileo, and Cassini spacecrafts, we summarize the sources as well as the dynamical and compositional properties of these nanoparticles. Future stream particle studies should focus on the connection between dust, moons, and the magnetosphere.

1 Introduction

After decades of space exploration, our knowledge about the outer solar system has grown tremendously. Cosmic dust—the microscopic solid bodies—behaves different from neutral and ionized gas and the grains dynamically interact with

H.-W. Hsu (✉)
Laboratory for Atmospheric and Space Physics, University of Colorado, Boulder, Colorado, USA
e-mail: sean.hsu@lasp.colorado.edu

H. Krüger
Max-Planck-Institut für Sonnensystemforschung, 37191 Katlenburg-Lindau, Germany

Max-Planck-Institut für Kernphysik, 69117 Heidelberg, Germany
e-mail: krueger@mps.mpg.de

F. Postberg
Max-Planck-Insitut für Kernphysik, 69117 Heidelberg, Germany

Institut für Geowissenschaften, Ruprecht-Karls Universität, Heidelberg, Germany
e-mail: frank.postberg@mpi-hd.mpg.de

I. Mann et al. (eds.), *Nanodust in the Solar System: Discoveries and Interpretations*, Astrophysics and Space Science Library 385, DOI 10.1007/978-3-642-27543-2_5,

each other. The electric charges carried on dust grains are constantly modified by the ambient plasma which consequently influences the dust dynamics. Due to erosion, heavy dust particles may act as extensive mass reservoirs and serve as sources of gas/plasma in the system. Moreover, the solid particles deliver specific information about their sources that cannot be obtained by other means. Studying the connection between dust particles and their environments therefore provides a unique opportunity to apply and examine the knowledge of various research fields to derive a comprehensive understanding.

Among various dust populations, nanoparticles are of particular interest due to their high mobility and their interaction with the electromagnetic environment. In the outer solar system, the two largest planet systems are found to be sources of nanodust particles. This type of nanoparticles is conventionally called "stream particles." Jovian stream particles were discovered in 1992 with the in situ dust detector on board the Ulysses spacecraft (Grün et al. 1992b, 1993) and were later studied with the dust detectors on board the Galileo (Grün et al. 1992a, 1996b, 1998) and Cassini spacecraft (Graps et al. 2001b; Postberg et al. 2006; Hsu et al. 2010b). The existence of Saturnian stream particles was confirmed by Cassini measurements in 2004 (Kempf et al. 2005a). These discoveries created a new field of research due to the rising attention to not only their extreme properties but also the implications concerning the interactions between dust, magnetosphere, and moons in the Jovian and Saturnian systems.

In fact many common features can be found between the magnetospheres of Jupiter and Saturn. Firstly, their magnetic field polarities are opposite to the Earth's magnetic field (Bagenal 1992), and, second, both systems host an active, *outgassing* satellite. At Jupiter the magnetospheric plasma is governed by the volcanically active moon Io (Bagenal 2007), while the tiny icy satellite Enceladus plays a similar but milder role in the Saturnian system (Hansen et al. 2006; Delamere et al. 2007). Due to the fast rotation and the strong magnetic field, the newly produced plasma is picked up to corotation with the planet and jointly triggers the magnetospheric circulation with the outward transportation.

Still, many differences exist between these two nanoparticle-emitting systems. Regarding the dust dynamics, the gravitational force that bounds particles to the system depends on the mass of the planet. The electromagnetic field and the charging conditions that determine the Lorentz force acting on the particle are regulated by the planet's magnetic field, magnetosphere structure, and the plasma environment. The aforementioned differences should leave distinguishable imprints on the dynamical properties of nanoparticles originating from these two systems. Moreover, materials emitted by Io and Enceladus show very different compositions (Lellouch et al. 1996; Waite et al. 2006). The chemical and the physical properties of Jovian and Saturnian stream particles therefore provide a unique window for a comparative study of these two magnetospheres.

To complement the specific discussions of nanoparticles originating from the Jovian and Saturnian systems in the following sections, we first introduce the general physics concepts relevant to the study of dust-stream particles. The dust stream measurements performed with Galileo in the Jovian system were also reviewed by Krüger et al. (2004).

1.1 Nanoparticle Ejection from a Planetary Magnetosphere

Once a nanoparticle obtains an electric charge, its charge-to-mass ratio can reach 10^3 or even $10^4\,\mathrm{C\,kg^{-1}}$ (Grün et al. 1998; Hsu et al. 2010a). The Lorentz force on a nanoparticle hence exceeds the gravity and dominates its dynamics (Zook et al. 1996). As a consequence, dust particles within proper size and charge ranges can gain enough energy to overcome the planet's gravity and escape. Driven by the planet's rotation, the corotation electric field ($\mathbf{E}_{\mathrm{co}}$) results in the corotation of the magnetospheric plasma and has great influence on the dynamics of charged dust particles. Although other mechanisms may also contribute to the ejection of nanoparticles from a planet's magnetosphere; here, we consider the corotation electric field to provide the nanoparticles with the energy to overcome the planet's gravitational attraction, which ultimately leads to the particle ejection.

The basic dynamical properties of these dust particles, i.e., their sizes and ejection speeds, can be estimated by considering the energy conservation law (Morfill et al. 1980b; Horányi et al. 1993b; Kempf 2007). The kinetic energy of a particle ejected from a Keplerian orbit can be written as

$$\frac{1}{2}m_{\mathrm{d}}v_{\mathrm{ex}}^2 = -\frac{GM_{\mathrm{P}}m_{\mathrm{d}}}{2r_0} + \int_{r_0}^{r_{\mathrm{ms}}} Q_d \cdot E_{\mathrm{co}}\,\mathrm{d}r. \tag{1}$$

Equation 1 shows on its left hand side the kinetic energy of an escaping dust particle with mass m_{d} and ejection speed v_{ex}. On the right hand side, the first term is the total energy of a dust particle in Keplerian motion, where G is the gravitational constant and M_{P} is the mass of the planet. The second term is the energy that a particle, with charge Q_d, obtains from the corotation electric field. The corotation electric field can be expressed as $\mathbf{E}_{\mathrm{co}} = -\mathbf{V}_{\mathbf{co}}\times\mathbf{B}$, where $\mathbf{V}_{\mathbf{co}} = \Omega\times r$ is the corotation velocity and $\mathbf{B}$ is the planetary magnetic field. Ω is the planet's rotation rate, r is the distance to the planet, r_{ms} is the size of the magnetosphere (or the location where $\mathbf{E}_{\mathrm{co}}$ breaks down), and r_0 is the equatorial distance where a particle starts to be accelerated by $\mathbf{E}_{\mathrm{co}}$.

Assuming a dipole magnetic field with a rigid corotating plasma and constant grain charges, from (1) the ejection speed of particles located in the magnetic equatorial region can be expressed as

$$v_{\mathrm{ex}}^2 = 2GM_{\mathrm{P}}r_0^{-1}\left(L^\star - \frac{1}{2}\right), \tag{2}$$

where $L^\star = L(1 - r_0/r_{\mathrm{ms}})$ and

$$L = \frac{Q_d}{m_\mathrm{d}} \frac{\Omega |B_0| R_\mathrm{P}^3}{GM_\mathrm{P}} , \tag{3}$$

is the ratio between the electromagnetic and the gravitational forces (Hamilton 1993a). R_P and $\mathbf{B_0}$ are the planet's radius and the equatorial magnetic field strength, respectively.

Two stream-particle-formation criteria can be read from (2). First, only particles with $L^\star > 1/2$ can be ejected as stream particles, which leads to the second implication: the fate of these charged particles is determined by their electric sign. At Jupiter and Saturn, only positively charged particles can gain energy from the outward pointing $\mathbf{E}_\mathrm{co}$ to become stream particles. The first criterion gives the upper size limit of stream particles:

$$a_\mathrm{max}{}^2 = \frac{6\varepsilon_0 \phi_d}{\rho_\mathrm{d}} \frac{\Omega \mathbf{B}}{GM_\mathrm{P}} \left(1 - \frac{r_0}{r_\mathrm{ms}}\right) , \tag{4}$$

where $\epsilon_0 = 8.854 \times 10^{-12}$ A s/(V m) is the vacuum permittivity, ϕ_d and ρ_d are the dust surface potential and the dust density, respectively.

Particles whose gyroradii are too small will be trapped by the magnetic field and their motion mimic plasma particles which can be described by the guiding center approximation. The theoretical lower size limit of stream particles can therefore be derived by using the Alfv́en criterion (Alfvén and Fälthammar 1963; Morfill et al. 1980b),

$$r_\mathrm{g} \cdot |\frac{\nabla \mathbf{B}}{\mathbf{B}}| < 0.1 , \tag{5}$$

where r_g is the particle gyroradius. Assuming a dipolar magnetic field, outside the synchronous orbit, (5) gives the lower size limit of stream particles:

$$a_\mathrm{min}^2 < \frac{\epsilon_0 \phi_\mathrm{d} B_0}{10 \varrho_\mathrm{d} \Omega} \left(\frac{R_\mathrm{P}}{r_0}\right)^3 \left(1 - \sqrt{\frac{GM_s}{\Omega^2 r_0^3}}\right)^{-1} . \tag{6}$$

Estimates derived from (3), (4), and (6) depend not only on the properties of the host planet but also on the amount of charge carried by the dust particle. It is reasonable to assume a constant dust charge for deriving the ejection size limits. However, as the devil lies in the details, we should be aware of the dust charge variations and understand its effects on the nanodust dynamics.

1.2 Charging Processes

The charging of dust particles is determined both by the environment and the dust properties (Whipple 1981). In interplanetary space where the solar wind plasma density is low, dust particles are charged positively due to the domination of the photo-electron emission induced by solar UV photons. To counterbalance the positive surface potential, more electrons are attracted to the particle until the equilibrium potential is reached (i.e., the net current becomes zero). As both the UV photon flux and the solar wind electron density decrease with the square of the heliocentric distance, the dust potential remains around +5V in interplanetary space (Mukai 1981; Zook et al. 1996).

In a planetary magnetosphere, where the plasma density is much higher, the most important dust-charging current is due to thermal electron collection. Capturing plasma electrons charges dust particles to negative potentials. In the opposite, positive currents such as photo-electron current, ion collection currents and, more importantly, the secondary electron current induced by hot electrons ionize dust particles and balance the negative current.

The photo-ionization and the secondary emission rates depend on both the composition and the particle size. Compared with metal, insulating materials generally emit fewer photo-electrons but eject more secondary electrons (Horányi 1996a). Due to the limited electron penetration length, radiation-excited electrons carried by the smaller dust particles are located closer to the surface, which allows more photo-electrons / secondary electrons to be emitted (Chow et al. 1993). The dust photon-absorption efficiency is conditional on the grain size, thus the strength of the photo-electron current varies among dust grains having different sizes (Mukai 1981; Sodha et al. 2011).

A spherical dust particle with a radius a (in meters) has a charge that corresponds to a surface potential (ϕ, in volts) of

$$Q_d = 4\pi \epsilon_0 a \phi . \tag{7}$$

To first order (if any size-dependent effect is neglected), dust particles will reach the same equilibrium potential after some time in a given plasma environment. As the current strength is proportional to the dust's surface area, from (7) we know that the time for a particle to reach the equilibrium potential is inversely proportional to the grain size. In other words, the charging time of a 10-nm particle would be hundred times longer than that of a micron-sized particle. One of the consequences of the different charging times is the divergent charging potentials between small and large dust particles.

Since the current intensity depends on the charging history, as well as on the temporal and spacial variations of the plasma, the charge states among particles with different sizes could be different (Meyer-Vernet 1982). On top of that, considering the quantized nature of electric charges, a nanoparticle with an electric potential of a few volts carries only few elementary charges. Their charging hence cannot be described by continuous currents and the stochastic effect must be taken into account (Cui and Goree 1994; Hsu et al. 2011b). Nanodust thus might be charged to opposite polarities and behave differently from larger dust grains. Their charge variations consequently affect their dynamics and bring further complexity to the system (see Schaffer and Burns 1995).

Adopting the classic two-body system theory, the charging currents can be calculated by considering the motion of ambient plasma particles in a centripetal electric field generated by a charged dust particle (Mott-Smith and Langmuir 1926). This is the so-called Orbital-Motion-Limited theory (OML theory) and it applies at the condition: $a \ll \lambda_D \ll D_d$, where λ_D is the Debye length and D_d is the inter-particle distance. Except for the extreme high dust-density conditions (i.e., the complex plasma), the OML criteria are fulfilled in most space environments. With the background plasma model, the charging of dust particles as well as the charging of spacecraft in various environments can be calculated (Whipple 1981; Horányi 1996a; Kempf et al. 2006 and the aforementioned references for more details).

After being released into the magnetosphere, the dynamical evolution of the stream particles begins with obtaining charges from the ambient plasma. Due to the stochastic charging process, each charging and discharging event acts as an orbital maneuver. The varying Lorentz force spreads nanoparticles into the magnetosphere. To recap, only positively charged particles within the proper size range can be ejected from Jupiter and Saturn. The rest wander in the magnetosphere and may eventually be eroded by plasma sputtering (see Sect. 3.4) or deposited onto nearby satellites.

1.3 Stream Particles in Interplanetary Space

Subsequent to escape from a planetary magnetosphere, the dynamics of stream particles is governed by the interplanetary magnetic field (IMF) carried by the outward-flowing solar wind. The relative motion between stream particles and the solar-wind plasma induces a co-moving electric field, $\mathbf{E}_c = (\mathbf{v_d} - \mathbf{v_{sw}}) \times \mathbf{B_{IMF}}$, where $\mathbf{v_d}$ and $\mathbf{v_{sw}}$ are the dust and the solar wind velocity vectors, and $\mathbf{B_{IMF}}$ is the IMF vector. Within the ecliptic plane, the direction of $\mathbf{E}_c$ predominantly points northward or southward (with respect to the ecliptic plane). As the direction and strength of $\mathbf{E}_c$ depend on IMF conditions, the stream particle flux varies with the solar wind structure (Hamilton and Burns 1993b). Solar-wind structures such as corotating interaction regions (CIRs) and probably coronal mass ejections (CMEs) are found to strongly influence stream particle dynamics (Kempf et al. 2005a; Flandes and Krüger 2007; Hsu et al. 2010a; Flandes et al. 2011). Within the enhanced IMF regions associated with CIRs, the IMF strength can increase by a factor of ten, and its direction flip by 180° twice during each solar rotation (Balogh et al. 1999; Crooker et al. 1999; Jackman et al. 2004). Compared to IMF variations, dust charge fluctuations play only a minor role in the stream-particle dynamics.

More details about observations carried out by the Ulysses, Galileo, and Cassini spacecrafts are presented in the following sections. The coupling of the solar wind structure and the stream particle detection pattern in the interplanetary space makes these nanoparticles distinctive from other dust populations. The stream particle measurements in interplanetary space provide not only important constraints on the dust dynamics, together with the composition information, but also specific information about their formation conditions (the "Dust Astronomy" concept, Grün et al. 2001b).

2 Dust Escaping from Jupiter's Magnetosphere

2.1 Jovian Dust Streams' Discovery and Grain Dynamics in Interplanetary Space

On 10 March 1992, about one month after the flyby of Ulysses at Jupiter, perhaps one of the most unexpected discoveries of the Ulysses mission was made: an intense stream of dust grains which was soon recognized to originate from the Jupiter system (Grün et al. 1993). During this burst, more than 350 impacts were detected within 26 h. This exceeded by more than two orders of magnitude the typical impact rates of interplanetary and interstellar dust seen before. A total of ten more dust streams were seen in the Ulysses data obtained within 2 AU from Jupiter (Baguhl et al. 1993; Grün et al. 1993, 2001a). The impact rate showing the dust streams as individual spikes is displayed in Fig. 1. The streams occurred at approximately monthly intervals (28 $\pm$ 3 days) which came as a complete surprise because no periodic phenomenon in interplanetary space was known before for small dust grains. Details of the stream measurements were summarized by Grün et al. (2001a).

For the following reasons, the Jovian system was the most likely source of the dust-stream particles:

- The streams were narrow and collimated, which required a relatively close-by source. Otherwise, they should have been dispersed in space and time.
- The streams were concentrated near Jupiter and the strongest one was detected closest to the planet.
- Before the Jupiter flyby, the streams approached Ulysses from directions almost opposite to the streams measured after flyby. Besides, all dust streams' moving directions were close to the line-of-sight direction to Jupiter.
- The observed periodicity suggested that all streams originated from a single source and seemed to rule out cometary or asteroidal origins of individual streams. Especially comet Shoemaker-Levy 9, before its tidal disruption in 1992, was considered a possible source (Grün et al. 1994). Only two of the eleven streams, however, were compatible with an origin from the comet.

In 1995, the Galileo dust measurements confirmed the dust streams (Fig. 1): dust "storms" with up to 10,000 impacts per day were recorded about half a year before Galileo's arrival at Jupiter within 0.5 AU from the planet (Grün et al. 1996a). These impact rates were at least two orders of magnitude larger than the rates measured with Ulysses at a similar distance from Jupiter. Given that both dust detectors had the same sensitivity for dust impacts and the same total sensor area (Grün et al. 1992a,b), this difference was most likely due to temporal variations in the

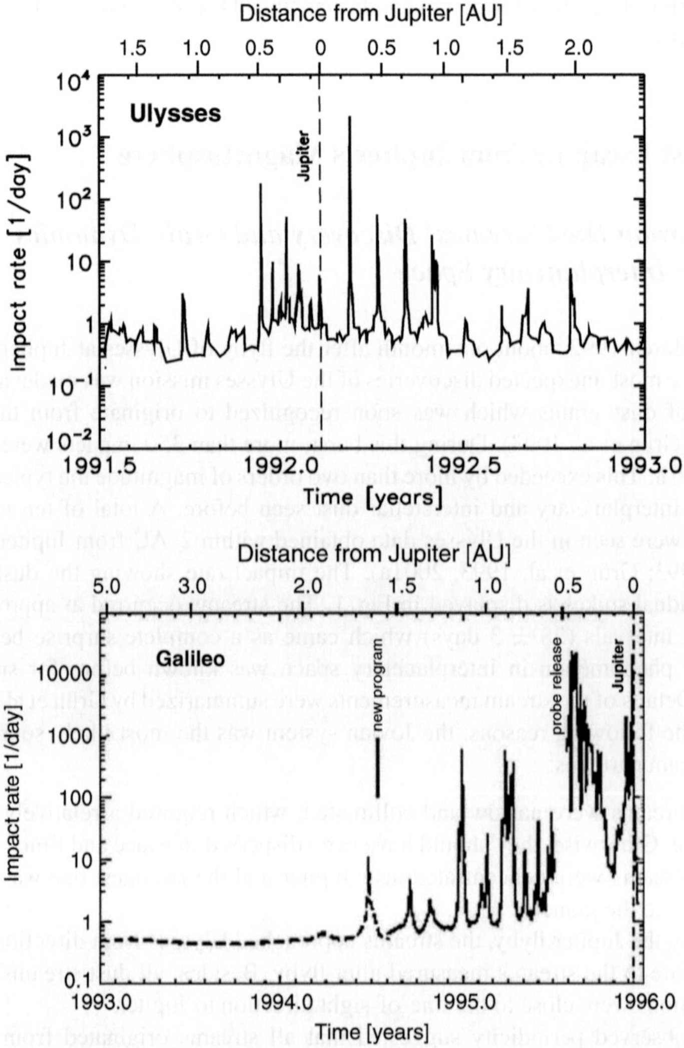

Fig. 1 Impact rates of dust particles measured with the Ulysses (*top*) and Galileo (*bottom*) in situ dust detectors in the outer interplanetary space. *Top:* around Ulysses' Jupiter flyby (*dashed vertical line*). *Bottom:* during Galileo's approach to Jupiter. The impact rates are sliding averages over six impacts. The distance from Jupiter is indicated at the top. Note that after Jupiter flyby the Ulysses spacecraft receded from the planet at about 35° jovigraphic latitude. Reprinted by permission from Macmillan Publishers Ltd: Nature (Grün et al. 1993, 1996a), copyright 1993, 1996

dust emission from the Jovian system and due to different detection geometries of the instruments for dust particles approaching from the Jupiter direction. During approach to Jupiter, the Galileo detector was facing toward the planet so that practically the entire sensor surface was exposed to the dust streams. In the Ulysses case, the sensor was pointing much further away from the planet's direction so that the effective sensing area for dust-stream particles was much smaller.

Later, during its orbital mission about Jupiter, Galileo continuously measured the dust streams in the Jovian system between December 1995 and September 2003 (Sect. 2.3.1; Krüger et al. 2005b). The streams measured in interplanetary space were the continuation of the streams detected within the magnetosphere (Grün et al. 1996b). The grains also showed a strong electromagnetic interaction with the Jovian magnetic field (Grün et al. 1997, 1998) and the majority of them originated from Jupiter's moon Io (Graps et al. 2000). The grains acquire their initial electric charge in the Io plasma torus before they are ejected from the Jovian system (Horányi et al. 1993a, 1997).

In February 2004, Ulysses had its second Jupiter flyby (at 0.8 AU distance from the planet) and again measured Jovian dust streams (Krüger et al. 2005a, 2006b; Flandes et al. 2011). A total of 28 streams were detected within about 3 AU from Jupiter. Frequency analysis revealed impact rate fluctuations with a 26-day periodicity, closely matching the solar rotation cycle. Near the Jovian equatorial plane the periodicity switched to about 13 days and the grain impact directions fluctuated with the orientation of the interplanetary magnetic field vector. This is consistent with theoretical considerations, as suggested by Hamilton and Burns (1993b), who explain the periodic impact pattern with recurrent changes of the IMF polarities: The periodic nature of the dust streams observed with Ulysses is most likely caused by the changes in the azimuthal component of the rotating solar-wind magnetic field, periodically accelerating these particles towards and away from the ecliptic plane.

2.2 *Masses and Speeds of the Jovian Stream Particles*

By applying the instrument calibration, estimates for masses and speeds of the stream particles were derived (Table 1). The calibrated values, however, were challenged by later investigations. It turned out that the impact speeds derived from the instrument calibration were much too low and, as a result, the calibrated grain masses were too large.

Although the particles' approach directions were close to the line-of-sight direction to Jupiter, the approach direction of most streams deviated too much from the Jupiter direction to be explained by gravitational forces alone. The deviation from the Jupiter direction was correlated with the magnitude and direction of the interplanetary magnetic field (especially its tangential component; Grün et al. 1996a; Zook et al. 1996). Strong nongravitational forces must have been acting on the grains to explain the measured impact directions.

In numerical simulations Zook et al. (1996) integrated the trajectories of many particles backward in time and away from Ulysses in various directions. Only particles that came close to Jupiter (< 0.05AU) were considered compatible with a Jovian origin. The dust particles were assumed to be electrically charged to $+5$V by

Table 1 Masses, speeds, and sizes of stream particles obtained from the instrument calibration compared with the values derived from numerical modeling. Radii are for spherical particles with $\rho = 1\,\mathrm{g\,cm^{-3}}$. Note that the values derived from dynamical modeling are widely accepted. From Krüger (2003a)

	Calibration (Grün et al. 1995)	Modeling (Zook et al. 1996)
Speed	$20\,\mathrm{km\,s^{-1}} \leq v \leq 56\,\mathrm{km\,s^{-1}}$	$v > 200\,\mathrm{km\,s^{-1}}$
Mass	$10^{-19}\,\mathrm{kg} \leq m \leq 9 \times 10^{-17}\,\mathrm{kg}$	$m \simeq 10^{-21}\,\mathrm{kg}$
Radius	$30\,\mathrm{nm} \leq a \leq 300\,\mathrm{nm}$	$5\,\mathrm{nm} \leq a \leq 10\,\mathrm{nm}$

a balance between solar photo-electron emission and neutralization by solar-wind electrons. Forces acting on the particles included the Sun's and Jupiter's gravity and the interaction with the IMF as measured by Ulysses (Balogh et al. 1993).

One important result from this analysis was that the particles were about 10^3 times less massive and 5–10 times faster than the values implied by the calibration of the dust instrument. Only particles faster than $200\,\mathrm{km\,s^{-1}}$ with masses of the order of 10^{-21} kg were compatible with an origin in the Jovian system (Table 1). The corresponding particle radii were only $a \simeq 5 - 10\,\mathrm{nm}$. Particles smaller than 5 nm were unable to travel from Jupiter to Ulysses. They were rather caught up by the solar-wind and swept away. Particles significantly larger than 25 nm were not compatible with the measured impact directions because they did not interact strongly enough with the interplanetary magnetic field. Later measurements by Cassini's dust detector allowed a determination of the average radius of the Jovian stream particles (see also Sect. 2.6): From the instrument's impact ionization mass spectra, Postberg et al. (2006) inferred an average particle size of about 6 nm.

For 10-nm particles, the Lorentz force exceeds gravity by more than a factor of 1,000 and the trajectories of such particles are totally dominated by the interaction with the interplanetary magnetic field. The periodic nature of the dust streams observed with Ulysses is most likely caused by the changes in the azimuthal component of the rotating solar-wind magnetic field, periodically accelerating these particles toward and away from the ecliptic plane (Hamilton and Burns 1993b).

With (2), (4), and (6) derived in Sect. 1.1, the size range and the ejection speed of Jovian stream particles can be estimated:

$$v_{\mathrm{ex}} \approx 59.5\,\mathrm{km\,s^{-1}} \left\{ 75.6 \cdot a_{nm}^{-1} \left[\frac{\phi_v}{\rho_{\mathrm{cgs}}} \right]^{1/2} \left(1 - \frac{r_0}{r_{\mathrm{ms}}} \right) - \frac{1}{2} \right\} \left[\frac{R_{\mathrm{P}}}{r_0} \right]^{1/2}, \quad (8)$$

$$a_{\max} \approx 107\,\mathrm{nm} \left\{ \frac{\phi_v}{\rho_{\mathrm{cgs}}} \left[1 - \frac{r_0}{r_{\mathrm{ms}}} \right] \right\}^{1/2}, \quad (9)$$

$$a_{\min} \approx 46\,\mathrm{nm} \left[\frac{\phi_v}{\rho_{\mathrm{cgs}}} \right]^{1/2} \left[\frac{R_{\mathrm{P}}}{r_0} \right]^{3/2} \left\{ 1 - 3.3 \left[\frac{R_{\mathrm{P}}}{r_0} \right]^{3/2} \right\}^{-1/2}, \quad (10)$$

where ϕ_v is the particle's potential in V and ρ_{cgs} is the density of a particle in $\mathrm{g\,cm^{-3}}$.

Assuming $\phi_v = +5\text{V}$, $\rho_{cgs} = 2.2\,\text{g cm}^{-3}$, $r_0 = 6\ \text{R}_\text{J}$, and $r_{ms} = 100\ \text{R}_\text{J}$, the stream particle size ranges from 5.4 nm to 156 nm. From (8), the ejection speed of a 10-nm particle is about 248 km s^{-1}. These values generally agree with the backward tracing results (Zook et al. 1996) but, due to oversimplified assumptions, are slightly different from the numerical simulation results.

Although the simulated impact directions explained the observations, Ulysses should have detected particles in between the dust streams when no impacts occurred. This indicates that, apart from interaction with the interplanetary magnetic field, there must be other processes modulating the particle trajectories (Grün et al. 2001a).

The analysis by Zook et al. (1996) demonstrated that the solar-wind magnetic field acts as a giant mass–velocity spectrometer for charged dust grains. In particular, masses and impact speeds of the grains derived from the instrument calibration (Grün et al. 1995) turned out to be invalid for the tiny Jovian dust-stream particles detected by Ulysses (Table 1). Masses and speeds of stream particles derived from modeling the Galileo measurements agreed very well with the values obtained for the streams detected by Ulysses (J. C. Liou and H. A. Zook, priv. comm.).

The coupling of the dust-stream particles to the IMF is illustrated in Fig. 2. For each dust stream detected in 2004, the plot shows the deviation δ of the average approach direction of the particles forming a dust stream as a function of the magnetic field strength measured at the location of Ulysses during the occurrence of that stream. The magnetic field vector is split into its radial, tangential, and normal components (B_R, B_T, B_N). Figure 2 shows a correlation of the deviation of the average dust stream impact direction onto the dust detector from the line-of-sight direction to Jupiter with the radial and the tangential components of the magnetic field vector while the normal component of the magnetic field vector shows no correlation. Thus, the radial and tangential components of the magnetic field vector affect the grain dynamics while the normal component has no effect on the grain dynamics, consistent with theoretical expectations. The detection of dust streams is correlated with the occurrence of CIRs, in which high-speed solar-wind plasma is compressed. The magnetic field strength and, hence, the Lorentz force acting on the grains can be 5–10 times larger during CIRs, leading to (1) strong deviations from the Jupiter direction; (2) grain acceleration and enhanced detectability; and (3) perhaps even destruction of larger grains. The dust streams at both Jupiter and Saturn are correlated with such events (Grün et al. 1993; Kempf et al. 2005a): Dust streams occur a few days after a CIR was detected by Ulysses and this time delay closely matches the particle flight time from Io to the spacecraft. Furthermore, the duration of a stream seems to be correlated with the duration of the previously detected CIR (Flandes and Krüger 2007; Flandes et al. 2011).

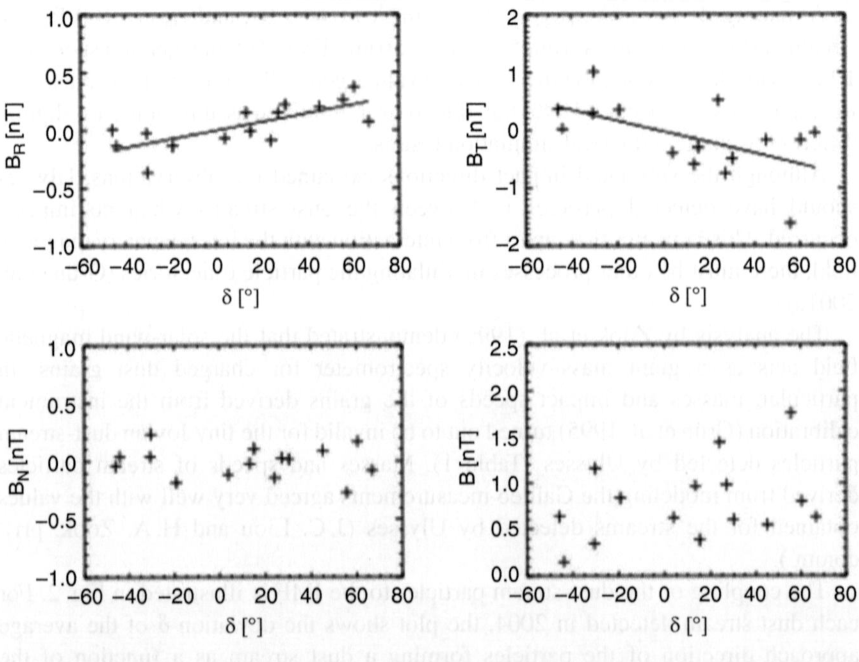

Fig. 2 Magnetic-field strength measured with Ulysses (radial, B_R, tangential, B_T, normal component, B_N and magnetic field magnitude $|B|$) vs. deviation δ of the average dust-stream impact direction (*spacecraft rotation angle*) onto the dust detector from the line-of-sight direction to Jupiter for 15 dust streams measured in 2004. In this time interval Ulysses traversed a large range in jovigraphic latitude ($-20° \leq \beta_J \leq +70°$) with only relatively little variation in ecliptic latitude ($-9° \leq \beta_{ecl} \leq +10°$). The solid lines are least squares fits to the data. Reprinted from Krüger et al. (2006b), Copyright 2006, with permission from Elsevier

2.3 *In Situ Dust Measurements Throughout the Jovian Magnetosphere*

Io is the most volcanically active celestial body in the Solar System. The most dramatic signs of Io's volcanism are the energetic plumes that rise to heights of hundreds of kilometers and cover the surface of the satellite with colorful tephra. When the plumes were discovered on images obtained with the Voyager spacecraft in 1979, it was suggested that tiny dust grains entrained in the plumes might be ejected into the circumjovian space by electromagnetic forces (Johnson et al. 1980; Morfill et al. 1980a). However, no observational proof for this mechanism was available at the time.

First indications that this electromagnetic ejection mechanism really works in nature came more than ten years later with the Ulysses spacecraft discovery of 10-nm dust particles emanating from the Jovian system (Grün et al. 1992c, 1993; Zook et al. 1996). Later Galileo measurements revealed strong electromagnetic interaction of the grains within the Jovian magnetosphere (Horányi et al. 1997; Grün et al. 1998) and confirmed that Io is the source for the majority of particles (Graps et al. 2000). Furthermore, the electromagnetically coupled dust grains may be used as a monitor of Io's volcanic plume activity and as probes of the plasma conditions in the Io plasma torus (Krüger et al. 2003b,c). The grains acquire their initial charge in the torus before they are ejected from the Jovian system. This way, the Jovian magnetosphere acts as a giant mass-velocity spectrometer and a dust accelerator for electromagnetically interacting nanodust. During its flyby in 2000/2001 the Cassini spacecraft was able to provide a compositional analysis of the dust streams. The grains' chemistry not only confirmed that Io is their source but also allowed insights into the condensation process of Io's volcanic plumes (Postberg et al. 2006).

2.3.1 Galileo Dust Streams Monitoring

The Galileo mission offered the unique opportunity for long-term monitoring of the dust environment of Jupiter. Figure 3 shows the flux of Jovian dust-stream particles measured from 1996 to 2002 superimposed upon Galileo's trajectory. In regions where no flux is shown, the dust streams were not detectable, because the detector was pointing in the wrong direction (most of the time when no flux is shown) or no dust data were transmitted to Earth. Figure 3 shows at least five interesting features:

- Fluxes of Io dust-stream particles are highly variable by approximately five orders of magnitude between 3×10^{-3} and $6 \times 10^{2}\,\mathrm{m^{-2}\,s^{-1}}$. The corresponding number density in the Jovian environment varies from 10 to $10^{6}\,\mathrm{km^{-3}}$. Number densities in the dust streams are roughly comparable with those found in dust clouds surrounding the Galilean moons, although the mass densities of dust in the clouds are an order of magnitude higher (Krüger et al. 2003d).
- The dust flux is usually higher close to Jupiter, as one would expect if the dust source is located in the inner Jovian system. The radial dependence of the dust flux measured with Galileo showed strong variations from orbit to orbit. This includes, in early 2001, a nearly perfect r^{-2} drop of the flux, as would be expected from simple particle dispersion in space. Taking all orbits since 1996 together, slopes were mostly in the range between r^{-1} and r^{-5} (Krüger et al. 2010). These variations may be due to variations in Io's dust production or the magnetosphere or both.
- In September 2000, the dust flux was about three orders of magnitude larger outside the magnetosphere ($\sim 10\,\mathrm{m^{-2}\,s^{-1}}$; jovicentric distance $\sim 280\,\mathrm{R_J}$) than within the magnetosphere. The reason for this "burst" may be increased dust emission from Io. Interestingly enough, a strong Io period is evident in periodograms from this time interval, whereas the 5-h and 10-h signature imposed by Jupiter is absent (Graps et al. 2001b).
- Such a high dust flux outside the magnetosphere was not always measured in this spatial region. A rather high flux ($\sim 1\,\mathrm{m^{-2}\,s^{-1}}$) occurred again in mid-2001, whereas other orbits showed a low flux ($0.01 - 0.1\,\mathrm{m^{-2}\,s^{-1}}$) at comparable locations during early 2001 and throughout 2002.

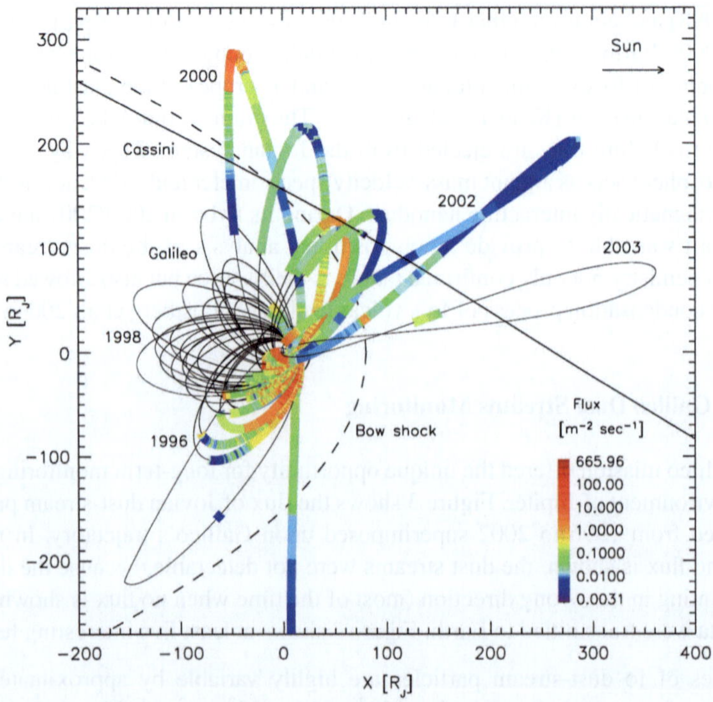

Fig. 3 Galileo's trajectory from 1995 to 2003 and Cassini's trajectory in December 2000 and January 2001, projected onto Jupiter's equatorial plane. The fluxes of Jovian dust-stream particles are superimposed (impact charge $Q_I \leq 10^{-13}$ C). For calculating the impact direction of the particles, a typical trajectory of a 10-nm grain was assumed (Grün et al. 1998). Galileo fluxes were corrected for instrument aging (Krüger et al. 2005b). Due to strong operational constraints, Cassini measured the dust streams during a short period of 12 h around Jupiter closest approach only. The rough location of the bow shock is indicated by a dashed parabola. The Sun is to the right. Adapted from Krüger (2003a)

- The data indicate a variation of the dust flux with Jovian local time: significantly higher fluxes were measured on the dawn and the dusk sides than on the noon side of Jupiter (Krüger et al. 2003c). This is in good agreement with modeling results for the dust streams (Horányi et al. 1997). The data—together with detailed modeling of the particle-charging and dynamics—may provide important information about the plasma conditions in the Io plasma torus.

2.3.2 Electromagnetically Interacting Jovian Dust Streams

In addition to the Ulysses measurements which revealed strong grain coupling to the interplanetary magnetic field in interplanetary space, the Galileo dust measurements within Jupiter's magnetosphere revealed strong electromagnetic interaction of the

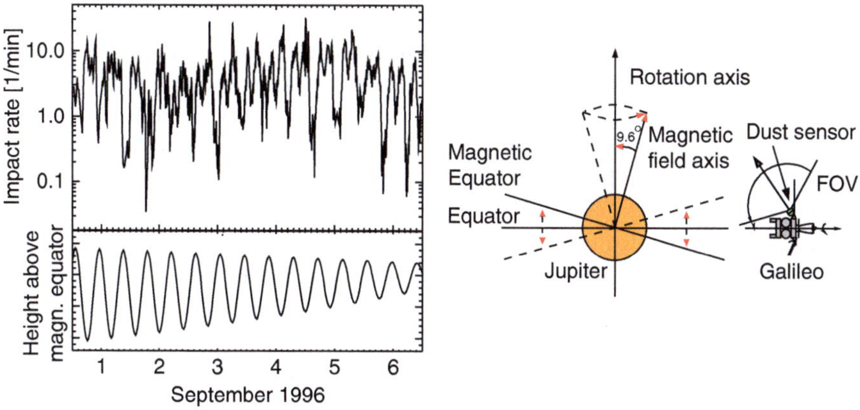

Fig. 4 *Left:* Phase relation between Galileo's position in Jupiter's magnetic field (here Galileo's distance from the magnetic equatorial plane, Z) and measured impact rate of stream particles (impact charge $Q_I \leq 10^{-13}$ C) during part of Galileo's G2 orbit in 1996. A dipole tilted at 9.6° with respect to Jupiter's rotation axis (*right*) pointing toward $\lambda_{III} = 202°$ was adopted for the magnetic field. In the time interval shown, the spacecraft approached Jupiter from 60 to 10 R_J jovicentric distance. From Krüger (2003a)

charged grains with the Jovian magnetosphere. Figure 4 shows an example of impact rates measured while Galileo was in the inner part of the magnetosphere. During this time, the rate fluctuated with 5-h and 10-h periodicities, compatible with Jupiter's rotation period, and the fluctuations reached up to two orders of magnitude. These fluctuations were correlated with the position of Galileo in the Jovian magnetic field. Due to the 9.6° tilt of Jupiter's magnetic axis with respect to the planet's rotation axis, the magnetic equator swept over the spacecraft in either upward- or downward direction every 5 h, shaping the fluctuations in the measured impact parameters (*right panel* in Fig. 4). In addition to the impact rate, the grain impact direction, the charge rise times, and the charge amplitudes showed similar fluctuations (Grün et al. 1998). With the Galileo dust detector, the electromagnetic interaction of charged dust particles with a planetary magnetosphere could be demonstrated for the first time. The source of the dust particles, however, could not be derived from measured impact rates and impact directions alone.

2.3.3 Io as a Source of the Jovian Dust Streams

In addition to the 5-h and 10-h periods, a modulation of the dust impact rate with Io's orbital period (42 h) could be recognized during some time intervals (Grün et al. 1998; Krüger et al. 1999a,b, 2001, 2006a, 2010) while at other times an Io modulation was missing. A frequency analysis of a two-year dataset shows Io's orbital frequency as a "carrier frequency" and primary source of the stream particles (Graps et al. 2000). Jupiter's magnetic field frequency modulates Io's frequency signal, giving rise to modulation side-lobe products seen around first order (10 h) and harmonic (5 h) Jupiter magnetic field frequencies. These modulation products confirm Io's role as a primary source of the Jovian dust streams. Io as a source can best explain the time-series analysis showing Io's orbit periodicity. Ultimately, a compositional analysis carried out by the Cassini Spacecraft (Postberg et al. 2006) indicated that at least 95% of the detected stream particles were of Io origin (Sect. 2.6).

An Io source is also compatible with the deduced particle sizes of $a \sim 10$ nm: photometric observations of Io's Loki plume obtained with Voyager imply a size range of 5–15 nm (Collins 1981), in agreement with numerical simulations (Zook et al. 1996). Theoretical considerations imply that particles of this size and below may be able to escape Io (Johnson et al. 1980; Ip 1996; Flandes and Maravilla 2004). Observations of the Pele plume constrained the grains to be smaller than 80 nm (Spencer et al. 1997), and showed the plume gas to be rich in S_2 and SO_2 (Spencer et al. 2000). Compositional measurements during the Cassini flyby in 2000/2001 showed that the escaping grains are heavily depleted in S_2 and SO_2 (Sect. 2.6). This led Postberg et al. (2006) to conclude that the majority of grains which fall back to the surface are an entirely different population than the escaping nanograins. It was predicted that the volatile, sulfur-rich particles condense a long time after emission of the hot gas. Recent observations with the New Horizons spacecraft indeed showed that most visible plume particles condensed within the plume rather than being ejected directly from the source (Spencer et al. 2007).

2.4 Particle Dynamics in the Jovian Magnetosphere

Dust grains escaping Io enter the cold plasma torus, where they become negatively charged and remain confined there. For sufficiently high electron temperatures, secondary electron production becomes important so that an incoming electron can generate more than one outgoing electron, and the net current can turn positive. Hence, grains visiting the outer hot regions of the plasma torus can change their sign of charge to positive.

The parameter that sets the size range for the grains escaping from the plasma torus is the secondary electron yield. Characteristic average surface potentials of particles in the cold plasma torus are ≈ -5V and $+5$V elsewhere in the torus (Horányi 2000). In the case of a positive potential, a 10-nm grain has lost about 30 elementary charges.

To a first-order approximation, Jupiter's magnetic field can be represented by a dipole magnetic field tilted by 9.6° with respect to the planet's rotation axis and rotating rigidly with Jupiter. The field strength decreases with r^{-3}. In the equatorial plane of Jupiter, the magnetic field vector points roughly perpendicular to the equatorial plane. Due to its rigid rotation, the magnetic field sweeps by a body with a speed linearly increasing with distance from Jupiter. Outside the "co-rotational distance" at 2.24 R_J, where the Keplerian orbital period equals the rotational period

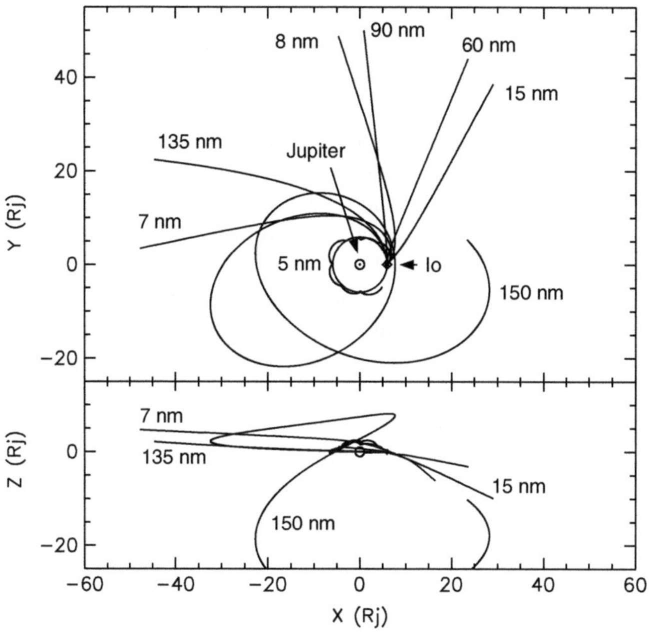

Fig. 5 Trajectories of dust-stream particles released from Io (surface potential +5 V). The numbers give approximate grain radii in nanometers. Particle trajectories are shown in the Jovian equatorial plane (X - Y) and in a perpendicular plane (X - Z). Adapted from Grün et al. (1998). See also Fig. 9

of Jupiter, positively charged particles feel an outward-directed Lorentz force. The co-rotational electric field points away from Jupiter and ejects positively charged grains in a certain size range out of the magnetosphere.

A window of particle sizes exists, for which grains can escape from traveling on bound orbits in Jupiter's magnetosphere. This is depicted in Fig. 5. For sufficiently large particles, the Lorentz force is a perturbation to the gravitational force and the particles follow Keplerian orbits. For small particles, the gravitational force is a perturbation to the Lorentz force. Therefore, the dust particles act as plasma ions and electrons, which gyrate about the planet's magnetic field lines. Between these two regimes there is a window $a_{\min} < a < a_{\max}$ of particle radii a, for which positively charged grains can escape from the Jovian magnetosphere, given by (9) and (10). The particles with the smallest radii in this window, $a_{\min}$, experience a weaker Lorentz force and larger gyroradii than those even smaller particles which are bound to the magnetic field lines. The particles with the largest grain sizes in this window, $a_{\max}$, barely have enough energy to escape from being gravitationally bound to Jupiter (Hamilton and Burns 1993b; Horányi 1996b; Burns et al. 2001). By allowing only particles in a certain size range to escape from the Jovian system, the planet's magnetosphere acts as a giant mass–velocity spectrometer for tiny dust grains.

For particles released from Io's orbital location, the window of particle sizes ejected from the magnetosphere is approximately 5 nm $\lesssim a \lesssim$ 160 nm (Fig. 5). This size window depends on the charging assumptions, especially the secondary electron emission coefficient.

The particles' ejection speeds from the magnetosphere are given by (8) and can be quite high for the smallest ejected dust-stream particles. For the particles eventually leaving the magnetosphere, the main acceleration occurs within a relatively short distance from the source (Hamilton and Burns 1993b; Grün et al. 1998). For example, a 10-nm particle with a surface potential of +5 V travels at a speed of roughly 200 km s^{-1} at 12 R_J, and reaches more than 300 km s^{-1} outside the Jovian magnetosphere at 130 R_J (Horányi et al. 1993b; Graps 2001a).

The Io plasma torus shows a dawn–dusk brightness asymmetry which is most likely the result of a large-scale, cross-tail electric field related to the global flow pattern in the plasma tail of Jupiter (Barbosa and Kivelson 1983; Ip and Goertz 1983). The dawn–dusk electric field shifts the entire plasma torus by 0.15 R_J in the dawn direction, resulting in a higher plasma temperature on the dusk side. The charge of a dust particle escaping the plasma torus is a function of Io's position in both magnetic and inertial coordinates, reflecting the plasma density and temperature variations along Io's orbit. The cross-tail electric field is small compared to the co-rotational electric field and its direct effect on the dynamics of the dust particles is small. However, the displacement of the torus leads to a strong dawn-to-dusk asymmetry in the plasma parameters, influencing the escape of the dust particles. Numerical simulations show that grains starting on the dawn side of Jupiter remain captured by the Io torus for some time, whereas particles released on the dusk side have a higher chance to escape (Horányi et al. 1997). This leads to a strong variation of the expected dust flux with Jovian local time even when the dust ejection rate of Io remains constant: higher fluxes are expected on the dawn and dusk sides of Jupiter than at noon.

This theoretical expectation from the modeling is impressively confirmed by six years of Galileo dust measurements at Jupiter (Krüger et al. 2003c). It shows that the plasma model, the description of the charging processes, and the dynamics of the grains used in the model reproduce the long-term behavior of the observations. In particular, the plasma conditions in the torus must be well represented, as they are crucial for dust-charging there. Models without the torus' dawn–dusk asymmetry do not show any local time dependence of the dust flux. Thus, dust-stream particles provide a probe of the plasma conditions in the Io torus.

To a first-order approximation, particles released from a source close to Jupiter's equatorial plane are accelerated along this same plane. The 9.6° tilt of Jupiter's magnetic field with respect to the planet's rotation axis, however, causes the particles to experience a significant out-of-plane component of the Lorentz acceleration. The result is that particles continuously released from Io move away from Jupiter in a warped dust sheet which is sometimes called "Jupiter's dusty ballerina skirt" (Fig. 9; Horányi et al. 1993b; Grün et al. 1998). Particles are separated according to their charge-to-mass ratio (corresponding to their size) and the time of formation.

2.5 *Cassini Flyby at Jupiter*

Aiming at the Saturnian system, the Cassini–Huygens spacecraft was launched on 15 October 1997 and flew by Jupiter on 30 December 2000 with a closest approach distance of 137 R_J. For the first time, the flux and directionality of stream particles could be measured simultaneously by two spacecraft at different locations. Furthermore, the Cassini dataset not only contributed to the study of the dust dynamics but also provided the first composition measurement of these tiny dust particles. The synergistic observation helped to connect the knowledge of stream-particle dynamics from interplanetary space back to their source deep inside the Jovian magnetosphere, while the chemical analysis shed light on the source of these tiny dust particles, namely Io's Pele-type volcanoes.

2.5.1 Simultaneous Measurements of Cassini and Galileo Spacecraft

As mentioned earlier, Io's dust production rate may reflect its volcanic activity and, therefore, stream-particle measurements can serve as a remote monitor of plume activity (Krüger et al. 2003b). Hence it is interesting to examine the results from two locations/spacecraft at the same time, not only for revealing the local effect (e.g., magnetic-field variation) but also for a better understanding of the global picture. However, different from the constantly rotating Galileo spacecraft, Cassini is a three-axis stabilized spacecraft. Such an arrangement allows the precise instrument-pointings required for swift object flybys, but is less favorable for long-term in situ monitoring. Moreover, to fulfill various scientific goals, the pointing profile of the Cassini spacecraft has been usually complex and not uniform. It is therefore difficult to compare Cassini dust measurements directly with Galileo's during the Cassini-Jupiter flyby. Nevertheless, a comparative study of these two measurements can still be made by using the Cassini datasets and the total dust production rate derived by Krüger et al. (2003b) from the Galileo measurements.

Figure 6 shows (from top to bottom) the measurements of the Cassini Cosmic Dust Analyzer (Cosmic Dust Analyser (CDA); the first two panels) together with the dust production rate derived from the Galileo measurements and also the field strength of the IMF measured by Cassini. The CDA impact directionality is expressed by two angles: γ_{CDA} is the angle between the CDA boresight and the source (Jupiter or Saturn in the next section). δ_{CDA} is the elevation angle between the CDA boresight and a reference plane. The ecliptic plane and the Saturnian ring plane are chosen as reference planes for Jovian and Saturnian observations, respectively. Two features can be seen when comparing the two spacecraft dust measurements in Fig. 6.

First, during these measurements, the Sun was active and thus the IMF conditions fluctuated throughout the observation. Regardless of the IMF conditions, stream particles were detected as long as the CDA pointed toward the Jupiter line of sight (LOS) direction (small γ_{CDA} in Fig. 6). Second, although the two spacecraft were at

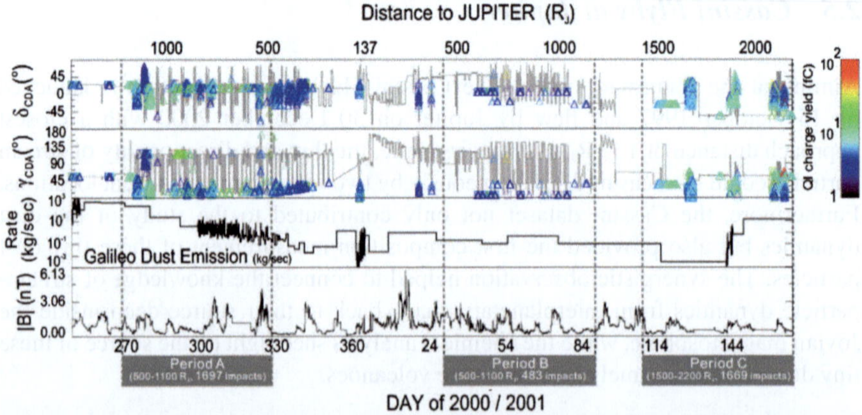

Fig. 6 Cassini's Jovian stream-particle observation during the Jupiter flyby. The pointing profile (*upper two panels*), the total dust production rate derived from Galileo measurements (Krüger et al. 2003b), and the IMF strength (*bottom panel*) are presented. Impacts on the Chemical Analyzing Target (CAT) and Impact Ionization Target (IIT) are shown as squares and triangles, respectively, and are color-coded with the CDA ion-grid signal (QI) to represent the impact strength. To compare with the Galileo results, three periods with corresponding number of CDA impacts and the spacecraft-Jupiter distance are marked along the lower horizontal axis. γ_{CDA} and δ_{CDA} are explained in the text. Reprinted with permission from Hsu et al. (2010b). Copyright 2010, American Institute of Physics

different locations and experienced different electromagnetic fields, the CDA data are similar to the Galileo total dust emission rate.

As shown in Fig. 6, though the Cassini spacecraft was at a similar jovicentric distance in periods A and B, the number of impacts is clearly higher during period A. This matches the declining dust flux measured with Galileo. On the right end of the plot, the intensive dust bursts detected by CDA in period C again coincided with the Galileo trend. The variation of CDA-measured flux does not agree with the longitudinal asymmetry proposed by Horányi et al. (1997) but does follow the time variation found by Galileo as described in Sect. 2.4 and Krüger et al. (2003b). Thus, measurements from these two spacecraft qualitatively agree with each other.

2.5.2 The Link Between the Production Rate of Stream Particle and the UV Emission of the Io Plasma Torus

Besides the dust measurements, the UV emission from Io's plasma torus as observed by Cassini UVIS during Cassini's flyby showed a significant time variability and was found to correlate with the Galileo dust production rate (Delamere et al. 2004). To match the UV observations, a plasma-chemistry model was developed by these authors. They suggested that the neutral gas produced from Io (presumed to be of volcanic origin) changed from 1.8 ton/s to 0.7 ton/s during the observation period.

Using the derived plasma properties, the charging condition of dust grains located inside the Io plasma torus can be calculated. Based on results from Delamere et al. (2004), the equilibrium potential of dust particles in the vicinity of Io's orbit is +1.9 V during the high dust/neutral gas production period (around day 275 of 2,000), but becomes −3.6 V during the low-production period (around day 14 of 2001; Hsu 2010). As mentioned earlier, only particles with positive charges can escape and become stream particles. The variation of dust-charging conditions fits Galileo's dust measurements (Krüger et al. 2003b) as well as CDA observations; it indicates a complicated connection between Ionian volcanic activity, neutral production, plasma chemistry, and the stream-particle activity. Further studies regarding time-variability of stream-particle activity are necessary.

Although it was only a distant flyby with short duration, this example demonstrates the scientific value of simultaneous measurements by two spacecraft. This is especially true for dust science as it is always not trivial to disentangle the information sealed in dust due to the complicated interaction between dust grains and their environment.

2.6 Composition and Formation of Jovian Stream Particles

The dust detector onboard the Cassini spacecraft, the CDA, is an impact-ionization dust detector with an integrated time-of-flight mass spectrometer (Srama et al. 2004), which allows the characterization of the elemental composition of the individual impinging dust grains. As described earlier in this chapter, measurements by the Ulysses and Galileo spacecraft provided important dynamical information of Jovian stream particles. Through the periodogram analysis of two-year Galileo stream particle measurements, Io's orbital period has been identified which suggests the moon as the major source of Jovian stream particles (Graps et al. 2000). CDA's compositional analysis helped to justify this finding. Moreover, the data narrowed down contributions from other dust sources and allowed insight into the formation process inside Io's volcanic plumes.

The surface of Io is covered by colorful volcanic ash that erupted from the moon's tidally heated interior. The variety of the surface colors indicates sulfur in its various molecular forms as well as specific silicates. As SO_2 is found to be the main constituent of Io's tenuous atmosphere and its volcanic plumes (Pearl et al. 1979), it was naturally assumed that the composition of particles of Ionian origin is similar.

At a Jupiter distance of more than 1 AU, long before Cassini's closest approach to Jupiter, CDA started to detect impacts of tiny and fast particles. Similar to previous measurements (as described in Sect. 2.2), most particles were detected from directions close to the Jupiter line-of-sight (LOS) and could therefore be identified as Jovian stream particles (Fig. 6). At the time of CDA catching the last Jovian stream particle, Cassini was more than 2.2 AU away from Jupiter with a total

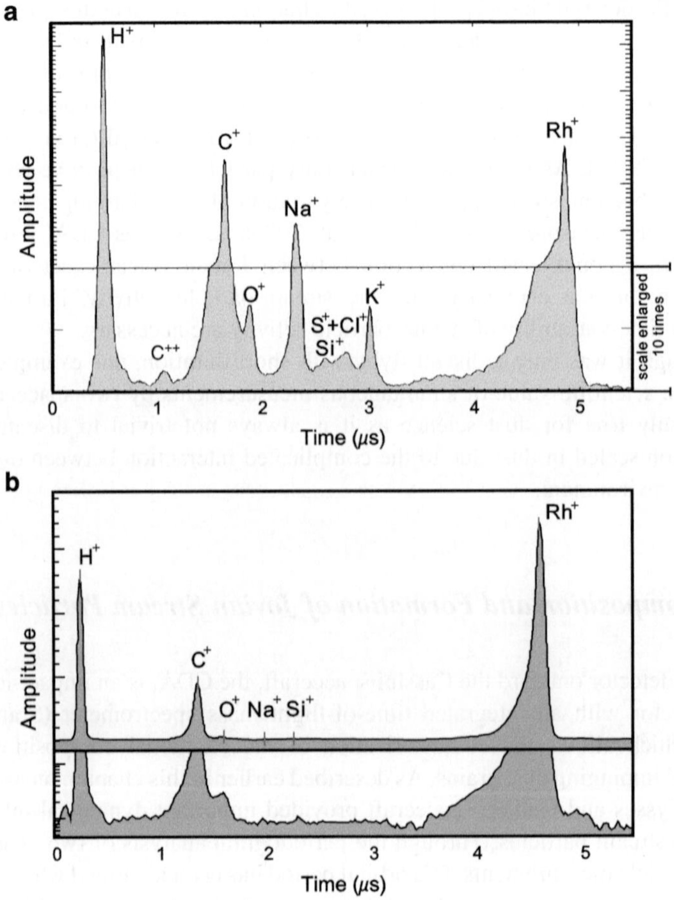

Fig. 7 CDA mass spectra of (**a**) Jovian and (**b**) Saturnian stream particles. Due to the high impact velocity and small impactor mass, stream-particle spectra are usually dominated by target and contaminant ions. (**a**) The mass spectrum is co-added from 30 Jovian stream particle spectra recorded before Cassini's Jupiter flyby in 2000. The Na^+, K^+, S+, Cl+, and partly Si^+ and O^+ lines are particle constituents. (**b**) Saturnian stream particle mass spectrum recorded in 2004. The lower panel is in logarithmic scale to show the weak lines. The main peaks in the spectrum are target (Rh^+) and target contamination ions (H^+, C^+)

of more than 7,000 impact events recorded, nearly 700 of which yielded a mass spectrum.

For impact mass spectra produced by hypervelocity nanoparticles, the understanding of impurities and contamination of the instrument target is vital (see Postberg et al. 2009a for more details). Figure 7 shows examples of CDA mass spectra. A statistical analysis of 287 CDA mass-spectra by Postberg et al. (2006) suggests that Jovian stream particles are mainly composed of NaCl while potassium-bearing components play a minor role. Sulfur or sulfurous components are other important but minor constituents. Trace amounts of silicates or rocky minerals are often identified, representing a minor contribution, possibly located in the grain cores.

The finding of NaCl and sulfurous compounds being part of most of the particles confirms Io as a source for the dust streams. NaCl is known to be the parent molecule for the vast majority of the Na and Cl species detected in Io's gas and plasma environment (Fegley and Zolotov 2000; Küppers and Schneider 2000; Schneider et al. 2000; Moses et al. 2002b; Lellouch et al. 2003). Furthermore, NaCl's atmospheric abundance is dependent on the volcanic activity on this moon (Lellouch et al. 2003; Mendillo et al. 2004, 2007). Thus, the results of Postberg et al. (2006) support Io being the source for the vast majority, maybe all, of the stream particles detected far outside the Jovian magnetosphere. However, a minor contribution from Jupiter's gossamer rings, with an upper limit of about 5% of the observed impacts (mainly contributing to the smallest of the observed stream particles), is consistent with the data. This estimate is made under the assumption that gossamer ring particles would not have an abundant alkali metal or sulfur component and mainly consist of rock-forming minerals.

Nevertheless, the result comes as a surprise since, compared with SO_2 and S_2, NaCl is only a minor component in Io's atmosphere, albeit with the highest concentration in the largest (the so called Pele-type) volcanic plumes (Lellouch et al. 2003). Obviously, the finding that NaCl is more abundant than sulfur in stream particles needs a mechanism that explains the inversion of these proportions. Stream particles consist of the most refractory materials found in the plume gases. Based on plume gas-chemistry models (Moses et al. 2002a,b; Schaefer and Fegley 2005a,b), Postberg et al. (2006) therefore suggested that Jovian stream particles are formed inside Pele-type plumes prior to the condensation of the very volatile SO_2. The minor sulfur contribution found in the grains' elemental composition is then mainly bound in refractory alkali salts, Na_2SO_4 and K_2SO_4. Stream particles probably build up primarily by accreting alkali-salt molecules around pyroclastic silicate cores. The less refractory sulfuric compounds, in particular SO_2, do not condense abundantly before the grains escape into Io's exosphere.

Besides the compositional information, the size and mass of the particles can also be inferred from the particle-forming ions in the spectra. The average mass of NaCl in Jovian stream particles is estimated to be about 1.5×10^{-21} kg (Postberg et al. 2006), equivalent to a particle radius of about 6 nm, varying between 3 and 12 nm.[1] Since NaCl is by far the most abundant component this number can be considered as a lower limit. This size limit agrees with numerical simulation results (Zook et al. 1996; Ip 1996; Horányi et al. 1997; Grün et al. 1998).

Observations by the New Horizons spacecraft confirm the hypothesis of Postberg et al. (2006) that most plume particles condense from the gas-phase close to the plume's shock front rather than being ejected from the source (Spencer et al. 2007).

[1]Due to a mixup of size and radius, the values given here are two times smaller than in the original publication.

The spectral analysis furthermore implies that there are no pure SO_2 condensates of stream particle size present in Pele-type plume tops and thus these condensates do not appear to play an important role within the first minutes after outgassing from the volcano's vent. Sulfur species like SO_2 or SO definitely condense upon contact with Io's cold surface. Condensation of the more volatile species probably also occurs in shocks or colder regions of the plume at greater horizontal distances from the vent. An alternative scenario would be that SO_2 condenses on particle surfaces as the final layer of the condensation cascade. Subsequent sputtering of the stream particles' surfaces on their voyage through the ionosphere and the plasma torus of Io would lead to quick dissociation and evaporation of the highly volatile SO_2 frost.

3 Stream Particles from Saturn

In 2004, when 1 AU of Saturn, the Cassini dust detector started to register faint impacts caused by fast and tiny dust particles (Kempf et al. 2005a). The signals and the solar-wind-associated variations of these impacts were found to be similar to those of the fast and tiny particles ejected from the Jovian system. Aside from detections during enhanced magnetic field regions, these particles were mainly detected from directions close to the Saturn LOS. Hence, these particles are referred to, and classified, as Saturnian stream particles. After Jupiter, the second large planet in the solar system—Saturn—is also identified as a source of fast, nanometer-sized dust particles (Kempf et al. 2005a).

In fact, this discovery was not a total surprise. Horányi (2000) used the plasma model derived from the Voyager data for dynamical simulations and proposed that in Saturn's magnetosphere, positively charged dust particles with proper sizes may gain enough energy to overcome gravity from the planet and escape. The particle sizes and ejection speeds for Saturnian particles were predicted to be smaller and slower compared to their Jovian counterparts.

Nevertheless, this important finding allows comparative studies to be made. The knowledge learned from Ulysses and Galileo measurements at Jupiter can then be applied and examined by studying the dynamics and composition of Saturnian stream particles. It also raises further questions, such as: What are the sizes and velocities of these particles? What is their composition? Where do they originate?

3.1 The Dynamics of Saturnian Dust-Stream Particles in Interplanetary Space

The discovery of Saturnian stream particles took place during Cassini's Saturn approach phase in 2004 (Kempf et al. 2005a). The flux and the impact signals of stream particles observed at Saturn were generally weaker than those at Jupiter.

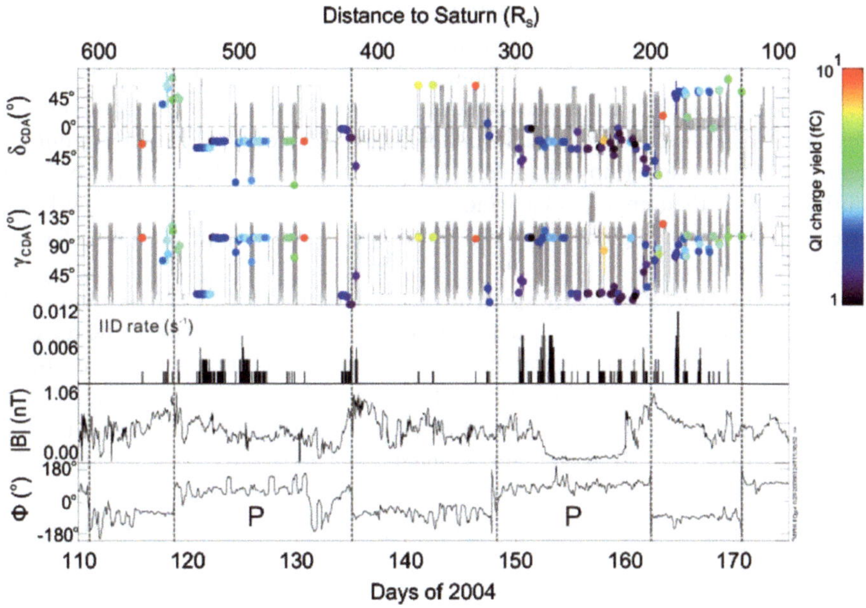

Fig. 8 CDA measurements of Saturnian stream particles before Saturn orbit insertion in 2004. The first two panels are similar to those in Fig. 6. The middle panel is the impact rate, and the bottom two panels show the IMF strength and the IMF azimuthal angle (Φ) defined from an observer-centric spherical coordinates system. Two IMF sectors can be clearly seen from the Φ angle panel. Most impacts were clustered in positive Φ periods (marked with P in the *bottom panel*), implicating the IMF modulation of the stream-particle dynamics

However, the detection pattern as well as the directionality of Saturnian stream particles showed a peculiar correlation to the IMF structure. The IMF conditions during this period were monitored by the Cassini magnetometer over more than 10 solar rotations (Jackman et al. 2004, 2005). Under declining solar activities, the IMF was found to be highly structured and consisted of compression and rarefaction regions, which were associated with CIRs (Jackman et al. 2004), a common solar-wind structure between 2 and 10 AU resulting from the interaction of locales containing slow and fast solar wind. The field direction agreed with the Parker spiral model and showed a two-sector structure during each solar rotation with the heliospheric current sheet (HCS, where the field direction reverses), embedded in the compression region (Jackman et al. 2008 see also Fig. 8). A clear IMF structure associated to CIRs can also be found in the Ulysses measurements close to Jupiter (Flandes and Krüger 2007).

The reason why these tiny particles are named "stream" particles is due to their interactive nature with the solar-wind magnetic field in interplanetary space. The "stream" phenomenon, or actually dust-impact bursts, is a combined effect of dust properties, IMF conditions, and the spacecraft location. Studying the interaction between the solar wind and Saturnian stream particles thus provides an opportunity for a comparative analysis of the properties, the physical and chemical processes, as well as the source of nanodust from two giant planet systems.

Within ∼500 R_S (0.2 AU) of Saturn, CDA observations showed a recurrent stream-particle impact pattern that correlated with CIR rarefaction and compression regions. Most impacts were detected either during enhanced IMF periods or during particular IMF sectors (those marked with P in Fig. 8). Impact bursts were observed during high-IMF-strength, solar-wind compression periods, while in low-IMF-strength, rarefaction regions only very faint impacts were detected from the Saturn LOS direction (Fig. 8).

According to the Parker IMF spiral model, at Saturn's orbit the IMF is dominated by its tangential component. Consequently, the electric field, induced by the relative motion between dust particles and the solar wind plasma, points either northward or southward (with respect to the ecliptic plane) depending on the IMF sector. Since the spacecraft approached the planet from its southern hemisphere, the clustering of impacts in only one IMF sector can be understood: apparently, only during one of the IMF sectors are particles bent southward toward the spacecraft, which results in the stream-particle detection. This indicated that the Saturnian stream particles are preferentially ejected when moving along the planet's ring plane (Hsu et al. 2010a see Fig. 9).

However, this scenario changed slightly in later orbits. As shown in Fig. 10, the faint Saturn LOS impact component was observed during both IMF polarities. One noticeable difference is that during this orbit (orbit A), the Cassini spacecraft was much closer to Saturn and its ring plane than it was during the Saturn approach phase. The directionality of these particles suggests that they are recently ejected from the system. For particles detected very close to the magnetosphere right after their ejection, the time they experience the solar wind magnetic field, no matter which IMF sector, is too short to have a significant effect on the dynamics of these particles. Hence, their original velocity vectors are roughly preserved and the impact pattern becomes nearly independent of the IMF sector. From this measurement, the stream particle ejection angle (with respect to the ring plane) can be estimated and is inferred to be 20–30° (Hsu et al. 2010b).

Another important feature presented in Fig. 10 is the evolution of the dynamical properties of stream particles as IMF conditions change. Following the changing IMF conditions, not only impact direction but also impact charge yield react almost instantaneously. Taking day 232 as a reference, the δ_{CDA} changes from negative to positive (particle's approach-direction changes from north to south) while impacts were detected from directions deviated from the Saturn LOS only (large γ_{CDA} angle). The longer Cassini stayed in the region with enhanced IMF, the higher the impact yields, which suggests that particles become faster. Following the decreasing IMF strength, the energetic impact flux dropped and the typical background Saturn LOS impacts resumed.

The observations shown in Fig. 10 agree with the result by Flandes and Krüger (2007) and Flandes et al. (2011), who reported a positive correlation between the duration of impact bursts and the length of high IMF-strength periods for Jovian dust streams. Furthermore, CDA measurements during Cassini's first three orbits about

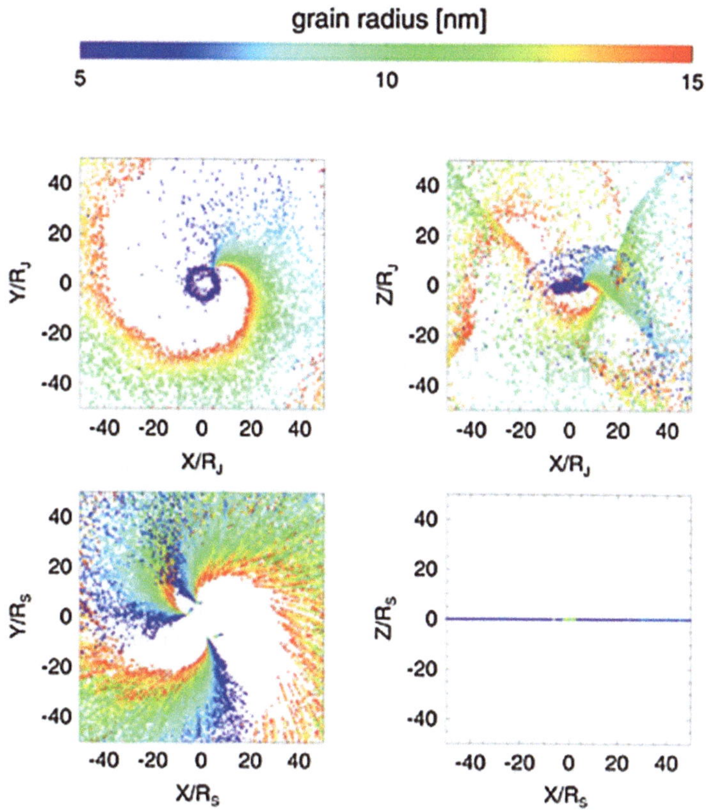

Fig. 9 Modeling the ejection of stream particles from Jupiter (*upper panels*) and Saturn (*lower panels*). Due to the tilted magnetic field, Jovian stream particles are ejected and form a dusty "ballerina skirt." In contrast, the ejection of Saturnian stream particles is well aligned with the ring plane. The color code represents particle sizes range from 5 to 100 nm. Reprinted with permission from Horányi (2000) and Horányi et al. (2004). Copyright 2000, American Institute of Physics. Copyright 2004, American Geophysical Union

Saturn show semiregular stream particle impact bursts with intervals that mimic the solar rotation period (Hsu et al. 2011a). Accordingly, Cassini measurements strongly support the idea that the local acceleration of stream particles within the enhanced IMF regions mainly accounts for the formation of stream particle impact bursts (i.e., the "dust stream" phenomenon) (Kempf et al. 2005a; Hsu et al. 2010a,b, 2011a).

3.2 *Masses and Speeds of the Saturnian Stream Particles*

CDA, like its predecessors on the Ulysses and Galileo spacecraft, is not calibrated for stream-particle measurements. The mass and the impact velocity of these fast

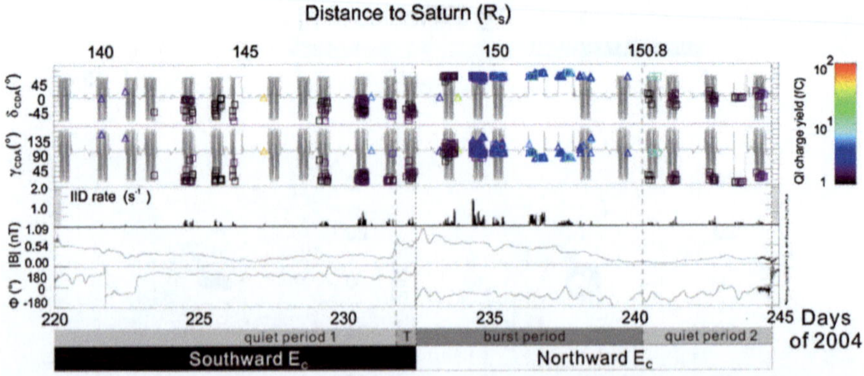

Fig. 10 Saturnian stream-particle observations between day 220 and day 245 of 2004. Panels are similar as Fig. 8. The IMF structure can be used as a reference frame to define three stream-particle-IMF interaction periods. The direction of the co-moving electric field (**E**$_c$) is also shown at the bottom. Note the correlation between the **E**$_c$ direction and δ_{CDA}. Reprinted with permission from Hsu et al. (2010b). Copyright 2010, American Institute of Physics

nanoparticles cannot be derived directly from the instrument impact signals, as described in Sect. 2.2. Determining their sizes and ejection speeds is important as these parameters provide information about the formation and acceleration process of these tiny particles (Kempf et al. 2005a; Maravilla and Flandes 2005; Hsu et al. 2011b).

Following Horányi (2000) and Kempf et al. (2005a) and (2), the theoretical ejection speed of Saturnian stream particles is

$$v_{ex} \approx 35.5\,\mathrm{km\,s^{-1}} \left\{ 23.3 \cdot a_{nm}^{-1} \left[\frac{\phi_v}{\rho_{cgs}}\right]^{1/2} \left(1 - \frac{r_0}{r_{ms}}\right) - \frac{1}{2} \right\} \left[\frac{R_P}{r_0}\right]^{1/2}. \tag{11}$$

From (4) and (6), the upper and lower size limits of Saturnian stream particles are

$$a_{max} \approx 33\,\mathrm{nm} \left\{ \frac{\phi_v}{\rho_{cgs}} \left[1 - \frac{r_0}{r_{ms}}\right] \right\}^{1/2}, \tag{12}$$

$$a_{min} \approx 11\,\mathrm{nm} \left[\frac{\phi_v}{\rho_{cgs}}\right]^{1/2} \left[\frac{R_P}{r_0}\right]^{3/2} \left\{ 1 - 2.5 \left[\frac{R_P}{r_0}\right]^{3/2} \right\}^{-1/2}. \tag{13}$$

Assuming $\phi_v = +2.3\mathrm{V}$, $\rho_{cgs} = 2.3\,\mathrm{g\,cm^{-3}}$, and $r_{ms} = 25\,\mathrm{R_S}$, the stream particle size ranges from 1.6 to 30 nm for $r_0 = 4\,\mathrm{R_S}$, and from 0.5 to 27 nm for $r_0 = 8\,\mathrm{R_S}$. From (11), the ejection speed of a 3-nm particle is around 106 and $60\,\mathrm{km\,s^{-1}}$ for $r_0 = 4$ and 8 $\mathrm{R_S}$, respectively.

Adopting the dust-charge-potential profile (the lower panel of Fig. 11), Kempf et al. (2005a) show the ejection size and velocity as functions of the distance from

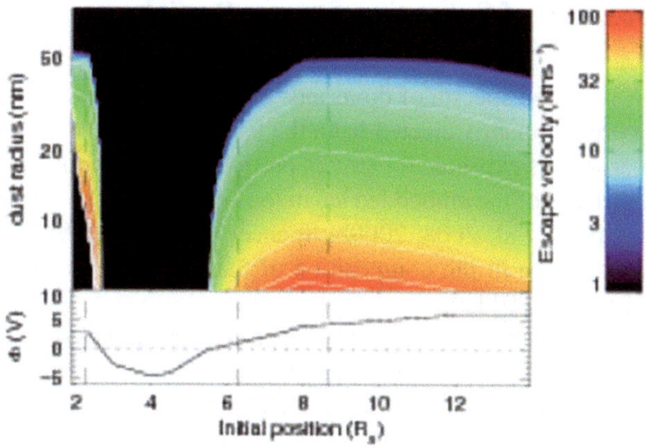

Fig. 11 Dependence of the Saturnian stream particle's dynamics on its initial position. The ejection speed and the corresponding particle size is shown in the top panel. The calculated dust potential profile is shown in the bottom panel. Reprinted by permission from Macmillan Publishers Ltd: Nature (Kempf et al. 2005a), Copyright 2005

Saturn (the upper panel of Fig. 11). Based on measurements within solar-wind rarefaction regions where the IMF strength and direction are stable, the sizes and the velocities of Saturnian stream particles are constrained (Hsu et al. 2010a,b). However, since nanoparticles are sensitive to the nature of the electromagnetic force, the best way to derive the dynamical properties of stream particles is by the backward-simulation (Zook et al. 1996).

Employing Cassini dust, IMF, and solar-wind speed measurements (Hill et al. 2004), a backward-tracing method, similar to Zook et al. (1996), has been applied for CDA Saturnian stream particle measurements (Hsu et al. 2011b). The result shows that Saturnian stream particles are particles with radii between 2 and 8 nm that have ejection speeds between 50 and 200 km s^{-1}, as indicated by theoretical modeling (Horányi 2000) and the observed impact signals (Kempf et al. 2005b; Hsu et al. 2011b). This confirms that the Saturnian stream particles are indeed smaller and slower than their Jovian counterparts.

More important, adopting the derived sizes and the velocities into (1), the region where the stream particles start to be accelerated outward can be located (the r_0 in (11), (12), and (13)). This "ejection region" is found to be at the outer E ring, roughly between the orbits of the icy satellites Dione and Rhea (Hsu et al. 2011b $r_0 \sim 8\,R_S$). However, since all the prominent dust reservoirs—such as the dense part of the E ring, the active moon Enceladus, and the main rings—are located in the inner Saturnian system, this information alone is not enough to constrain the original dust source.

3.3 Composition of Saturnian Stream Particles

Cassini's CDA recorded spectra of thousands of Saturnian stream-particle impacts. Since they are predominantly ejected from the E-ring region, a composition similar to the E-ring particles is expected for stream particles. The composition of the E ring as well as the surfaces of the embedded ice moons is dominated by water ice (Hillier et al. 2007; Postberg et al. 2008, 2009b; Spencer et al. 2009). Surprisingly, water plays only a minor role in the composition of Saturnian stream particles (Kempf et al. 2005b; Hsu et al. 2011b).

Because the grains that hit the CDA target at extreme velocities are tiny, the impact produces many more ions from target material and contaminants (Rh^+, C^+, H^+) than from the dust particle itself (Fig. 7). The ion signal from the particle often is barely above the noise level or not present (Postberg et al. 2009a).

About 80% of the registered Saturnian stream particles show ion yields below 2 fC and are too small to produce highly significant particle mass lines (Hsu et al. 2011b). However, if the ultraweak spectral signals with ion yields below 2 fC are co-added, they show the same characteristic particle mass lines as the larger impacts. The similar proportion of mass line intensities indicates that the fraction of larger particles with distinct particle mass lines is indeed representative for the whole ensemble (Hsu et al. 2011b).

Si^+ is the most reliably identified particle species (Kempf et al. 2005b; Hsu et al. 2011b Fig. 12b). Unlike Jovian stream particles (Postberg et al. 2006), the Saturnian counterparts show a clear trend of increasing Si^+ occurrence with increasing total ion yield. Often the ratio of Si^+ to O^+ is in good agreement with silicate as a main particle constituent.

In most cases water ice is likely not the bulk component of stream particles as it is observed for E-ring particles. Only a varying and usually minor water-ice purity of less than about 50% (and maybe far less) agrees with most spectra. However, about 5% of the spectra are outliers with a significantly increased abundance of O^+ ions (Hsu et al. 2011b Fig. 12a). This finding strongly indicates the occasional dominance of water ice.

In most cases, the faint and rare sodium mass lines correspond to a putative low target-contamination (Postberg et al. 2009a). This is particularly relevant since alkali ions play a major role in the spectra of E-ring grains (Postberg et al. 2009b) which are (to a greater degree) produced by the active venting of the moon Enceladus. About 6% of E-ring ice grains contain sodium salts at a percent level which makes Na^+ the most abundant ion in these spectra (due to its low ionization energy). For Saturnian stream particles a strong depletion of alkali compounds coincides with depleted water ice (Hsu et al. 2011b).

Organic compounds are also suspected to be a minor constituent of E-ring grains (Postberg et al. 2008). In the impact speed regime of stream particles, hydrocarbons likely decompose completely into H^+ and C^+ (Srama et al. 2009). Since these ions also form strong mass lines from CDA target contamination (Postberg et al.

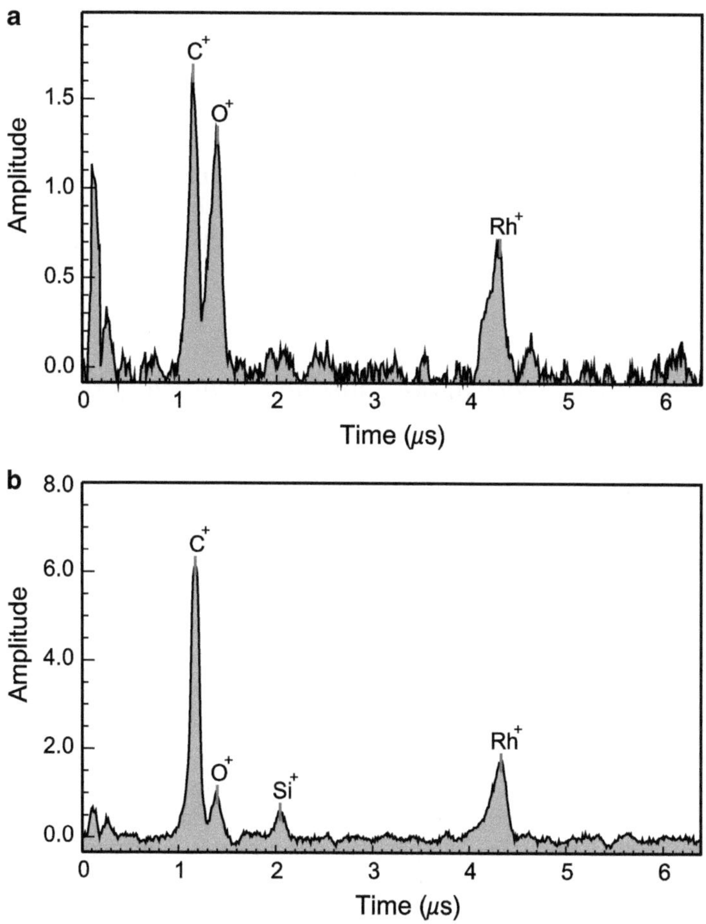

Fig. 12 The CDA mass spectra of Saturnian stream particles. (**a**) Example of a particularly oxygen-rich spectrum. These rare cases likely mark an impact of a stream particle with abundant water ice. Mass lines from carbon (C^+) and rhodium (Rh^+) stem from the instrument's target. Since the spectrum shows no silicon mass line, it is in best agreement with a pure water-ice grain. (**b**) Spectrum of an impact from a relatively large Saturnian stream particle (ion yield = 6 fC) with unusually distinct particle mass lines and typical composition. Since there is no excess of oxygen, the spectrum suggests a silicate particle with little or no water ice. From Hsu et al. (2011b). Copyright 2011 American Geophysical Union. Reproduced/modified by permission of American Geophysical Union

2009a), no conclusions can be drawn about the contribution of organic compounds to Saturnian stream particles.

Kempf et al. (2005b) also suggest that nitrogen-bearing compounds might play a role for the composition of stream particles because signatures appear occasionally at 14 (N^+) and 18 amu (NH_4^+). According to a revised analysis (Hsu et al. 2011b) H_2O^+ is responsible for the rare and faint 18-amu mass lines rather than NH_4^+.

However, the mass line at 14 amu agrees with (but does not prove) an occasional contribution of a nitrogen-bearing compound.

In summary, Si^+ is the most abundant particle compound identifiable in stream-particle spectra. Water and alkali signatures are clearly depleted with respect to the E-ring's composition. There seem to be only a small subpopulation of stream particles which consists predominantly of water ice. This is particularly puzzling, as the stream particles probably are derived from the ice plumes of Saturn's active moon Enceladus. A plausible explanation for the distinct compositional discrepancy follows.

3.4 Particle Dynamics in the Saturnian Magnetosphere

Motivated by the composition results, a dynamical model similar to the one by Horányi (2000) was developed to simulate the nanoparticle ejection process in Saturn's magnetosphere (Hsu et al. 2011b). That model includes the difference between water ice and the silicateous material in regard to their charging and the plasma sputtering properties. Moreover, the charging of nanoparticles is described by a Poisson process when modeling the stochastic behavior.

As mentioned in Sect. 1.1, only positively charged particles can become stream particles. This is an important issue especially for Saturnian stream particles. Right outside the dense Io plasma torus, the potential of Jovian dust particles turns positive immediately due to secondary electron emission. Contrary to the Jovian case, at Saturn the neutral gas density exceeds the plasma density by more than an order of magnitude across the magnetosphere (Delamere et al. 2007). With the buffer of the neutral gas, the plasma temperature is suppressed especially in the vicinity of Enceladus. As a consequence, this cold and dense plasma in the inner Saturnian system builds up a negative potential barrier that prohibits the ejection of nanoparticles (Fig. 11). However, the amount of charge carried by a nanoparticle is so small that most of the time the potential of nanoparticles deviates significantly from the equilibrium condition. As shown in Fig. 13, even within a usually negative charging environment a nanoparticle can still be occasionally charged positively. The stochastic charge variation of nanoparticles hence provides a diffusion mechanism that eases this negative potential barrier and can eventually lead to the ejection of stream particles.

Considering the stochastic charging process, the dynamical evolution of nanoparticles in Saturn's magnetosphere is modeled by Hsu et al. (2011b). Figure 14 shows an example trajectory of a 3-nm particle. Combining simulations for water ice and silicateous particles, the ejection size–velocity relation is shown in Fig. 15a. Four curves (with symbols) represent different combinations of parameters. In addition to water-ice and silicateous particles, two additional sets adopting the enhanced secondary electron emission, as suggested by Chow et al. (1993), are also simulated to examine the size-dependent secondary electron emission scenario. Nevertheless, four curves in Fig. 15a fit well to the backward-tracing simulation results shown in

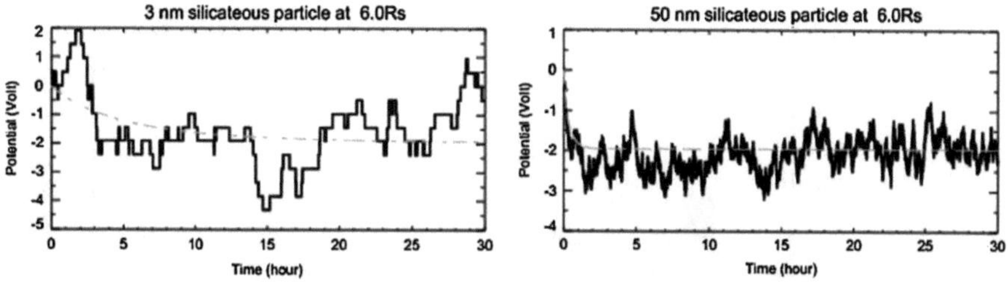

Fig. 13 Comparison of stochastic charge variation on 3-nm and 50-nm particles. The potential calculated from continuous charging currents is marked by the gray dash line as a reference. The potential of the 50-nm particle follows the equilibrium potential within ± 1V. The potential of the 3-nm particle, on the contrary, is highly variable and can even reach the opposite polarity. Note the different y axis labelling. From Hsu et al. (2011b). Copyright 2011 American Geophysical Union. Reproduced/modified by permission of American Geophysical Union

the background contour. Although the simulations with higher secondary electron yield show higher ejection speed, the charging property difference seems not the key factor that causes the aforementioned discrepancy in composition.

Before the discovery of Enceladus' active plume, one of the mechanisms proposed to explain the OH observation was plasma sputtering erosion of E-ring dust grains (Jurac et al. 2001), this is a passive feedback from the dusty rings to the magnetosphere. The dust mass is transformed to neutral gas and ultimately becomes part of the magnetospheric plasma. It therefore determines the lifetime of dust particles (Johnson et al. 2008). The lifetime of micron-sized, icy E-ring dust grains due to plasma-sputtering is estimated to be 50 years (Jurac et al. 2001).

Taking this into account, ratios of the ejection time[2] and the sputtering lifetime for particles with different sizes and parameters are shown in Fig. 15b. We see that the ejection time of icy nanoparticles is close to, or even shorter than, the sputtering lifetime. To the contrary, due to their higher plasma-sputtering resistance (Tielens et al. 1994), silicateous particles are more likely to survive erosion and to be ejected as stream particles. This result explains the composition discrepancy between Saturnian stream particles and the constituents of the E ring. Moreover, it implies that Saturn's main rings and the nanoparticle population detected by the Cassini Plasma Spectrometer (CAPS) within Enceladus' plume (Jones et al. 2009; see Chap. 2) cannot be the major source of Saturnian stream particles. In conjunction with the CDA mass-spectra analysis (Sect. 3.3) and the backward-simulation result, Hsu et al. (2011b) suggest that the silicateous impurities released from the larger E-ring grains through sputtering erosion at the outer E ring are the major source of Saturnian stream particles. While Enceladus has been found to be the dominant source of E-ring dust grains (Spahn et al. 2006), the ultimate source of Saturnian stream particles is thus also suggested to be Enceladus (Hsu et al. 2011b).

[2]The time between the release of a particle into the magnetosphere and its ejection from the system.

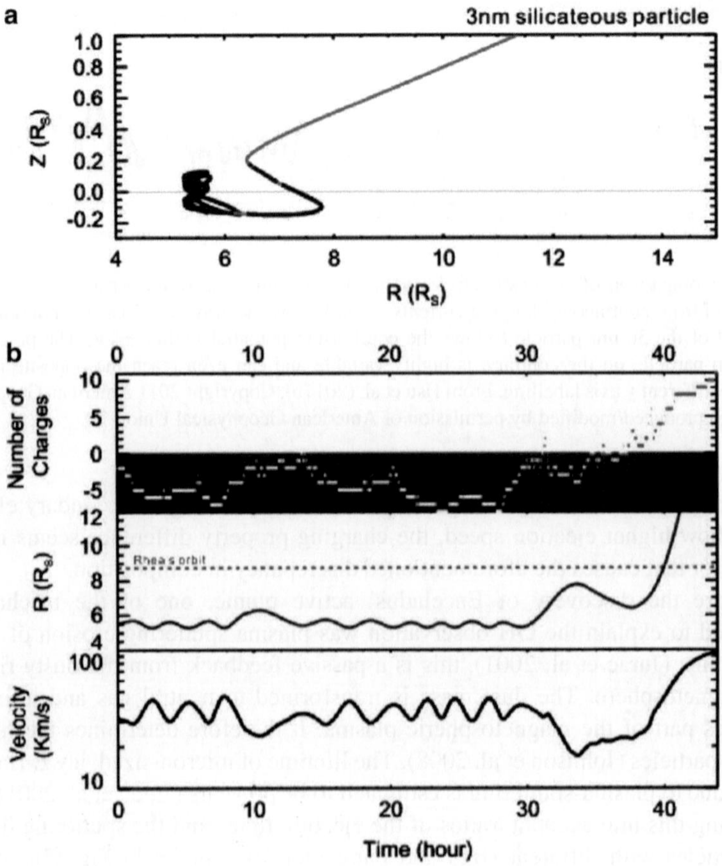

Fig. 14 The evolution path of a 3-nm silicateous grain. (**a**) The particle trajectory in the meridian plane. The colors represent the charge polarity: black—negative and neutral, gray—positive. (**b**) The particle charges, the planetocentric distance, and the velocity are shown with respect to time. Note the changes of particle velocity during the positive charge epochs. From Hsu et al. (2011b). Copyright 2011 American Geophysical Union. Reproduced/modified by permission of American Geophysical Union

The detection of silicateous material from this tiny but active moon yields further constraints to the history of Enceladus (Postberg et al., manuscript in preparation).

4 Summary

In this chapter we have reviewed the discovery and the ensuing studies of Jovian and Saturnian stream particles. The name "stream particle" stems from the "dust stream" phenomenon, which is in fact impact bursts resulting from the acceleration

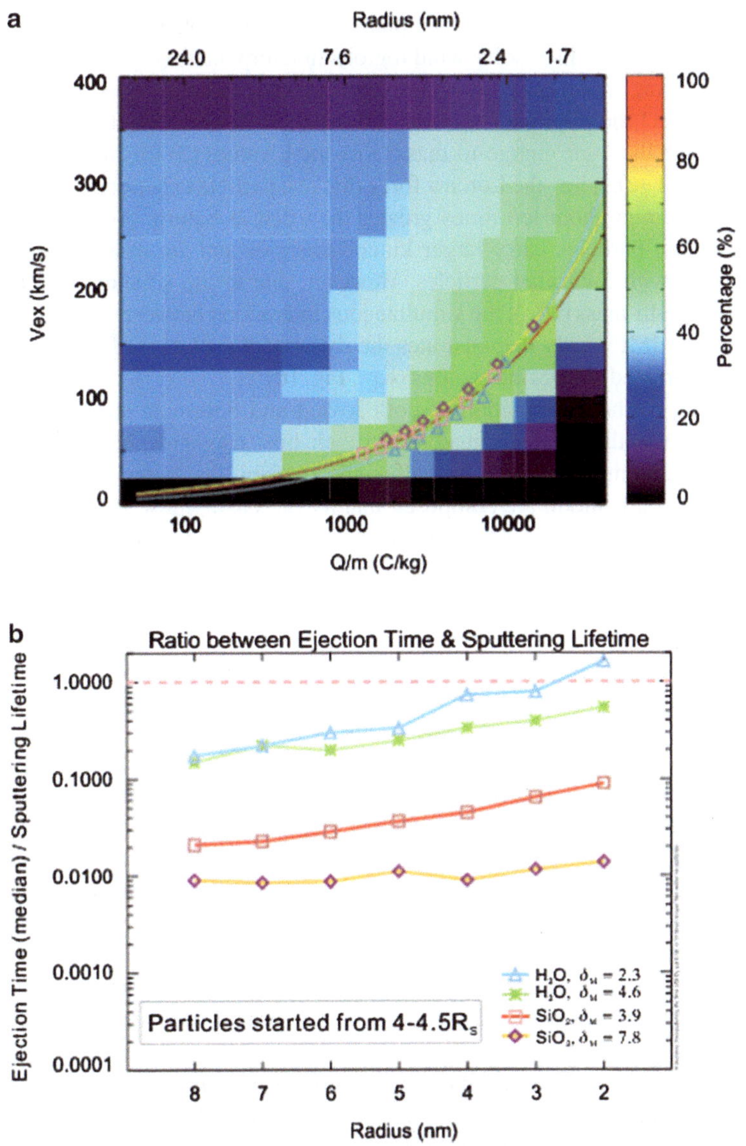

Fig. 15 (**a**) Comparison of stream particles' dynamical properties derived from the backward simulation (background colors) and the ejection model (*lines and symbols*). (**b**) The ratio between the ejection time and the sputtering lifetime as a function of particle radius. Particles with smaller secondary electron yield are more difficult to eject. The sputtering lifetime for SiO_2 particles is assumed to be ten times longer than that for water-ice particles (Tielens et al. 1994). Adapted from Hsu et al. (2011b). Copyright 2011 American Geophysical Union. Reproduced/modified by permission of American Geophysical Union

and ejection of nanoparticles from a planetary magnetosphere and the subsequent interaction with high IMF solar-wind regions in interplanetary space.

Before being modified by the IMF, the initial dynamical properties of stream particles reflect the physical condition differences between Jupiter and Saturn. For particles with the same charge-to-mass ratio, the L values (3) for Jupiter and Saturn have a ratio of $\sim$0.1, i.e., the Lorentz force that one particle experiences in the Jovian magnetosphere is about ten times greater than that at Saturn. As a consequence, Jovian stream particles carry larger kinetic energies and are easier to observe in comparison with Saturnian particles. However, due to the smaller sizes (i.e., the larger charge-to-mass ratio), the simultaneous interaction between Saturnian stream particles and IMF nicely demonstrates the dynamical evolution of stream particles inside compressed solar wind regions (e.g., Fig. 10).

Intriguingly, the composition of both Jovian and Saturnian stream particles is not the same as the prevalent material in their host magnetospheres. Despite Io's volcanic gas being dominated by sulfur and oxygen, the Jovian stream particles released from this moon are composed mainly of NaCl, whose high condensation temperature indicates the gas chemistry in the ionian volcanic plumes where these particles form. Silicateous Saturnian stream particles, on the other hand, are dynamically old particles, which have survived plasma sputtering, and provide independent constraints to the interior of Enceladus. The amount of information sealed in these solid particles is clearly disproportionate to their tiny sizes.

Nevertheless, many interesting aspects remain and require further study. Regarding Saturnian stream particles, due to the limitation of CDA observation, the temporal and spatial variation of the magnetosphere is not included in previous work and should be considered for future studies. For their Jovian counterparts, both Galileo and Cassini observations show that stream-particle activities vary with Io's volcanic output. Hypervelocity impacts of nanoparticles from Io may have a "gardening" effect on the surfaces of Europa and Ganymede. The connection between neutral gas, plasma, and nanoparticles during different Io activity phases remains unclear. Furthermore, the existence of stream particles from Jupiter and Saturn implies that the ejection of nanoparticles might be a common phenomenon for planetary systems with dust sources and a proper electromagnetic environment.

As the Cassini mission continues, upcoming measurements will surely provide new insights and complement the results presented here. Moreover, with improved sensitivity, dust particles can also be measured with other instruments, like plasma wave instruments (e.g., Cassini RPWS (Meyer-Vernet et al. 2009a) and Stereo SWAVE (Meyer-Vernet et al. 2009b), see Meyer-Vernet and Zaslavsky (2012)) and the plasma spectrometer (e.g., Cassini CAPS, see Jones et al. (2009); Jones (2012)). Synergetic development of theoretical and laboratory research should continue for supporting the future outer planet missions.

References

Alfvén, H. and Fälthammar, C. G.: 1963, *Cosmical Electrodynamics, Fundamental Principles*, Oxford University Press, New York, 2nd edition

Bagenal, F.: 1992, *Ann. Rev. Earth. Planet. Sci.* **20**, 289

Bagenal, F.: 2007, *Jour. of A. Solar-Terrestrial Phy.* **69**, 387

Baguhl, M., Grün, E., Linkert, G., Linkert, D., and Siddique, N.: 1993, *Planet. Space Sci.* **41**, 1085

Balogh, A., Erdos, G., Forsyth, R. J., and Smith, E. J.: 1993, *Geophys. Res. Lett.* **20**, 2331

Balogh, A., Bothmer, V., Crooker, N. U., Forsyth, R. J., Gloeckler, G., Hewish, A., Hilchenbach, M., Kallenbach, R., Klecker, B., Linker, J. A., Lucek, E., Mann, G., Marsch, E., Posner, A., Richardson, I. G., Schmidt, J. M., Scholer, M., Wang, Y., Wimmer-Schweingruber, R. F., Aellig, M. R., Bochsler, P., Hefti, S., and Mikić, Z.: 1999, *Space Sci. Rev.* **89**, 141

Barbosa, D. D. and Kivelson, M. G.: 1983, *Geophys. Res. Lett.* **10**, 210

Burns, J. A., Hamilton, D. P., and Showalter, M. R.: 2001, in E. Grün, B. A. S. Gustafson, S. F. Dermott, and H. Fechtig (eds.), *Interplanetary Dust*, pp 641–725, Springer Verlag, Berlin Heidelberg New York

Chow, V., Mendis, D., and Rosenberg, M.: 1993, *J. Geophys. Res.* **98**, 19 065

Collins, S. A.: 1981, *J. Geophys. Res.* **86**, 8621

Crooker, N. U., Gosling, J. T., Bothmer, V., Forsyth, R. J., Gazis, P. R., Hewish, A., Horbury, T. S., Intriligator, D. S., Jokipii, J. R., Kóta, J., Lazarus, A. J., Lee, M. A., Lucek, E., Marsch, E., Posner, A., Richardson, I. G., Roelof, E. C., Schmidt, J. M., Siscoe, G. L., Tsurutani, B. T., and Wimmer-Schweingruber, R. F.: 1999, *Space Sci. Rev.* **89**, 179

Cui, C. and Goree, J.: 1994, *IEEE Trans. Plasma Sci.* **22**, 151

Delamere, P. A., Steffl, A., and Bagenal, F.: 2004, *J. Geophys. Res. (Space Physics)* **109**, 10216

Delamere, P. A., Bagenal, F., Dols, V., and Ray, L. C.: 2007, *Geophys. Res. Lett.* **34**, 9105

Fegley, B. and Zolotov, M. Y.: 2000, *Icarus* **148**, 193

Flandes, A. and Maravilla, D.: 2004, *Adv. Space Res.* **34**, 2251

Flandes, A. and Krüger, H.: 2007, in H. Krüger and A. L. Graps (ed.), *Dust in Planetary Systems*, pp 87–90, ESA SP-643

Flandes, A., Krüger, H., Hamilton, D. P., Valdés-Galicia, J. F., Spilker, L., and Caballero, R.: 2011, *Planet. Space Sci.*, **59**, 1455

Graps, A. L., Grün, E., Svedhem, H., Krüger, H., Horányi, M., Heck, A., and Lammers, S.: 2000, *Nature* **405**, 48

Graps, A. L.: 2001a, *Ph.D. thesis*, Ruprecht-Karls-Universität Heidelberg

Graps, A. L., Grün, E., Krüger, H., Horányi, M., and Svedhem, H.: 2001b, in *Proceedings of the Meteoroids 2001 Conference, ESA SP-495*, pp 601–608

Grün, E., Fechtig, H., Hanner, M. S., Kissel, J., Lindblad, B. A., Linkert, D., Maas, D., Morfill, G. E., and Zook, H. A.: 1992a, *Space Sci. Rev.* **60**, 317

Grün, E., Fechtig, H., Kissel, J., Linkert, D., Maas, D., McDonnell, J. A. M., Morfill, G. E., Schwehm, G. H., Zook, H. A., and Giese, R. H.: 1992b, *Astron. and Astrophys. Suppl. Ser.* **92**, 411

Grün, E., Zook, H. A., Baguhl, M., Fechtig, H., Hanner, M. S., Kissel, J., Lindblad, B. A., Linkert, D., Linkert, G., and Mann, I. B.: 1992c, *Science* **257**, 1550

Grün, E., Zook, H. A., Baguhl, M., Balogh, A., Bame, S. J., Fechtig, H., Forsyth, R., Hanner, M. S., Horányi, M., Kissel, J., Lindblad, B. A., Linkert, D., Linkert, G., Mann, I., McDonnell, J. A. M., Morfill, G. E., Phillips, J. L., Polanskey, C., Schwehm, G. H., Siddique, N., Staubach, P., Svestka, J., and Taylor, A.: 1993, *Nature* **362**, 428

Grün, E., Hamilton, D., Baguhl, M., Riemann, R., Horányi, M., and Polanskey, C.: 1994, *Geophys. Res. Lett.* **21**, 1035

Grün, E., Baguhl, M., Hamilton, D. P., Kissel, J., Linkert, D., Linkert, G., and Riemann, R.: 1995, *Planet. Space Sci.* **43**, 941

Grün, E., Baguhl, M., Hamilton, D. P., Riemann, R., Zook, H. A., Dermott, S. F., Fechtig, H., Gustafson, B. A., Hanner, M. S., Horányi, M., Khurana, K. K., Kissel, J., Kivelson, M.,

Lindblad, B. A., Linkert, D., Linkert, G., Mann, I., McDonnell, J. A. M., Morfill, G. E., Polanskey, C., Schwehm, G. H., and Srama, R.: 1996a, *Nature* **381**, 395
Grün, E., Hamilton, D. P., Riemann, R., Dermott, S. F., Fechtig, H., Gustafson, B. A., Hanner, M. S., Heck, A., Horányi, M., Kissel, J., Kivelson, M., Krüger, H., Lindblad, B. A., Linkert, D., Linkert, G., Mann, I., McDonnell, J. A. M., Morfill, G. E., Polanskey, C., Schwehm, G. H., Srama, R., and Zook, H. A.: 1996b, *Science* **274**, 399
Grün, E., Krüger, P., Dermott, S. F., Fechtig, H., Graps, A. L., Gustafson, B. A., Hamilton, D. P., Heck, A., Horányi, M., Kissel, J., Lindblad, B. A., Linkert, D., Linkert, G., Mann, I., McDonnell, J. A. M., Morfill, G. E., Polanskey, C., Schwehm, G. H., Srama, R., and Zook, H. A.: 1997, *Geophys. Res. Lett.* **24**, 2171
Grün, E., Krüger, H., Graps, A., Hamilton, D. P., Heck, A., Linkert, G., Zook, H., Dermott, S. F., Fechtig, H., Gustafson, B., Hanner, M., Horányi, M., Kissel, J., Lindblad, B., Linkert, G., Mann, I., McDonnell, J. A. M., Morfill, G. E., Polanskey, C., Schwehm, G. H., and Srama, R.: 1998, *J. Geophys. Res.* **103**, 20011
Grün, E., Krüger, H., and Landgraf, M.: 2001a, in A. Balogh, R. Marsden, and E. Smith (eds.), *The Heliosphere at Solar Minimum: The Ulysses Perspective*, pp 373–404, Springer Praxis
Grün, E., Kempf, S., Krüger, H., Landgraf, M., and Srama, R.: 2001b, in B. Warmbein (ed.), *Proceedings of the Meteoroids 2001 Conference, 6–10 August 2001, Kiruna, Sweden*, ESA SP-495, pp 651–662, ESA Publications Division, Noordwijk
Hamilton, D. P.: 1993a, *Icarus* **101**, 244
Hamilton, D. P. and Burns, J. A.: 1993b, *Nature* **364**, 695
Hansen, C. J., Esposito, L., Stewart, A. I. F., Colwell, J., Hendrix, A., Pryor, W., Shemansky, D., and West, R.: 2006, *Science* **311**, 1422
Hill, M. E., Hamilton, D. C., Gloeckler, G., Krimigis, S. M., and Mitchell, D. G.: 2004, *AGU Fall Meeting Abstracts* p. A1412
Hillier, J. K., Green, S. F., McBride, N., Schwanethal, J. P., Postberg, F., Srama, R., Kempf, S., Moragas-Klostermeyer, G., McDonnell, J. A. M., and Grün, E.: 2007, *Mon. Not. R. Astron. Soc.* **377**, 1588
Horányi, M., Morfill, G. E., and Grün, E.: 1993a, *Nature* **363**, 144
Horányi, M., Morfill, G. E., and Grün, E.: 1993b, *J. Geophys. Res.* **98**, 21245
Horányi, M.: 1996a, *Ann. Rev. Astron. Astrophys.* **34**, 383
Horányi, M.: 1996b, in T. W. Rettig and J. M. Hahn (eds.), *Completing the Inventory of the Solar System*, ASP Conference Series, pp 129–136, Astronomical Society of the Pacific
Horányi, M., Grün, E., and Heck, A.: 1997, *Geophys. Res. Lett.* **24**, 2175
Horányi, M.: 2000, *Phys. Plasmas* **7(10)**, 3847
Horányi, M., Hartquist, T. W., Havnes, O., Mendis, D. A., and Morfill, G. E.: 2004, *Rev. Geophys.* **42**, RG4002
Hsu, H., Kempf, S., and Jackman, C. M.: 2010a, *Icarus* **206**, 653
Hsu, H., Kempf, S., Postberg, F., Srama, R., Jackman, C. M., Moragas-Klostermeyer, G., Helfert, S., and Grün, E.: 2010b, *Twelfth International Solar Wind Conference, AIP Conference Proceedings* **1216**, 510
Hsu, H. W.: 2010, *Ph.D. thesis*, Ruprecht-Karls-Universität Heidelberg
Hsu, H.-W., Kempf, S., Postberg, F., Trieloff, M., Burton, M., Roy, M., Moragas-Klostermeyer, G., and Srama, R.: 2011, *J. Geophys. Res.* **116**, 08213
Hsu, H.-W., Postberg, F., Kempf, S., Trieloff, M., Burton, M., Roy, M., Moragas-Klostermeyer, G., and Srama, R.: 2011, *J. Geophys. Res.* **116**, 09215
Ip, W. H. and Goertz, C. K.: 1983, *Nature* **302**, 232
Ip, W. H.: 1996, *Geophys. Res. Lett.* **24**, 3671
Jackman, C. M., Achilleos, N., Bunce, E. J., Cowley, S. W. H., Dougherty, M. K., Jones, G. H., Milan, S. E., and Smith, E. J.: 2004, *J. Geophys. Res.* **109(A18)**, 11203
Jackman, C. M., Achilleos, N., Bunce, E. J., Cowley, S. W. H., and Milan, S. E.: 2005, *Adv. Space Res.* **36**, 2120
Jackman, C. M., Forsyth, R. J., and Dougherty, M. K.: 2008, *J. Geophys. Res. (Space Physics)* **113**, 8114

Johnson, T. V., Morfill, G. E., and Grün, E.: 1980, *Geophys. Res. Lett.* **7**, 305
Johnson, R. E., Famá, M., Liu, M., Baragiola, R. A., Sittler, E. C., and Smith, H. T.: 2008, *Planet. Space Sci.* **56**, 1238
Jones, G. H., Arridge, C. S., Coates, A. J., Lewis, G. R., Kanani, S., Wellbrock, A., Young, D. T., Crary, F. J., Tokar, R. L., Wilson, R. J., Hill, T. W., Johnson, R. E., Mitchell, D. G., Schmidt, J., Kempf, S., Beckmann, U., Russell, C. T., Jia, Y. D., Dougherty, M. K., Waite, J. H., and Magee, B. A.: 2009, *Geophys. Res. Lett.* **36**, 16204
Jones. Ge. H.: 2012, Nanodust Measurements by the Cassini Plasma Spectrometer. In: Mann, I., Meyer-Vernet, N., Czechowski, A., (eds.) Nanodust in the Solar System: Discoveries and Interpretations, XXX–XXX
Jurac, S., Johnson, R. E., and Richardson, J. D.: 2001, *Icarus* **149**, 384
Kempf, S., Srama, R., Horanyi, M., Burton, M. E., Helfert, S., Moragas-Klostermeyer, G., Roy, M., and Grün, E.: 2005a, *Nature* **433**, 289
Kempf, S., Srama, R., Postberg, F., Burton, M., Green, S. F., Helfert, S., Hillier, J. K., McBride, N., McDonnell, J. A. M., Moragas-Klostermeyer, G., Roy, M., and Grün, E.: 2005b, *Science* **307**, 1274
Kempf, S., Beckmann, U., Srama, R., Horanyi, M., Auer, S., and Grün, E.: 2006, *Planet. Space Sci.* **54**, 999
Kempf, S.: 2007, *Saturnian Dust: Rings, Ice Volcanoes, and Streams*, Habilitation Thesis, Technische Universität Braunschweig
Krüger, H., Grün, E., Graps, A. L., and Lammers, S.: 1999a, in E. M. J. Büchner, I. Axford and V. Vasyliunas (eds.), *Proceedings of the VII. International Conference on Plasma Astrophysics and Space Physics, held in Lindau in May 1998*, Vol. 264, pp 247–256, Kluwer Academic Publishers
Krüger, H., Grün, E., Heck, A., and Lammers, S.: 1999b, *Planet. Space Sci.* **47**, 1015
Krüger, H., Grün, E., Graps, A. L., Bindschadler, D. L., Dermott, S. F., Fechtig, H., Gustafson, B. A., Hamilton, D. P., Hanner, M. S., Horányi, M., Kissel, J., Lindblad, B., Linkert, D., Linkert, G., Mann, I., McDonnell, J. A. M., Morfill, G. E., Polanskey, C., Schwehm, G. H., Srama, R., and Zook, H. A.: 2001, *Planet. Space Sci.* **49**, 1285
Krüger, H.: 2003a, *Jupiter's Dust Disc, An Astrophysical Laboratory*, Shaker Verlag Aachen, ISBN 3–8322–2224–3, Habilitation Thesis, Ruprecht-Karls-Universität Heidelberg
Krüger, H., Geissler, P., Horányi, M., Graps, A. L., Kempf, S., Srama, R., Moragas-Klostermeyer, G., Moissl, R., Johnson, T. V., and Grün, E.: 2003b, *Geophys. Res. Lett.* **30**, 2101
Krüger, H., Horányi, M., and Grün, E.: 2003c, *Geophys. Res. Lett.* **30**, 1058
Krüger, H., Krivov, A. V., Sremčević, M., and Grün, E.: 2003d, *Icarus* **164**, 170
Krüger, H., Horányi, M., Krivov, A. V., and Graps, A. L.: 2004, Jovian dust: streams, clouds and rings, pp 219–240, *Jupiter. The planet, satellites and magnetosphere.* Edited by Bagenal, F., Dowling, T. E., McKinnon, W. B., Cambridge planetary science, Vol. 1, Cambridge, UK: Cambridge University Press, ISBN 0–521–81808–7, 2004
Krüger, H., Forsyth, R. J., Graps, A. L., and Grün, E.: 2005a, in Boufendi, L., Mikikian, M. and Shukla, P. K., AIP conference proceedings (ed.), *New Vistas in Dusty Plasmas*, pp 157–160
Krüger, H., Grün, E., Linkert, D., Linkert, G., and Moissl, R.: 2005b, *Planet. Space Sci.* **53**, 1109
Krüger, H., Bindschadler, D., Dermott, S. F., Graps, A. L., Grün, E., Gustafson, B. A., Hamilton, D. P., Hanner, M. S., Horányi, M., Kissel, J., Lindblad, B., Linkert, D., Linkert, G., Mann, I., McDonnell, J. A. M., Moissl, R., Morfill, G. E., Polanskey, C., Schwehm, G. H., Srama, R., and Zook, H. A.: 2006a, *Planet. Space Sci.* **54**, 879
Krüger, H., Graps, A. L., Hamilton, D. P., Flandes, A., Forsyth, R. J., Horányi, M., and Grün, E.: 2006b, *Planet. Space Sci.* **54**, 919
Krüger, H., Bindschadler, D., Dermott, S. F., Graps, A. L., Grün, E., Gustafson, B. A., Hamilton, D. P., Hanner, M. S., Horányi, M., Kissel, J., Linkert, D., Linkert, G., Mann, I., McDonnell, J. A. M., Moissl, R., Morfill, G. E., Polanskey, C., Roy, M., Schwehm, G. H., and Srama, R.: 2010, *Planet. Space Sci.* **58**, 965
Küppers, M. and Schneider, N. M.: 2000, *Geophys. Res. Lett.* **27**, 513

Lellouch, E., Strobel, D. F., Belton, M. J. S., Summers, M. E., Paubert, G., and Moreno, R.: 1996, *Astrophys. J. Lett.* **459**, L107

Lellouch, E., Paubert, G., Moses, J. I., Schneider, N. M., and Strobel, D. F.: 2003, *Nature* **421**, 45

Maravilla, D. and Flandes, A.: 2005, *Geophys. Res. Lett.* **32**, 6202

Mendillo, M., Wilson, J., Spencer, J., and Stansberry, J.: 2004, *Icarus* **170**, 430

Mendillo, M., Laurent, S., Wilson, J., Baumgardner, J., Konrad, J., and Karl, W. C.: 2007, *Nature* **448**, 330

Meyer-Vernet, N.: 1982, *Astron. Astrophys.* **105**, 98

Meyer-Vernet, N., Lecacheux, A., Kaiser, M. L., and Gurnett, D. A.: 2009a, *Geophys. Res. Lett.* **36**, L03103

Meyer-Vernet, N., Maksimovic, M., Czechowski, A., Mann, I., Zouganelis, I., Goetz, K., Kaiser, M. L., St. Cyr, O. C., Bougeret, J.-L., and Bale, S. D.: 2009b, *Solar Phys.* **256**, 463

Meyer-Vernet N., and Zaslavsky A.: 2012, In Situ Detection of Interplanetary and Jovian Nanodust with Radio and Plasma Wave Instruments. In: Mann, I., Meyer-Vernet, N., Czechowski, A., (eds.) Nanodust in the Solar System: Discoveries and Interpretations, XXX–XXX

Morfill, G. E., Grün, E., and Johnson, T. V.: 1980a, *Planet. Space Sci.* **28**, 1101

Morfill, G. E., Grün, E., and Johnson, T. V.: 1980b, *Planet. Space Sci.* **28**, 1087

Moses, J. I., Zolotov, M. Y., and Fegley, B.: 2002a, *Icarus* **156**, 107

Moses, J. I., Zolotov, M. Y., and Fegley, B.: 2002b, *Icarus* **156**, 76

Mott-Smith, H. M. and Langmuir, I.: 1926, *Phys. Rev.* **28**, 727

Mukai, T.: 1981, *Astron. Astrophys.* **99**, 1

Pearl, J., Hanel, R., Kunde, V., Maguire, W., Fox, K., Gupta, S., Ponnamperuma, C., and Raulin, F.: 1979, *Nature* **280**, 755

Postberg, F., Kempf, S., Green, S. F., Hillier, J. K., McBride, N., and Grün, E.: 2006, *Icarus* **183**, 122

Postberg, F., Kempf, S., Hillier, J. K., Srama, R., Green, S. F., McBride, N., and Grün, E.: 2008, *Icarus* **193**, 438

Postberg, F., Kempf, S., Rost, D., Stephan, T., Srama, R., Trieloff, M., Mocker, A., and Goerlich, M.: 2009a, *Planet. Space Sci.* **57**, 1359

Postberg, F., Kempf, S., Schmidt, J., Brilliantov, N., Beinsen, A., Abel, B., Buck, U., and Srama, R.: 2009b, *Nature* **459**, 1098

Schaefer, L. and Fegley, B.: 2005a, *Icarus* **173**, 454

Schaefer, L. and Fegley, Jr., B.: 2005b, *Astrophys. J.* **618**, 1079

Schaffer, L. and Burns, J. A.: 1995, *J. Geophys. Res.* **100**, 213

Schneider, N. M., Park, A. H., and Kuppers, M. E.: 2000, in *Bulletin of the American Astronomical Society*, p. 1057

Sodha, M. S., Mishra, S. K., and Misra, S.: 2011, *J. App. Phys.* **109(1)**, 013303

Spahn, F., Schmidt, J., Albers, N., Hörning, M., Makuch, M., Seiß, M., Kempf, S., Srama, R., Dikarev, V., Helfert, S., Moragas-Klostermeyer, G., Krivov, A. V., Sremčević, M., Tuzzolino, A. J., Economou, T., and Grün, E.: 2006, *Science* **311**, 1416

Spencer, J. R., Sartoretti, P., Bellester, G. E., McEwen, A. S., Clarke, J. T., and McGrath, M. A.: 1997, *Geophys. Res. Lett.* **24(20)**, 2471

Spencer, J. R., Jessup, K. L., McGrath, M. A., Ballester, G. E., and Yelle, R.: 2000, *Science* **288**, 1208

Spencer, J. R., Stern, S. A., Cheng, A. F., Weaver, H. A., Reuter, D. C., Retherford, K., Lunsford, A., Moore, J. M., Abramov, O., Lopes, R. M. C., Perry, J. E., Kamp, L., Showalter, M., Jessup, K. L., Marchis, F., Schenk, P. M., and Dumas, C.: 2007, *Science* **318**, 240

Spencer, J. R., Barr, A. C., Esposito, L. W., Helfenstein, P., Ingersoll, A. P., Jaumann, R., McKay, C. P., Nimmo, F., and Waite, J. H.: 2009, Enceladus: An Active Cryovolcanic Satellite, Ch. 21, pp 683–724, *Saturn from Cassini-Huygens.* Edited by M. K.Dougherty, L. W. Esposito, and S. M. Krimigis, Springer

Srama, R., Ahrens, T. J. Altobelli, N., Auer, S., Bradley, J. G. Burton, M., Dikarev, V. V., Economou, T., Fechtig, H., Görlich, M., Grande, M., Graps, A. L., Grün, E., Havnes, O., Helfert, S., Horányi, M., Igenbergs, E., Jeßberger, E. K., Johnson, T. V., Kempf, S., Krivov,

A. V., Krüger, H., Moragas-Klostermeyer, G., Lamy, P., Landgraf, M., Linkert, D., Linkert, G., Lura, F., Mocker-Ahlreep, A., McDonnell, J. A. M., Möhlmann, D., Morfill, G. E., Müller, M., Roy, M., Schäfer, G., Schlotzhauer, G. H., Schwehm, G. H., Spahn, F., Stübig, M., Svestka, J., Tschernjawski, V., Tuzzolino, A. J., Wäsch, R., and Zook, H. A.: 2004, *Space Sci. Rev.* **114**, 465

Srama, R., Woiwode, W., Postberg, F., Armes, S., Fujii, S., Dupin, D., Ormond-Prout, J., Sternovsky, Z., Kempf, S., Moragas-Klostermeyer, G., Mocker, A., and Grün, E.: 2009, *Rapid Communications in Mass Spectrometry* **23**, 3895

Tielens, A. G. G. M., McKee, C. F., Seab, C. G., and Hollenbach, D. J.: 1994, *Astrophys. J.* **431**, 321–340

Waite, J. H., Combi, M. R., Ip, W.-H., Cravens, T. E., McNutt, R. L., Kasprzak, W., Yelle, R., Luhmann, J., Niemann, H., Gell, D., Magee, B., Fletcher, G., Lunine, J., and Tseng, W.-L.: 2006, *Science* **311**, 1419

Whipple, E. C.: 1981, *Reports on Prog. in Phys.* **44**, 1197

Zook, H. A., Grün, E., Baguhl, M., Hamilton, D. P., Linkert, G., Linkert, D., Liou, J.-C., Forsyth, R., and Phillips, J. L.: 1996, *Science* **274**, 1501

Nanodust Measurements by the Cassini Plasma Spectrometer

Geraint H. Jones

Abstract The Cassini Plasma Spectrometer, CAPS, is an instrument aboard the Cassini orbiter primarily designed to detect thermal plasma throughout the Saturn system. The instrument has achieved this goal very successfully, and, as presented here, added to its immensely valuable scientific return by unexpectedly detecting positively and negatively charged nanodust in the upper atmosphere of Saturn's largest moon Titan and in the plume of material ejected from the south pole of the icy moon Enceladus. Here, an overview is given of these observations, the sources of these particles, and the implications that their presence has for atmospheric chemistry at Titan, and the moon–magnetosphere interaction that takes place at Enceladus.

1 Introduction

The Cassini–Huygens spacecraft arrived at Saturn on July 1, 2004, carrying a suite of instruments designed to investigate in unprecedented detail the planet, its rings and moons, and the environment within which they reside. The Cassini orbiter payload includes an instrument dedicated to the observation of dust: the Cosmic Dust Analyzer, CDA (Srama et al. 2004), a highly successful experiment designed to observe grains with masses between 10^{-19} and 10^{-9} kg. Other instruments aboard the orbiter also provide information on the dust environment in the Saturn system: the Radio and Plasma Wave Experiment (Gurnett et al. 2004) detects impacts on the spacecraft of micron-scale grains. The Magnetospheric Imaging Instrument, MIMI

G.H. Jones (✉)
Mullard Space Science Laboratory, Department of Space and Climate Physics,
University College London, Holmbury St. Mary, Dorking, Surrey RH5 6NT, UK

The Centre for Planetary Sciences at UCL/Birkbeck, Gower Street, London WC1E 6BT, UK
e-mail: ghj@mssl.ucl.ac.uk

I. Mann et al. (eds.), *Nanodust in the Solar System: Discoveries and Interpretations*, Astrophysics and Space Science Library 385, DOI 10.1007/978-3-642-27543-2_6,

(Krimigis et al. 2004), can also detect dust through the discharges between the plates of its INCA instrument (Jones et al. 2008) and with the LEMMS sensor (N. Krupp, personal communication 2010). At the time of this writing, the characteristics of the dust that MIMI detects have not been tightly constrained. Here, we describe serendipitous detections of charged nanodust by another Cassini instrument, the Cassini Plasma Spectrometer, CAPS (Young et al. 2004).

The Saturnian system is a rich environment for the study of cosmic dust. The planet's main ring system is an obvious source of small particles that result from collisions between larger ring grains. This population has yet to be sampled directly, but the planet is also girdled by diffuse rings dominated by particles of radii $<100\ \mu$m. These low optical depth rings, referred to as D, G, and E, are environments where inter-grain collisions are rare, and their low-mass particles are susceptible to non-gravitational forces. The D and G rings reside on the inner and outer boundaries of the main ring system, whereas the E ring extends between at least 3.0 and 11.6 Saturn radii from the planet, where the planet's radius = 60,268 km. Although long believed to be the largest ring in the solar system, the E ring's extent is actually surpassed by a ring associated with the outer moon Phoebe, discovered remotely, extending from at least 128–207 Saturn radii from the planet (Verbiscer et al. 2009).

The E ring's population of grains is primarily fed by the ejection of particles by the moon Enceladus, which we discuss further below. As the Cassini orbiter spends much of its time within the E ring, CDA has sampled it extensively, revealing its spatial structure and composition, and has observed the reverse in sign of its constituent particles' potential at approximately 8 Saturn radii (Kempf et al. 2008, 2010; Postberg et al. 2008). Other concentrations of dust also exist. Discovered remotely through their effects on magnetospheric particles (Roussos et al. 2008), and later observed remotely, the two small moons Methone and Pallene are embedded in arcs of dust grains. Faint rings are also associated with Janus/Epimetheus and Pallene (Hedman et al. 2009). The reader is referred to the review of Horányi et al. (2009) of diffuse rings at Saturn.

Saturn's moons are being continuously bombarded by interplanetary dust, which creates a dynamic cloud of ejecta permanently surrounding all the icy satellites, a phenomenon first observed at the Galilean moons (Krüger et al. 1999). Quantifying the characteristics of these clouds provides valuable constraints on the influx of interplanetary material (Spahn et al. 2006; Jones et al. 2008). Dust populations are known to absorb magnetospheric particles, the efficiency of which depends on the absorbing dust and the incident particles. Broad, energetic (>20 keV) electron depletions observed downstream of the icy moon Rhea in 2005 and 2007 were challenging to explain, and a debris disk with possible embedded discrete rings was proposed as the explanation for the depletions (Jones et al. 2008). The existence of this proposed debris disk is however very doubtful as all attempts to observe it remotely have been unsuccessful (Tiscareno et al. 2010).

Saturn possesses the second-largest magnetosphere in the solar system. The planet's moons and rings are direct and indirect sources of nanodust that can be accelerated out of this magnetosphere (Kempf et al. 2005). The sources and evolution of these particles are discussed in this volume by Hsu, Krüger, and Postberg.

Here, an overview is given of the charged nanograin observations made by CAPS to date at the moon Titan, which spends most of its time inside the magnetosphere, and Enceladus, which permanently resides in the planet's inner magnetosphere.

2 The Detection of Nanodust by the CAPS Instrument

CAPS comprises three sensors: the electron spectrometer, ELS; the ion mass spectrometer, IMS; and the ion beam spectrometer, IBS. ELS detects electrons of energies from 0.6 eV/e to 28.8 keV/e, IMS positive ions from 1 eV/e to 50.3 keV, and IBS positive ions from 1 eV to 49.8 keV. ELS and IMS detect particles arriving at Cassini over a "fan" measuring 160° across, by 5° and 8°, respectively. IBS covers a different field of view, comprising three crossed angular passbands measuring 150° by 1.4°. Although designed for the detection of electrons and positive ions, all three have detected charged nanodust in the Saturn system.

The three sensors are all mounted on an actuator, which for much of the time sweeps the sensors' fields of view across approximately half the sky over a period of $\sim$3 min, allowing much larger coverage of the thermal plasma environment in Cassini's vicinity than would otherwise be possible on this three-axis stabilized spacecraft. Actuation is sometimes halted for satellite encounters, which removes the aliasing effects that result from actuation, in return for a loss in solid angle coverage by the three CAPS sensors.

As outlined below, all three sensors, when oriented in the local ram direction, are capable of detecting charged nanodust and measuring the particles' energy to charge ratio. The grains have to be electrically charged to be detected by CAPS: positively charged to be detected by IMS and IBS, and negatively charged for detection by ELS. When particles of the correct charge sense enter a sensor, they are deflected by an electrostatic analyser towards the respective sensor's detector, which is a microchannel plate in the case of IMS and ELS, while IBS uses channel electron multipliers. The electrostatic analyser of each sensor sweeps through a range of detectable energy per charge values, and if a charged grain possesses the correct energy per charge ratio at the moment they enter the sensor, they will reach the detector and be recorded.

It is important to note that IMS, IBS, and ELS measure energy per charge (E/q), rather than mass per charge (M/q). However, the former can be converted into the latter if the arrival speed of the particles is known. As the particles reported on here are known to be associated with moons being encountered by Cassini, it is a relatively safe assumption that the detected particles have speeds in the spacecraft frame very similar to the relative speeds of the spacecraft and those moons. Some uncertainties in the particles' exact speeds arise due to the existence of winds in the ionosphere of Titan (Crary et al. 2009), though likely to be on the scale of a few hundred ms^{-1} compared to the encounter speed of $\sim$6 km s^{-1}, and possibly much lower than this for the very highest mass particles. Even if winds are significant, their effects on arrival speeds at Cassini can be minimal if close to perpendicular to the spacecraft's trajectory through the ionosphere. At Enceladus, an uncertainty in the exact speed of the particles arises from the velocity at which the grains are ejected from the surface fissures at that moon's south pole. Their speeds will be less than the inferred plume gas velocity of >600 m s^{-1} (Hansen et al. 2008), but at the inferred low M/q values, likely larger than the moon's escape velocity of 207 m s^{-1}. These speeds are relatively low compared to most encounter velocities, but their possible effects on inferences regarding particles' M/q values should always be taken into consideration. M/q values quoted here all assume that detected particles are at rest with respect to the encountered moon.

The detected particles' mass per charge ratio, M/q, can be calculated from their measured energy per charge, E/q, by

$$(M/q) = 2(E/q)v^{-2}, \quad (1)$$

where v is the particle arrival speed at the spacecraft. A conversion to mass can only be made if the particle charge is that of a single electron. This is likely to be true for the very smallest particles discussed here, but may not be for larger grains, e.g. Kempf et al. (2006). We therefore refer to inferred particle parameters in terms of M/q, and any sizes quoted assume a single charge. As the conversion of E/q to M/q is based on particles' kinetic energy, the upper limit of detection of charged nanograins by CAPS is a function of encounter speed. All Titan encounters are very similar in relative velocity, but Enceladus speeds vary significantly, meaning that M/q values detectable during the slowest encounters will be beyond the range of the instrument for faster flybys.

Finally, we note that the primary detection of charged particles by CAPS is not thought to be strongly affected by the very high impact speed of the detected particles themselves. This is an important factor to consider, as, for example, measurements with sounding rockets with impact velocities of only $\sim$1 km s^{-1} show signals being influenced by impact generated charges (Havnes and Nsheim 2007). Due to the nature of the sampling by CAPS, only particles of the correct mass per charge ratio for a given velocity will be channelled to the detector in any of its sensors, i.e., during the Titan and Enceladus encounters, it is not thought that the CAPS sensors are detecting a significant number of charged particles resulting from impacts on or near the instrument apertures, or indeed inside the instruments themselves. Although particle fragmentation inside the instrument could result in fragmentation products being detected at a low level, it is believed that direct detection of the charged particles native to the sampled environments dominates the observed signatures.

3 Nanoparticles in Titan's Atmosphere

The first evidence of nanoparticle detection in the CAPS data came from relatively low altitude encounters with Saturn's largest moon Titan. This 2,575-km-radius satellite possesses a thick atmosphere, and Cassini has encountered it often, not only because of great scientific interest in this body, but also because its high mass allows its use to adjust the orbital trajectory of the spacecraft about the Saturnian system effectively. Depending on the scientific focus of each Titan encounter, as well as spacecraft safety concerns that arise when passing at high speed through the upper atmosphere, the spacecraft is placed in different orientations. CAPS too is placed in a different operational mode for each Titan encounter: it can be actuating, most often over a small angular range that encompasses when possible the local ram direction from which atmospheric particles arrive at the spacecraft, or can be fixed, pointing when possible to continuously cover the local ram direction, where particles close to being at rest in Titan's frame will directly enter the CAPS instrument's apertures.

During early Titan encounters where CAPS was actuating, a population of relatively high energy per charge particles was detected by ELS when the instrument was pointing in the ram direction. The unusual signatures in the ram direction were detected when Cassini was within Titan's ionosphere (Coates et al. 2007; Waite et al. 2007).

As is also the case for the other CAPS sensors, ELS records particles' energy per charge, E/q. The conversion of E/q to mass per charge, M/q, makes use of a simple technique that has now been used in other contexts to estimate the mass to charge ratio of particles detected by the instrument's three sensors. As the particles were detected in the ionosphere, it is safe to assume that their velocity with respect to Titan is very small compared to Cassini's speed relative to the moon. The speed of the spacecraft with respect to Titan is very well established for each encounter, and is always approximately 6 km s^{-1}. An ELS spectrogram from one such encounter is shown in Fig. 1, demonstrating how high fluxes of particles are detected when ELS is swept over the arrival direction of cold ionospheric plasma.

Conversion of E/q to M/q reveals an M/q value of up to 13,800 amu/q (Coates et al. 2010a,b), compared to maximum observed masses of positively charged ions of around 1,000 amu/q (Coates et al. 2010a). Based on the likely chemical composition and physical structure of these negatively charged organic aerosols, they are estimated to have sizes of around 10–30 nm (Coates et al. 2007). The high M/q value may be a lower limit for the particle mass. Representative parameters for Titan's ionosphere—$T_e \sim 1{,}000$ K, $n_e \sim 10^3$cm^{-3}—indicate a Debye length $\lambda_D \sim 0.07$ m. This suggests that particles of this size—radius a = 30 nm—will acquire a potential $\phi \sim -0.25$ V and hence a charge $Q = 4\pi\epsilon_0 a\phi\exp(-a/\lambda_D)$ of typically 5 electron charges. A 10,000-amu/q signature may therefore correspond to a 50,000 amu mass multiply charged particle (Coates et al. 2007).

As outlined by Waite et al. (2007), it is suspected that these very massive negatively charged molecular ions are the result of a series of chemical reactions and physical processes that lead to their growth from simple molecules such as

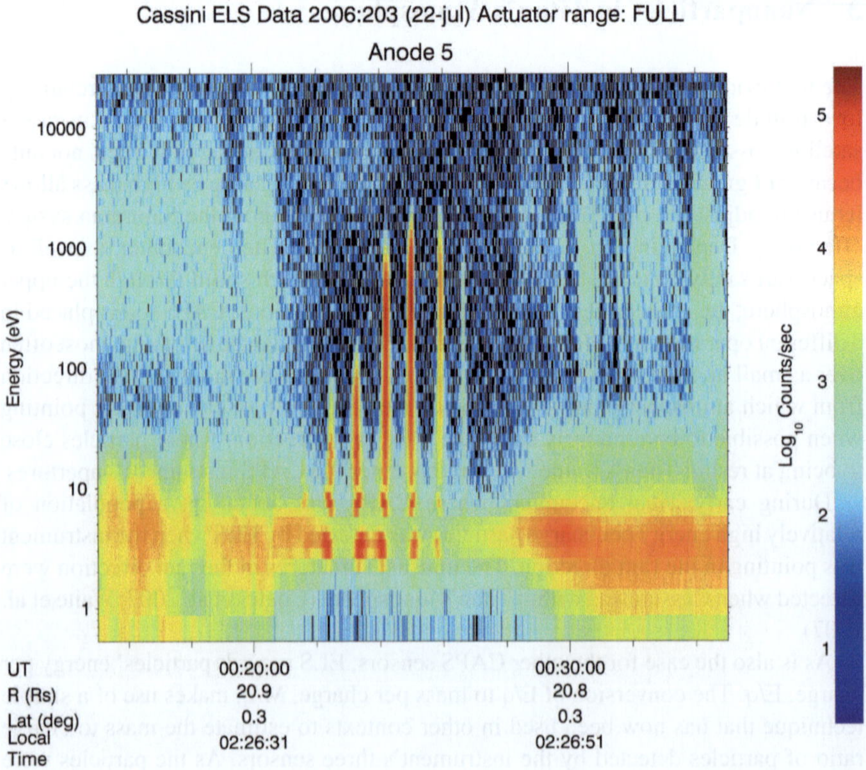

Fig. 1 Detection of charged nanoparticles in Titan's ionosphere. This spectrogram shows data from one CAPS-ELS anode, and covers a period of 20 min during Cassini's T16 encounter with the moon in July 2006. The vertical axis is energy per charge of particles, and colour coding shows the rate at which particles were detected per second. The count rates range from 10 (blue) to 100,000 (red) counts per second. The instrument was actuating during this encounter; when the detection fan encompassed the local ram direction, negative ions in Titan's ionosphere were detected, causing the vertical "spikes." The highest mass detected during this encounter was ~13,800 amu/q. Other features of the spectrogram are due to electrons: ionospheric electrons below ~30 eV and magnetospheric electrons above that energy

CH_4 and N_2, through larger, more complex positively charged ions of masses up to a few hundred amu, to the likely conglomerate negatively charged nanograins detected by CAPS-ELS. These nanodust particles settle into the lower levels of Titan's atmosphere constituting the organic aerosol population termed tholins. As these particles' fates appear to primarily lie in the lower atmosphere—a regime outside the scope of this work—we refrain from an exhaustive description of the implications of the nanodust's presence in Titan's ionosphere, instead of directing the interested reader to the works of Coates et al. (2007), Coates et al. (2009), Coates et al. (2010a), and Michael et al. (2011).

However, one possible implication for the presence of these charged grains could be that they act as a source of nanodust in the wider Saturnian system. Under ideal conditions, when the local electric field is directed towards the nadir, it is conceivable that negatively charged grains could be accelerated out of Titan's ionosphere onto an escape trajectory, placing them in a Saturnian orbit, thus contributing to the lower-mass end of the nanograin population that may eventually be ejected from the planet's magnetosphere.

4 Nanodust in the Enceladus Plume

Enceladus is a highly unusual satellite. During Cassini's first close encounter with the moon in February 2005, it became apparent from magnetometer data that there was a significant interaction between this icy moon and Saturn's magnetosphere, within which it continually resides. Closer encounters revealed more about the nature of the interaction, which was found to be caused by the presence of a significant plume of gas and solid particles emanating from Enceladus's south polar region (Dougherty et al. 2006; Spencer et al. 2006; Porco et al. 2006). The moon has a radius of 252.1 km and orbits at 3.95 Saturn radii from the planet. The primary components of the plume are water vapour (Hansen et al. 2006; Waite et al. 2006) and icy grains (Spahn et al. 2006; Porco et al. 2006). The plume gas emerges, largely from surface fissures, at >600 m s^{-1} (Hansen et al. 2008). Solid grains are entrained in this flow and are largely collimated into discrete jets. Most grains' velocities are believed to be lower than Enceladus's 207 m s^{-1} escape velocity, therefore the majority of these solid particles fall back to the surface.

During close Enceladus encounters E3 (March 12, 2008), E5 (October 9, 2008), E7 (November 2, 2009), E10 (May 18, 2010), and E13 (December 21, 2011), the CAPS sensors' apertures were oriented to encompass the ram direction, allowing direct sampling of plume material moving slowly relative to Enceladus. During the first of these encounters, IMS, ELS, and IBS unexpectedly recorded extremely high $>\sim 1$ keV/q count rates within the plume (Fig. 2). The IMS and ELS observations were reported upon by Jones et al. (2009). Populations of high energy magnetospheric ions and electrons were quickly ruled out as the cause of these signatures as they were only detected by CAPS anodes nearest the local ram direction.

The signatures were therefore almost certainly due to massive ions in the Enceladus plume. To estimate their masses, as is the case at Titan, as ions enter CAPS sensors at known velocities, their measured kinetic energy per charge can be converted to a mass-to-charge ratio (M/q). The E3 encounter, with its encounter speed of 14.4 km s^{-1}, turned out to have recorded negative particles with M/q values of up to 26,600 amu/q, and up to 46,500 amu/q for positive grains, implying that these particles are charged nanodust (Fig. 3). As it is the relative speed of spacecraft and plume material that determines the conversion factor between E/q and M/q, lower speeds allow a greater M/q

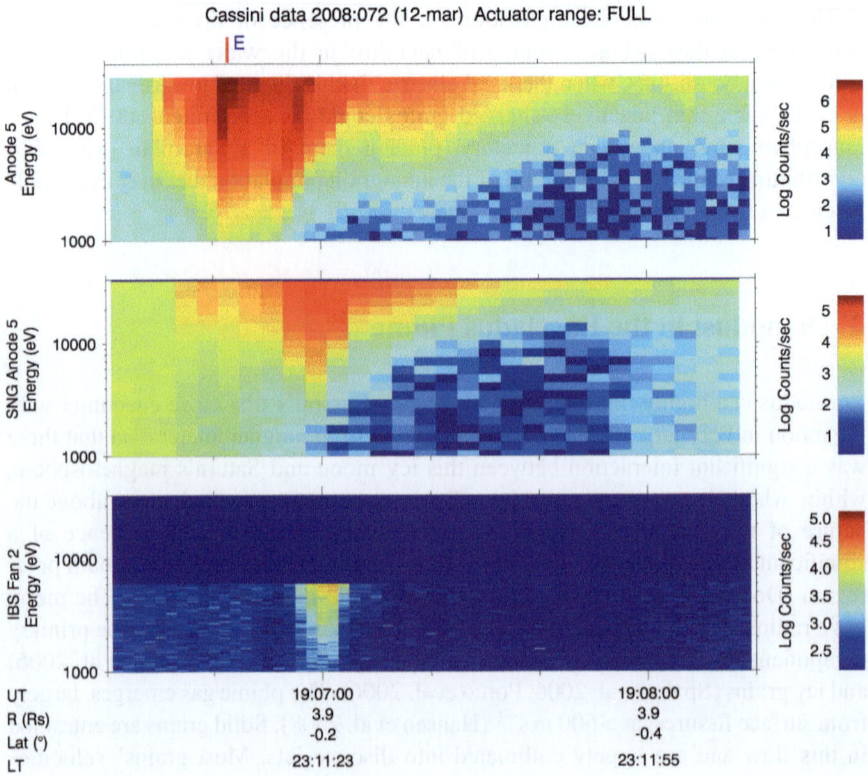

Fig. 2 Nanograin detections by CAPS ELS, IMS, and IBS, respectively, covering a 2 min period during the E3 encounter in March 2008, while Cassini was crossing the Enceladus plume. The count rates cover logarithmic scales from blue to red. Fine structure due to the crossing of individual jets is observable in all three datasets, and there is a clear discrepancy between the timing of detections in the negatively charged (ELS) and positively charged (IMS and IBS) grains. The three panels share the same E/q scale, where available covering 1–37 KeV in E/q, or 930–34,439 amu/q, if the particles are assumed to be at rest with respect to Enceladus

range to be covered by each CAPS sensor, at the expense of M/q resolution. The encounters with suitable CAPS pointing involved near-constant encounter speeds ranging from 6.2 to 17.7 km s^{-1}. Encounter E10, with its 6.2 km s^{-1}, speed potentially allowed the detection of negative grains with masses up to 144,600 amu/q, compared to the fastest encounter E5, where its 17.7 km s^{-1} speed allowed only masses per charge of up to only 17,700 amu/q to be detected.

Data from the E3 and E5 encounters remain the most striking obtained to date, showing very high fluxes up to the top of the sensors' E/q ranges (Jones et al. 2009); at lower energies, low mass positive and negative ions originating in the plume gas

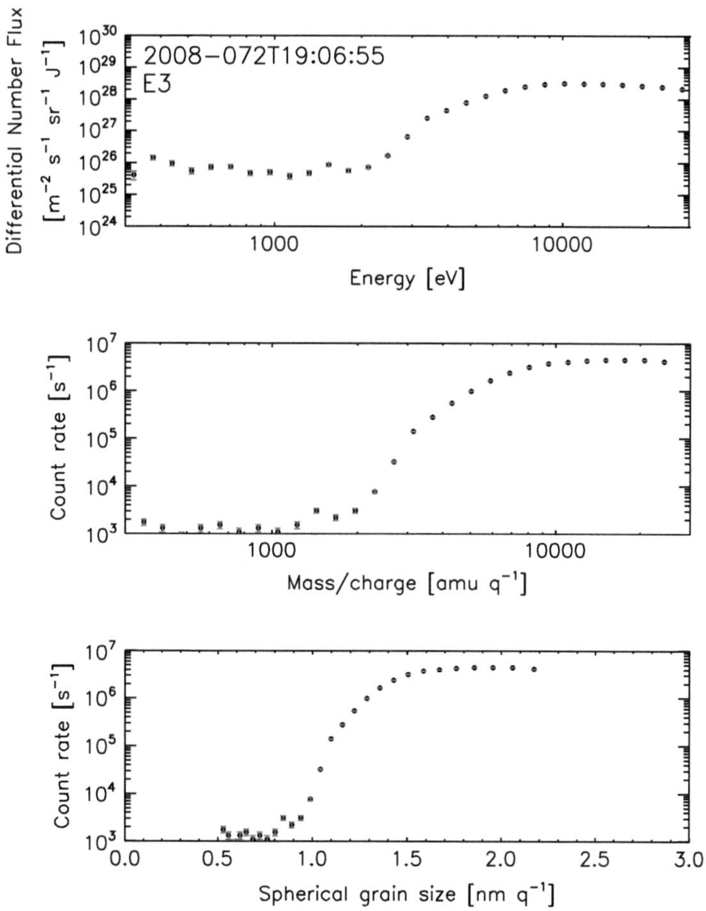

Fig. 3 Negatively charged nanograin spectrum obtained during Enceladus encounter E3. The top panel shows the original E/q spectrum, in units of differential number flux. The second panel shows the distribution having converted from E/q to M/q. The third panel shows the size per charge distribution of these grains assuming that the particles are solid water-ice spheres (courtesy C. S. Arridge)

itself were also detected (Tokar et al. 2009; Coates et al. 2010b). The $>\sim 500$ eV/e signatures during these encounters imply that ELS detected negative ions of M/q from ~400 up to 26,600 and 17,600 amu/q, respectively, and IMS detected positive ions up to 46,500 and 30,700 amu/q. The grains' charge state is undetermined, but if they are singly charged and have the density of solid water ice, they measure up to >2-nm radius, i.e., orders of magnitude smaller than plume grains detected by other instrumentation (Spahn et al. 2006). The derivation of the number density of the nanodust is currently ongoing; a direct calculation of this parameter from the raw data is not possible due to the many instrumental effects that need to be taken into account, including, for example, the effects of ELS microchannel plate detection efficiencies for high mass particles.

The source of these grains is the condensation of water vapour that emerges at Enceladus's south polar region at surface fractures. Modelling of this process (Schmidt et al. 2008) predicts the formation of grains on scales from nm to tens of mm, with the smallest attaining velocities close to the gas speed.

As shown in Fig. 2, although some differences between signatures observed by the different elements of CAPS are due to the sensors' time resolution, there is a clear mismatch between the timing of fine scale structures observed in the positive and negative grain data. In E3 IMS data, one clear, relatively broad flux peak is observed, preceded by a minor isolated peak. In ELS, two main peaks are seen: one detected over ELS's entire energy range and all anodes, possibly resulting from grain fragmentation inside ELS itself, and a second, more complex signature where the peak energy shifts over time, indicating negative grains' spatial separation.

The temporal, and therefore spatial, differences between positive and negative fine-scale structures are believed to be due to the effects of electromagnetic forces on the nanodust. Initially, the motional electric field associated with the flow of magnetospheric plasma past Enceladus will accelerate negative grains towards Saturn and positive grains away from the planet. As reported by Jones et al. (2009), at a qualitative level, this appears to agree with the jet source locations catalogued by Spitale and Porco (2007). Detailed modelling of the processes moulding the shape of the jets is ongoing (Arridge et al. 2010), but it seems that the electromagnetic fields and gravitational drift present in the vicinity of Enceladus; essentially allow the plume region to act as a mass spectrometer, splitting the particles by mass and charge state. The gyroradii of the grains detected by CAPS can reach scales of a few hundred km, i.e., larger than the radius of Enceladus itself. Jets that are initially collimated flows of particles are likely to be splayed into sheets of grains. When Cassini crosses one of these highly modified jets, it may only encounter particles with a very limited range of M/q at a given location, as may have occurred during encounter E10 (Fig. 4). A detailed knowledge of the flow speeds throughout the plume region is needed for accurate modelling of the grain trajectories: as Cassini's sampling of this region will be limited, plume plasma environment simulations are required for the trajectory models.

One puzzle arising from the CAPS observations is the colocation of positively and negatively charged nanometre particles. The overriding charging process near Enceladus is caused by the plume material's immersion in Saturn's corotational magnetospheric plasma, resulting in negative potentials (Kempf et al. 2006). The co-existence of both populations at first appears anomalous. At nanometre scales, however, the charging process can be stochastic, depending on local plasma parameters. In addition, during their subsurface formation in a collisional environment (Schmidt et al. 2008), grains could undergo triboelectric charging: those that condense within the vent even when of the same composition can acquire opposite charges. Smaller particles tend to charge negative, and larger ones positive, e.g., Duff and Lacks (2008). Most entrained nanodust particles are likely to have been accelerated to near-gas velocities.

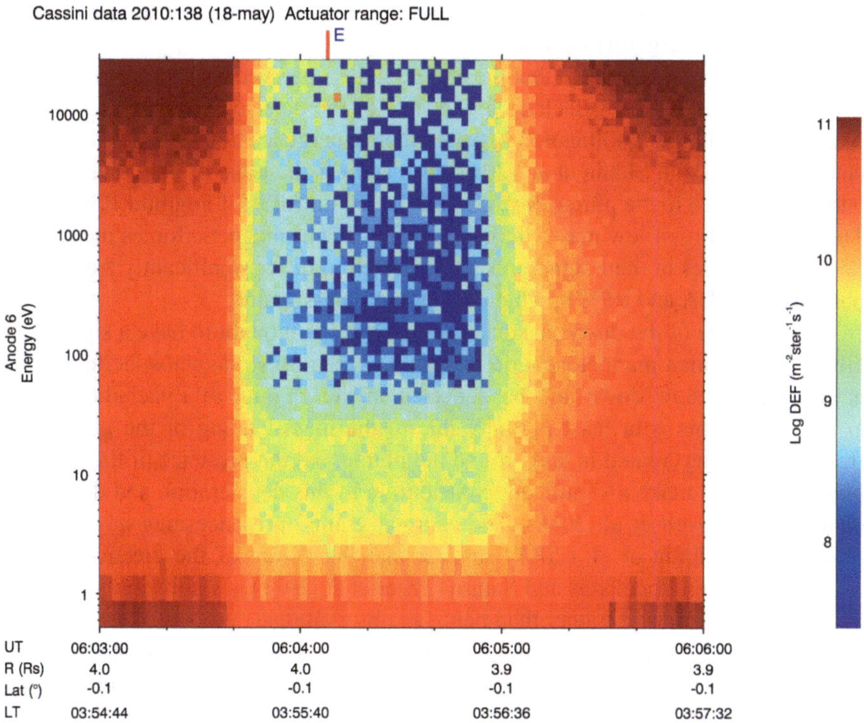

Fig. 4 Detection of nanodust by CAPS-ELS during the E10 encounter in May 2010. The broad decrease in electron fluxes is due to Enceladus itself shielding Cassini from magnetospheric energetic particles that form a background in the ELS data. Unlike the E3 encounter deep in the plume, this more distant crossing at 434-km minimum distance from the centre of the moon appears to have encountered a significantly lower nanograin flux and with over a limited M/q range. The nanograin detections, which were also registered by another anode, indicating that several particles were observed, appear here in red above 10,000 eV at 06:04:10. If these particle were singly charged, their masses were in the range of ~50,000–70,000 amu

Overall, during both E3 and E5 encounters, CAPS detected negative particles to much lower kinetic energies than for positive particles; if a proxy for lower masses, this supports the picture of triboelectric charging occurring within vents. Although sunlight will have little effect in the near-Enceladus environment, particles' charge state changes could vary once exposed to the plume and the corotational plasma flow, where plasma parameters can differ significantly. This may have implications for the behaviour of particles under the influence of electromagnetic forces, as their charge state changes, possibly reversing polarity, grains' paths can be affected significantly. Grains initially charged positive and deflected away from Saturn may charge negative, causing them to then be accelerated in the opposite direction. The positively charged population that can be accelerated outwards by the magnetospheric corotational electric field very likely forms a component of the dust streams observed outside Saturn's magnetosphere (Kempf et al. 2005; Hsu, Krüger, and Postberg, this volume).

Farrell et al. (2010) argued that the charge states observed by IMS and ELS may not reflect the initial triboelectric charge state of the detected particles, due to the likelihood of charge modification between their emergence at the surface and their detection at Cassini. Shafiq et al. (2011) also address the issue of grain charging of particles exposed to the plume environment, based on RPWS Langmuir Probe data. The susceptibility of low mass charged grains to electromagnetic forces means that the fine structures in their signatures are expected to differ significantly from those observed by CDA and RPWS, which detect high mass grains.

The presence of the dusty plasma envelope itself is thought to have a significant effect on Saturnian magnetospheric plasma flow. Saturn's magnetospheric plasma is observed to slow down and be deflected in the vicinity of Enceladus (Tokar et al. 2006). This is at least partially due to the mass-loading of the plasma by fresh pickup ions created in and around the plume, but the presence of the charged grains may also make a significant contribution to this deceleration and deflection of the flow (Farrell et al. 2010). The charged grains' presence may have further implications: Simon et al. (2011) suggest that the nature of the electrodynamic coupling between Enceladus and Saturn (Pryor et al. 2011) is significantly affected by the grains' presence. Rather than being an observational curiosity, it appears that this population of particles spanning the mass range between heavy molecules and micron-scale dust has important effects on several processes occurring in Saturn's inner magnetosphere.

5 Summary and Discussion

Cassini's CAPS instrument has unexpectedly proven itself to be a valuable source of data on the population of charged nanodust in two key contexts in the Saturn system. With the knowledge gained from these data, it is possible to include charged nanograin detection as a design criterion for future planetary exploration instruments. One key element that is missing from the CAPS measurements is a determination of charge state. All three components of CAPS measure the energy per charge of detected particles, not mass per charge, meaning that despite theoretical considerations indicating that single charge states are most likely, some uncertainty will persist in the masses of the detected grains. Charge state determination would therefore be a valuable addition to similar future instrumentation, and indeed invaluable where such an instrument is designed with the determination of nanograin parameters as a primary goal. A determination of the arrival direction of nanograins with a precision of a few degrees or less would also strongly benefit the study of nanograins using future plasma spectrometers, as it will allow a more precise measurement of particles' speeds with respect to the spacecraft.

At the time of writing, Cassini has performed E13, its 14th targeted encounter with Enceladus, a series of flybys that began with E0 in February 2005. Nine more encounters are planned for the remainder of the spacecraft's mission, ending with E22 in December 2015. CAPS will be oriented in the ram direction for several of the nine remaining encounters, and we therefore expect to obtain further valuable data on the charged nanograin populations in the moon's south polar plume. One puzzling question is why ELS seems to be more sensitive to the detection of negatively charged grains than IMS. For example, during E7, ELS clearly detected the presence of negative nanodust, while IMS made no positive detection of the equivalent positively charged population. For E7, IBS was not in a mode to make the complementary positively charged nanograin observations, so it remains to be seen whether parts of the plume can indeed be dominated by nanodust of a particular charge sense during certain periods, or whether instead this observation was at least partially due to an instrumental effect.

Titan too will be encountered many times during the remainder of Cassini's mission, ending with T126 on April 22, 2017. The detailed plans for these encounters are to be finalized, but it is certain that CAPS will be oriented in the ram direction, either fixed or actuating across it, for many of these encounters, adding to our knowledge of negatively charged aerosols high in Titan's atmosphere.

There may be data already gathered by CAPS recording the presence of charged nanodust elsewhere in the system. With our understanding of the nanograin signatures in CAPS data increasing, searches for signatures of these particles are planned. At the planned end of the mission, during April–November 2017, the Cassini orbiter will follow "proximal" orbits where periapsis occurs between Saturn's cloudtops and the main ring system. This should be an excellent opportunity to detect charged nanodust to the north and south of the main ring system.

Acknowledgements The Cassini Plasma Spectrometer team was designed, built, and operated by an international team led at the Southwest Research Institute, San Antonio, Texas (Principal Investigator: F. J. Crary). The author is supported by a UK Science and Technology Facilities Council Advanced Fellowship, and is grateful for useful discussions with C. S. Arridge, A. J. Coates, and A. Wellbrock.

References

Arridge, C. S., Jones, G. H., Crary, F. J., Young, D. T., Kanani, S.: 2010, *EPSC Abstracts* **5**, EPSC2010–691.

Coates, A. J., Crary, F. J., Lewis, G. R., Young, D. T., Waite Jr., J. H., Sittler Jr., E. C.: 2007, *Geophys. Res. Lett.* **34**, L22103.

Coates, A. J., Wellbrock, A., Lewis, G. R., Jones, G. H., Young, D. T., Crary, F. J., Waite Jr., J. H.:2009, *Planet. Space Sci.* **57**, 1866.

Coates, A. J., et al.:2010a, *Faraday Disc.* **147**, 293.

Coates, A. J., et al.: 2010b, *Icarus* **206**, 618.

Crary, F. J., et al.:2009, *Planet. Sp. Sci.* **57**, 1895.

Dougherty, M. K., et al.:2006, *Science* **311**, 1406.

Duff, N., Lacks, D. J.: 2008, *J. Electrost.* **66**, 51.

Farrell, W. M., et al.:2010, *Geophys. Res. Lett.* **37**, L20202.

Gurnett, D. A., et al.: 2004, *Sp. Sci. Rev.* **114**, 395.
Hansen, C. J., et al.: 2006, *Science* **311**, 1422.
Hansen, C. J., et al.: 2008, *Nature* **456**, 477.
Havnes, O., Nsheim, L. I.: 2007, *Ann. Geophys.* **25** 623.
Hedman, M. M., et al.: 2009, *Icarus* **199** 378.
Horányi, M., Burns, J. A., Hedman, M. M., Jones, G. H., Kempf, S.: 2009, In: Dougherty, M. K., Esposito, L., Krimigis, S. M. (eds.) *Saturn from Cassini-Huygens*, Springer Verlag, 511.
Jones, G. H., et al.: 2009, *Geophys. Res. Lett.* **36**, L16204.
Jones, G. H., et al.: 2008, *Science* **319**, 1380.
Kempf, S., et al.: 2005, *Nature* **433**, 289.
Kempf, S., et al.: 2006, *Planet. Space Sci.* **54**, 999.
Kempf, S., et al.: 2008, *Icarus* **193**, 420.
Kempf, S., Beckmann, U., Schmidt, J.:2010 *Icarus* **206** 446.
Krimigis, S. M., et al.: 2004, *Sp. Sci. Rev.* **114**, 233.
Krüger, H., et al.:1999, *Nature* **399**, 558.
Michael, M., Tripathi, S. N., Arya, P., Coates, A., Wellbrock, A., Young, D. T.: 2011, *Planet. Sp. Sci.* **59**, 880.
Porco, C. C., et al., *Science* **311**, 1393.
Postberg, F., et al., *Icarus* **193**, 438.
Pryor, W., et al.: 2011, *Nature* **472**, 331.
Roussos, E., et al.: 2008, *Icarus* **193**, 455.
Schmidt, J., Brilliantov, N., Spahn, F., Kempf, S.: 2008, *Nature* **451**, 685.
Shafiq, M., Wahlund, J.-E., Morooka, M. W., Kurth, W. S., Farrell, W. M.: 2011, *Planet. Sp. Sci.* **59**, 17.
Simon, S., Saur, J., Kriegel, H., Neubauer, F.M., Motschmann, U., Dougherty, M.K., Influence of negatively charged plume grains and hemisphere coupling currents on the structure of Enceladus' Alfvén wings: Analytical modeling of Cassini magnetometer observations. Journal of Geophysical Research (Space Physics), **116**(A15):4221 2011.
Spahn, F., et al.: 2006, *Science* **311**, 1416.
Spencer, J. R., et al.: 2006, *Science* **311**, 1401.
Spitale, J. N., Porco, C. C.: 2007, *Nature* **449**, 695.
Srama, R., et al.: 2004, *Sp. Sci. Rev.* **114**, 465.
Tiscareno, M. S., Burns, J. A., Cuzzi, J. N., Hedman, M. M.:2010, *Geophys. Res. Lett.* **37**, L14205.
Tokar, R. L., et al.: 2006, *Science* **311**, 1409.
Tokar, R. L., et al.: 2009, *Geophys. Res. Lett.* **36**, L13203.
Verbiscer, A. J., Skrutskie, M. F., Hamilton, D. P.:2009, *Nature* **461** 1098.
Waite, J. H., et al.: 2006, *Science* **311**, 1419.
Waite, J. H., et al.: 2007, *Science* **316**, 870.
Young, D. T., et al.: 2004, *Sp. Sci. Rev.* **114**, 1.

In Situ Detection of Interplanetary and Jovian Nanodust with Radio and Plasma Wave Instruments

Nicole Meyer-Vernet and Arnaud Zaslavsky

Abstract Radio and plasma wave instruments in space can detect cosmic dust over a wide range of sizes via impact ionisation. Such measurements were performed on a number of spacecraft in various environments, using instruments that were generally not designed to do so, and have been recently extended to nanodust. The technique is based on analysis of the electric pulses induced by the plasma clouds produced by impact ionisation of fast dust particles. Nanodust can be detected in this way despite their small mass because (1) their large charge-to-mass ratio enables them to be accelerated to high speeds, and (2) the amplitude of the induced electric pulses increases much faster with speed than with mass. As a result, the impacts of nanodust produce signals as high as those of larger and slower grains. This chapter describes the basic principles of such measurements, the underlying physics, the applications to the recent discovery of interplanetary nanodust near Earth orbit with STEREO/WAVES, and to the detection of Jovian nanodust with Cassini/RPWS. Finally, we give some perspectives for wave instruments as dust detectors.

1 Introduction

Even though this was not immediately recognised, the first in situ detection of fast nanodust in space took place 20 years ago, when the Ulysses cosmic dust analyser detected streams of particles ejected by Jupiter that were initially identified as 0.2-μm grains moving at about 50 km/s (Grün et al. 1992). A few years later, these streams were recognised as made instead of fast nanodust, $\sim 10^3$ times less massive and moving 5–10 times faster than previously reported, i.e., outside the calibration range of the instrument (Zook et al. 1996). This pioneering result opened the way to

N. Meyer-Vernet (✉) · A. Zaslavsky
LESIA - Observatoire de Paris, CNRS, UPMC, Université Paris Diderot, 92190 Meudon, France
e-mail: nicole.meyer@obspm.fr; arnaud.zaslavsky@obspm.fr

I. Mann et al. (eds.), *Nanodust in the Solar System: Discoveries and Interpretations*, Astrophysics and Space Science Library 385, DOI 10.1007/978-3-642-27543-2_7,

extensive studies of nanodust produced by outer planets' satellites and rings, whose electric charge enables them to be ejected by the electric field of the corotating magnetosphere and further accelerated by the magnetised solar wind (Johnson et al. 1980; Horányi et al. 1997; Hsu et al. 2012).

That nanodust could also be produced in the inner heliosphere and be accelerated to high speeds by the solar wind was suggested a few years ago (Mann et al. 2007). Nevertheless, when the STEREO/WAVES instrument detected serendipitously voltage pulses of amplitude corresponding to impacts of such fast nanodust, with a rate similar to that expected from extrapolation of the interplanetary dust model (Meyer-Vernet et al. 2009a), this came as a surprise since conventional dust detectors had not detected such interplanetary nanodust (Grün et al. 2001).

In fact, this capability of wave instruments to measure dust should not have been surprising since nearly 30 years ago, the first in situ measurement of microdust in the E and G rings of Saturn was performed serendipitously by the radio (Aubier et al. 1983) and the plasma wave (Gurnett et al. 1983) instruments on the spacecraft Voyager, despite the fact that neither the radio (Warwick et al. 1982) nor the plasma wave (Scarf et al. 1982) instrument was designed to do so. These pioneering results opened the way to microdust measurements with wave instruments in various environments, including other planetary environments and comets (see reviews by Oberc 1996; Meyer-Vernet 2001).

The capability of wave experiments to measure nanodust was confirmed by the detection of nanodust near Jupiter by the Cassini/RPWS instrument (Meyer-Vernet et al. 2009b), and the STEREO pioneering result was subsequently confirmed and expanded by a detailed analysis based on an independent dataset acquired by a different subsystem of the STEREO/WAVES instrument (Zaslavsky et al. 2012). On the theoretical front, detailed calculations of the nanodust dynamics confirmed their ejection from the inner heliosphere and their acceleration in the solar wind to several hundred of kilometres per second at 1 AU (Czechowski and Mann 2010).

In this chapter, we summarise the basic principles of dust detection with a wave instrument and their extension to nanodust (Sect. 2), and the main results obtained for interplanetary nanodust near 1 AU with STEREO (Sect. 3), and for Jovian nanodust with Cassini (Sect. 4). In Sect. 5, we give some perspectives for wave instruments, which are complementary to traditional dust detectors since they have a much greater collecting area and are much less reliant on a specific spacecraft attitude. Unless otherwise stated, we use the International System of units.

2 Basics of In Situ Dust Detection with a Wave Instrument

The traditional use of wave instruments is the observation of electromagnetic waves, whose propagation from large distances enables measurements of distant objects by radio techniques. It was soon realised that these instruments could also be used at lower frequencies for in situ measurements, by detecting intense plasma waves produced by instabilities. A crucial step was reached by showing that, since

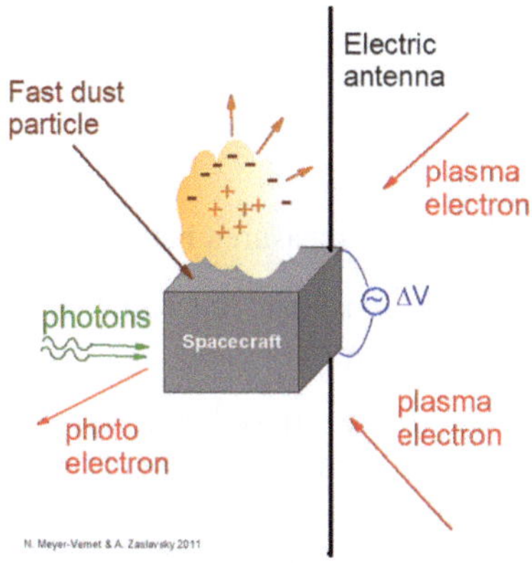

Fig. 1 Principle of in situ measurements of plasma and dust with a wave instrument. Plasma particles passing-by the antennas (as well as impacting and ejected particles) produce a quasi-thermal electrostatic noise whose power spectrum reveals the plasma density, temperature, and other properties. Dust impacts at fast speed produce partial ionisation of the dust and target, yielding an expanding plasma cloud. This produces voltage pulses whose analysis reveals some dust properties

electrostatic waves are closely coupled to plasma particles, a sensitive wave receiver in space can also measure in situ several bulk properties of *stable* plasmas (Meyer-Vernet 1979). This is because the motion of the charged particles around the antenna, as well as the impacts or emission (Fig. 1), produces a quasi-thermal noise whose analysis reveals their density, temperature, and possible nonthermal properties. This has led to the technique of quasi-thermal noise spectroscopy, which has been successfully used for plasma measurements in various space environments (Meyer-Vernet et al. 1998).

But electric antennas are not only sensitive and accurate plasma detectors of equivalent cross-section much greater than their physical size. They can also detect dust, since impacts of high speed dust particles vaporise and partially ionise them as well as some material of the impact craters, producing plasma clouds whose electric field reveals some dust properties. As a result, electric antennas can also be used as sensitive dust detectors of large detecting area since it may be the whole spacecraft surface.

2.1 What Do Radio and Plasma Wave Instruments Detect

Wave instruments used for in situ measurements of plasma and dust deliver a voltage[1] using two basic systems:

[1]We do not consider here magnetic field measurements.

- Electric antennas made of conductive booms,[2] which are used in two main ways:
  - In monopole mode, the voltage is measured between one antenna boom and the spacecraft conductive structure.
  - In dipole mode, the voltage is measured between two antenna booms.
- Electronic analysers which transform the signal into quantities suitable for analysis, and deliver two main types of data:
  - Time-integrated power spectra, equivalent to Fourier transforms of the auto-correlation function of the measured voltage; this part of the instrument is called a frequency receiver.
  - Broadband voltage waveforms made of time series data captured at a very high rate; this part of the instrument is called a time domain sampler (TDS).

The power spectrum delivered by frequency receivers corresponds to

$$V_f^2 = 2\int_{-\infty}^{+\infty} \mathrm{d}\tau \, \mathrm{e}^{\mathrm{i}\omega\tau} \langle V(t)V(t+\tau)\rangle \tag{1}$$

(where the frequency $f = \omega/2\pi$). Since this involves a time integration, frequency receivers are not adapted to study short individual events, even though in modern instruments such as STEREO/WAVES and the high-frequency (hf) receiver of Cassini/RPWS, the integration is calculated over short times (typically <1 s). Furthermore, in order to cover a large dynamic range, the STEREO low-frequency receiver (LFR) is equipped with an automatic gain control (AGC) which adjusts the gain according to the input level; an adequate response thus requires the signal to be stationary or to be made of a large number of individual events during the acquisition time. In contrast, TDSs are adapted to study individual events since the time series are rather long (typically 131 ms on STEREO) and acquired at a very high rate (typically 8 μs on STEREO). However, they involve so huge a quantity of data that a selection of the telemetered periods must be made on board; in complement, other types of data are telemetered as for example the peak signal within some given time periods, or histograms (Bougeret et al. 2008).

Figure 2 shows an example of spectrograms acquired by the low and high frequency receivers on STEREO A and B. The spectra are displayed as frequency vs. time with relative intensity above background (in dB) scaled in grey. They show solar type III bursts—a type of solar emission for the study of which the instrument was designed, and unexpected voltage pulses attributed to nanodust impacts that will be discussed in Sect. 2.3. Likewise, the TDS was designed to study Langmuir wave packets as the one shown in Fig. 3 or other types of plasma instabilities, but

[2]We do not consider antennas made of spheres because they must be mounted on booms, which complicates considerably the analysis (Manning, 1998), so that they are rarely used.

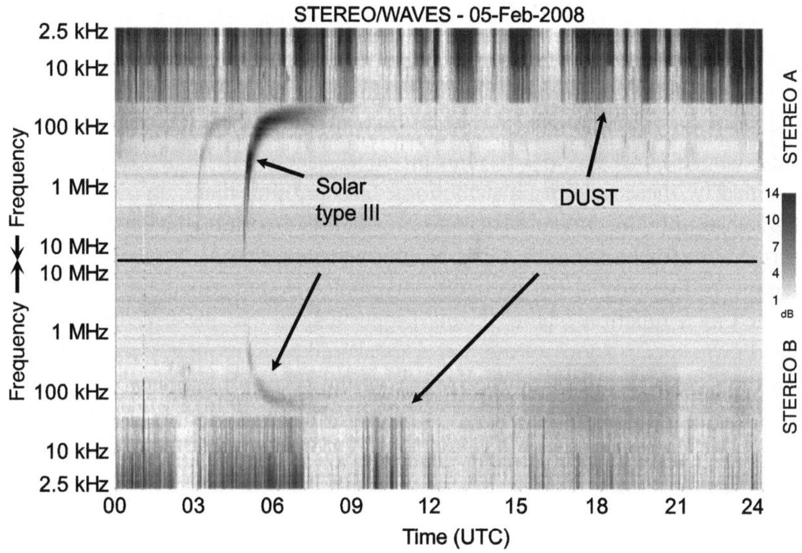

Fig. 2 Typical spectrograms displayed as frequency (2.5 kHz–20 MHz) vs. time (24 h) with relative intensity scaled in *grey*, from the low and high frequency receivers with the X–Y dipole on STEREO A and B (then separated by 45° longitude), showing solar type III bursts and voltage pulses produced by nanodust impacts. The discontinuities between frequency bands are due to differences in integration times. Both spacecraft see the same solar type III, albeit with different intensities, due to the directivity of the source of electromagnetic waves and to scattering by coronal and solar wind plasma. In contrast, they see different dust pulses since the measurement is local

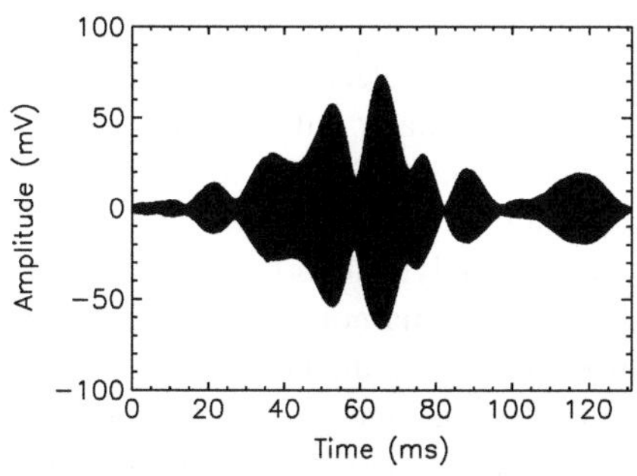

Fig. 3 Voltage waveform from the TDS measured on 31:01:2007 with the STEREO B/X-antenna, showing a Langmuir wave packet

it turned out to measure voltage pulses as shown in Fig. 7 that will be discussed in Sect. 2.3.

Since the detection of dust by wave instruments builds on many concepts introduced for plasma detection, we first remind them briefly.

2.2 How Do Passive Wave Instruments Measure the Ambient Plasma

A stable plasma is characterised by the velocity distributions of the particle species, whose quasi-thermal motion produces electric field fluctuations entirely determined by these velocity distributions. On the other hand, an electric antenna is characterised by its current distribution, which is determined by its geometry for a short dipole (of half-length $L \ll \lambda$, the wavelength). As a result, the voltage power spectral density at the ports of a given electric antenna can be theoretically calculated as a function of the plasma properties, so that these properties can be deduced from spectroscopy of the power measured by a frequency receiver.

For timescales smaller than the inverse of the plasma frequency ($f_p \propto n^{1/2}$, n being the ambient electron density), the electric field of a charged particle moving slower than the electron thermal speed v_{th} is screened at distances greater than the Debye length $L_D \propto (T/n)^{1/2}$ (T being the temperature). Thermal electrons moving within this distance from the antenna thus produce voltage pulses of timescale $\sim L_D/v_{th} \sim 1/2\pi f_p$. From the properties of Fourier transforms, this yields a flat noise spectrum at frequencies $< f_p$. On the other hand, faster charged particles (as well as some perturbations) produce Langmuir waves of frequencies $\simeq f_p$ and wavelengths $> L_D$ ($\sim$10 m in the solar wind). This produces a quasi-thermal noise spectral peak at f_p if the antenna length $L > L_D$.

Fitting the measured spectra to a theoretical calculation thus yields the electron density and temperature as well as other plasma properties, so that the electric antenna serves as a plasma detector of equivalent cross section $\sim 2L \times L_D$—typically several hundred square metres in the solar wind, which generally exceeds the antenna physical cross section by more than three orders of magnitude. An additional contribution comes from electrons collected by the antenna surface (or ejected by it), which produce voltage pulses of rise time $\tau_e \sim \mathrm{Min}(L, L_D)/v_{th}$) (decreasing at a much longer timescale); from the properties of Fourier transforms, this yields a power spectrum $\propto f^{-2}$ at frequencies $< 1/2\pi\tau_e$. Basic theoretical expressions and approximate analytical formulas are given by Meyer-Vernet and Perche (1989). Note that this plasma noise cannot be measured by TDSs since it involves the superposition of a huge number of extremely small signals.[3]

Finally, we note that the instantaneous voltage measured by each wire antenna boom shorter than the electromagnetic wavelength is equal to the average potential V_A along its length. However, the receiver of impedance Z_R detects $V_R = \Gamma V_A$, where $\Gamma = Z_R/(Z_R + Z_A) \simeq C_A/(C_A + C_{stray})$, since the impedances are mainly capacitive at the frequencies considered. The receiver impedance is essentially due to the stray capacitance C_{stray}, whereas the antenna impedance Z_A, mainly due to

[3]Each passing electron produces a voltage pulse of order of magnitude $e/4\pi\varepsilon_0 L_D$ (where e is the electron charge), which amounts to $\sim 10^{-10}$ V in the solar wind at 1 AU. The number of such pulses per second $\sim 2nv_{th}LL_D > 2 \times 10^{15}\ \mathrm{s}^{-1}$ in the solar wind at 1 AU if $L > L_D$.

the antenna capacitance C_A, can be calculated from the plasma properties (Meyer-Vernet and Perche 1989).

2.3 How Do Wave Instruments Measure the Ambient Dust Flux

In situ measurements by wave instruments are based on the detection of electrostatic fields produced by electric charges. For dust grains, this may be the charge carried by the grains, or the much greater charge produced by high speed impact ionisation.

2.3.1 Nonimpacting Dust Grains

Dust particles in a plasma carry an electric charge due to photoelectron emission, collection of plasma particles, and secondary emission. Their motion thus produces electrostatic field variations, so that an electric antenna can in principle detect the noise produced by dust grains passing by the antenna (and/or collected by its surface). This noise can be observed with a frequency receiver if the dust charge and concentration are high enough (Meyer-Vernet 2001). Contrary to the plasma thermal noise, this "dust thermal noise" can also be observed with a sensitive TDS for highly charged dust grains, with an equivalent cross section $\sim 2L \times L_D$—similar to that for plasma detection (typically $>200\,\mathrm{m}^2$ in the solar wind), as proposed by Meuris et al. (1996).

However, when the dust grains move faster than a few km/s, the sound speed in bulk matter, another effect becomes dominant: impact ionisation.

2.3.2 Impact Ionisation

A dust grain impacting a solid target such as a spacecraft or its appendages produces a strong shock compression which vaporises and ionises the dust as well as some material of the impact crater. This material then expands into the low-pressure ambient medium, cooling and partially recombining (Drapatz and Michel, 1974). The residual ionisation of the expanding plasma cloudlet can be used to detect the grain. In practice, one measures the charge Q carried by the electrons and/or the ions by separating and recollecting them, and deduces the grain mass m and speed v from laboratory calibrations of the timescales of the signals (to deduce v), and of the relationship $Q(m, v)$ (to deduce m). This is the principle of classical impact ionisation dust detectors (Auer, 2001).

The relationship $Q(m, v)$ depends on the material of both the grain and the target as well as on the impact angle; despite extensive theoretical calculations and simulations (Kissel and Krüger 1987; Hornung and Kissel 1994), it remains largely empirical (Krüger, 1996). In order of magnitude, we have $Q \propto mv^{3.5}$ (Dietzel et al. 1973), with a typical relationship

$$Q \simeq 0.7 m^{1.02} v^{3.48} , \quad (2)$$

with Q in Cb, m in kg, v in $\mathrm{km\,s^{-1}}$ (McBride and McDonnell, 1999). The three coefficients in (2) depend on mass, speed, angle of incidence, as well as grain and target composition (Göller and Grün 1989), so that Q may differ from this relationship by one order of magnitude.

An important consequence of (2) is that a 10-nm grain moving at 300 km/s in the solar wind, as predicted by dynamics (Mann et al. 2007; Czechowski and Mann 2012), should produce the same impact charge as a grain ~500 times more massive impacting at 50 km/s. Note, however, that such empirical relationships are based on simulations involving grains of mass greater than about 10^{-18} kg and speed smaller than about 80 km/s (Auer, 2001), i.e., outside the range of fast nanodust.

Some comparisons may be instructive. First, let us compare the amount of material involved in the charge Q given in (2) to that contained in the grain itself. The mass m_Q of the ions—assumed to be singly ionised—involved in Q is Am_pQ/e (m_p being the proton mass and A the average ion atomic mass), so that

$$m_Q/m \simeq (AQ/m)(m_p/e) \sim 10^{-8} \times Am^{0.02} v^{3.48} , \quad (3)$$

For $v \simeq 300$ km/s, this yields $m_Q/m \sim A$ over the whole grain mass range, so that for $A > 1$ a significant part of the impact charge comes from the target's material.

Second, let us compare the mass m_{crater} involved in the impact crater to the grain mass m. From empirical expressions for dust impacts on various targets (McBride and McDonnell, 1999), we get

$$m_{\mathrm{crater}}/m \sim m^{0.056} v^{2.42} , \quad (4)$$

in order of magnitude,[4] with m in kg and v in $\mathrm{km\,s^{-1}}$. This ratio exceeds unity in most practical cases. For a 10-nm grain ($m = 10^{-20}$ kg) impacting at $v = 300$ km/s, (4) yields $m_{\mathrm{crater}}/m \sim 10^5$, i.e., the crater is of size nearly 1 μm. This means that most of the ejected mass comes from the target, presumably in the form of debris, and since (3) yields $m_Q/m \geq 1$, the total ejected mass m_{crater} exceeds by a large amount the ionised mass m_Q.

Third, let us compare the impact charge Q to the charge q carried by the dust grain itself before impact. For a grain of radius r, we have $q \simeq 4\pi\epsilon_0 r\Phi$, with $\Phi \simeq$

[4] The ejected mass is very dependent on the materials involved, and is expected to increase as the dynamic yield strength and the mass density of the target material decrease (Shanbing et al. 1994). For deriving (4), we have assumed a crater volume $\sim\epsilon^3$, where ϵ is the penetration distance, a target yield stress of the same order of magnitude as that of aluminium, and a mass density $\sim 2.5 \times 10^3\ \mathrm{kg\,m^{-3}}$ for the target and the grain.

5–10 V in the interplanetary medium (see for example Meyer-Vernet 2007). With a grain mass density of $2.5 \times 10^3\,\mathrm{kg\,m^{-3}}$, this yields

$$Q/q \sim 2 \times 10^{10} m^{0.69} v^{3.48}\,, \qquad (5)$$

so that $Q/q \gg 1$ in almost all practical cases. Hence, for dust grains of impact speed greater than a few km/s, impact ionisation is generally the dominant mechanism for detection by wave instruments. For a 10-nm grain impacting at 300 km/s, (5) yields $Q/q \sim 10^5$.

However, in the absence of laboratory simulations and calculations for fast nanodust, the reliability of (2), as well as (3)–(5), for such particles remains open to question. Surface effects play an increasing role as size decreases; for example in a 1-nm grain, a significant proportion of the atoms lie at the surface. Furthermore, since the ejected neutral material, in the form of solid or liquid debris and/or gases, exceeds considerably the ionised material, it may affect the evolution of the system. The expected disintegration of the liquid phase into a large number of small droplets before vaporisation, complicates considerably the simulation (Hornung et al. 2000). These difficulties are especially important in the case of STEREO since detailed data on the material of the blankets covering the spacecraft are not easily available, and the published spacecraft description (Driesman et al. 2008) contains errors in the technical properties of the blankets. Finally, energy conservation yields a speed limit for (2) to hold. Since the energy required to produce free charges by vaporisation and ionisation, $\simeq 10\,\mathrm{eV}$ for each of the Q/e ions, cannot exceed the grain kinetic energy,[5] the validity of (2) requires at least $v < 2 \times 10^3$ km/s.

Be that as it may, it is noteworthy that the initial identification of Jovian dust streams by traditional detectors, which was based on their calibration, has been subsequently modified from dynamical arguments by multiplying the mass by 10^{-3} and the speed by 5–10 (Zook et al. 1996). Since $10^{-3} \simeq 7^{-3.5}$, this suggests that the $Q \propto mv^{3.5}$ law still holds for these grains.

Finally, let us note that a fast nanodust impact represents a huge incident power since for a grain of radius r and mass density ρ, the incident kinetic energy $\rho(4\pi r^3/3)v^2/2$ comes over the surface $\sim\pi r^2$ during a time $\sim r/v$, which yields a power $P \sim \rho v^3$. For $\rho \sim 2.5 \times 10^3\,\mathrm{kg\,m^{-3}}$ and $v \sim 300$ km/s, this yields $P \sim 10^{20}\,\mathrm{W/m^2}$—a huge power, greater by several orders of magnitude than that involved in laboratory simulations.

2.3.3 The Impact Plasma Cloud

Let us study the properties of the plasma cloud in the simple case when it behaves independently of its environment. This is expected to hold when two conditions

[5] The actual limit is expected to be smaller since part of the energy is also used for vaporisation, formation of debris, and kinetic energy of the expelled material.

are met: first its density should significantly exceed that of the surrounding plasma; second, the energy of its particles (in eV) should significantly exceed the electric potential of the spacecraft and antennas. We also neglect the neutral gas component of the impact plasma cloud as well as the solid or liquid debris, and assume that the cloud contains Q/e ions (singly ionised positively) and the same number of electrons, is uniform, of spherical shape, and expanding at constant speed v_E. As the radius increases with time t as $R \sim v_E t$, the electron number density decreases as $n \simeq 3Q/(4\pi R^3 e) \propto t^{-3}$. The maximum radius R_{max} and lifetime t_{max} of the cloud are reached when its electron density has decreased to the ambient value n_a, i.e.

$$R_{max} \sim (3Q/4\pi e n_a)^{1/3}, \tag{6}$$

$$t_{max} \sim R_{max}/v_E. \tag{7}$$

An important quantity is the ratio between the cloud's proper Debye length $L_D \propto (T/n)^{1/2}$ and its radius R, which controls charge separation

$$L_D/R \simeq \left(\frac{4\pi\epsilon_0 R T_{(eV)}}{3Q}\right)^{1/2}. \tag{8}$$

If $L_D/R \ll 1$, the cloud is quasi-neutral except in a small outer shell of width L_D which contains a charge $\sim Q \times L_D/R$ producing an electric potential

$$\phi_c \sim (Q L_D/R)/4\pi\epsilon_0 R. \tag{9}$$

The evolution of the cloud's temperature T as it expands depends on its collisional state. The electron and ion mean free path is determined by Coulomb collisions, whose effective cross section is determined by the radius r_L at which the Coulomb potential of a plasma particle $\sim e/4\pi\epsilon_0 r_L$ equals its kinetic energy $\propto T$; this yields a free path $\propto T^2$; taking into account large distance particle encounters up to distance L_D increases the effective cross section by a factor $\lambda_0 = \ln(L_D/r_L)$. The cloud's electrons and ions become collisionally decoupled when the mean free path becomes greater than the cloud's radius, which therefore takes place when the radius is equal to $R_0 \simeq (e/4\pi\epsilon_0 T_{0(eV)})(3\lambda_0 Q/4e)^{1/2}$ (Pantellini et al. 2012a). Here $T_{0(eV)} = k_B T_0/e$ where T_0 is the temperature of the cloud at this time, and λ_0 lies typically between 2 and 10. One deduces from (8) that at the beginning of the collisionless regime $L_{D0}/R_0 \simeq (\lambda_0 e/12Q)^{1/4} \ll 1$ since $Q/e \gg 1$ and $\lambda_0 > 1$.

This problem has been recently studied by Pantellini et al. (2012a) with a N-body simulation. As the cloud expands, the problem becomes self-similar, so that the ratio L_D/R remains constant; therefore, according to (8), the electron temperature decreases[6] as $T \propto 1/R$. An important consequence is that since $L_{D0}/R_0 \ll 1$, we

[6]This can be understood as follows. Since most of the electrons are trapped in the cloud, we have $T_{(eV)} \sim \phi_c$, the cloud's potential. Since $L_D \propto (T/n)^{1/2}$ and $n \propto R^{-3}$, we have

also have from (8) $L_D/R \ll 1$ during the collisionless expansion, i.e., the cloud is quasi-neutral at collisional decoupling and remains so as it expands further. For a 10-nm grain impacting at 300 km/s in the solar wind at 1 AU where $n_a \sim 5 \times 10^6\,\mathrm{m}^{-3}$, (2) and (6) yield $Q \sim 10^{-12}$ C and $R_{max} \sim 1$ m, so that $R_0 \sim 10^{-5}/T_{0(eV)}$, $L_D/R \sim 0.02$, and $\phi_c \sim 2\times10^{-4}$ V at distance R_{max}. Therefore the collisionless regime starts at a very small value of the cloud's radius, the cloud is quasi-neutral except in a thin outer shell, and it produces a very small electric potential (Pantellini et al. 2012a).

2.3.4 Application to Wave Instruments

The charge in the expanding impact plasma cloud can in principle be detected by wave instruments in a number of ways:

1. Direct recollection of the cloud's charges by the spacecraft or antennas (Aubier et al. 1983; Meyer-Vernet 1985).
2. Perturbation of the current balance of the spacecraft or antennas (Oberc et al. 1990).
3. Direct detection of the electrostatic field produced by charge separation in the impact plasma cloud (Oberc 1994, 1996).
4. Detection of electromagnetic radiation produced by charge oscillations in the plasma cloud (Foschini 1998).

Direct Charge Recollection

In the solar wind, the floating potential of the spacecraft and antennas is mainly determined by the balance of photoelectron emission and collection of ambient electrons. Since an uncharged surface generally ejects many more photoelectrons than it collects ambient solar wind electrons (whereas ambient ions contribute much less), the equilibrium potential is positive and of the order of magnitude of a few times the photoelectron temperature expressed in eV, $T_{ph} \sim 3$ eV, i.e. $\Phi \sim 5$–10 V, in order to bind the photoelectrons sufficiently to make their flux balance that of the ambient electrons. For nanodust, R_{max} is much smaller than the solar wind Debye length,[7] and of the order of magnitude of the Debye length of the photoelectrons ejected by the sunlit surfaces.[8] Hence the impact plasma cloud lies in the electric field produced by the spacecraft charge. Since the spacecraft (or

---

$\phi_c \propto (T/R)^{1/2}$, whence from (9) $T \simeq 1/R$. In contrast, an adiabatic behaviour would yield instead $T \propto n^{2/3} \propto R^{-2}$.

[7] The solar wind density $n_a \sim 5 \times 10^6\,\mathrm{m}^{-3} \propto d^{-2}$, so that from (6) $R_{max} \propto d^{2/3}$, and $L_{Da} \simeq 10 \times d^{\alpha}$ m where d is the heliocentric distance in AU and $1/2 < \alpha < 1$ (Meyer-Vernet 2007). Hence, $L_{Da}/R_{max} \sim 10$ for $Q \sim 10^{-12}$ C, with a variation as $Q^{-1/3}$ and a weak variation with d.

[8] With an ejected photoelectron current $I_{ph0} \simeq 5 \times 10^{-5}\,\mathrm{Am}^{-2}$ and $T_{ph} \simeq 3$ eV, the photoelectron Debye length $L_{Dph} \simeq (\epsilon_0/I_{ph0})^{1/2}(T_{ph(eV)})^{3/4}(e/m_e)^{1/4} \sim 1$ m at 1 AU, with $L_{Dph} \propto d$ (Meyer-Vernet 2007).

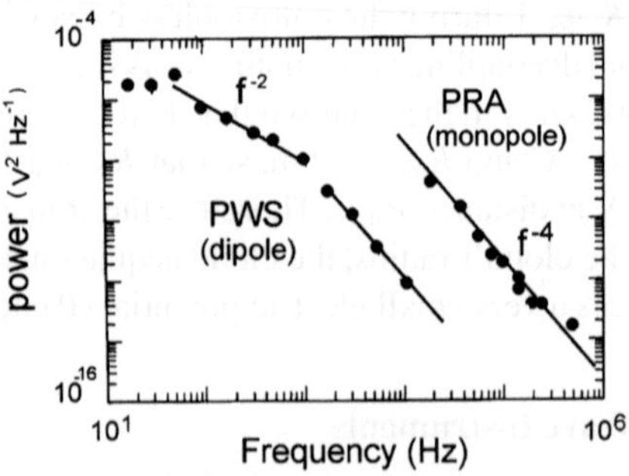

Fig. 4 Voltage power spectrum measured in Saturn G ring by the (high frequency) radio (PRA) and (lower frequency) plasma wave (PWS) instruments on Voyager 2, with respectively monopole and dipole antennas. At similar frequencies, the power is higher by four orders of magnitude on the monopole than on the dipole. Adapted from Mann et al. (2011)

antenna) electric potential is much greater than the small potential of the quasi-neutral cloud estimated in Sect. 2.3.3, and exceeds the kinetic energy (in eV) of the cloud's electrons, they are easily recollected. The spacecraft receives most of the dust impacts and of the recollected electrons because of its larger cross section compared to the antenna, so that an impact makes its voltage vary by $-Q/C_{sc}$, C_{sc} being the spacecraft capacitance.

Each monopole sees the difference between its own voltage (which barely changes except if the impact takes place close to it—see item 2) and that of the spacecraft. Therefore, each impact produces by this mechanism a *positive* voltage pulse of similar amplitude on all the *monopole* antennas, i.e., taking the receiver gain into account

$$\delta V_1 \sim \Gamma Q/C_{sc}. \tag{10}$$

On the other hand, since a dipole antenna sees the difference in voltage between two monopoles, it detects a much smaller signal, produced by circuit imbalances (the so-called common-mode rejection) and/or by weak recollection by the antennas themselves. This difficulty in calibrating the dipole mode for direct charge recollection is at the origin of the conflicting results obtained on Voyager in planetary rings by the radio (Aubier et al. 1983) and the plasma wave (Gurnett et al. 1983) instruments, which used respectively monopole and dipole configurations (see Fig. 4 and discussions by Oberc 1994; Meyer-Vernet 1985, 2001).

The impact voltage pulse has a short rise time τ_r determined by collection of the cloud's electrons, and a much longer decay time τ_d, determined by the discharge of the spacecraft via the ambient plasma and photoelectron currents. Such a pulse has the generic squared Fourier transform (Meyer-Vernet 1985)

$$V^2_{f\,(\mathrm{pulse})} \simeq \frac{2\delta V^2/\omega^2}{(1+\omega^2\tau_r^2)(1+1/\omega^2\tau_d^2)}. \tag{11}$$

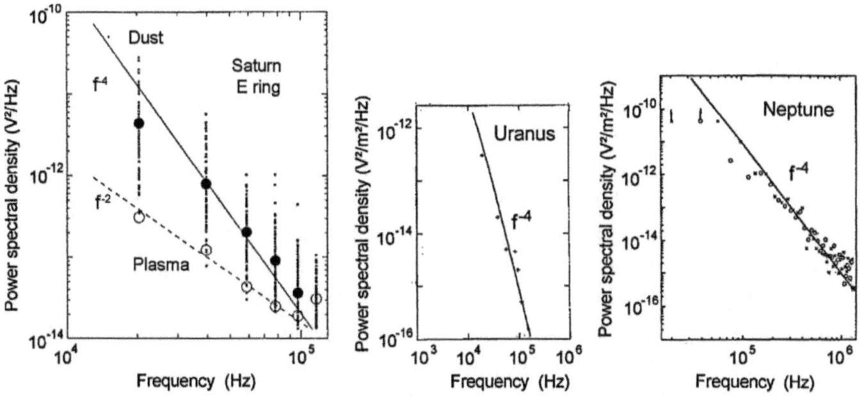

Fig. 5 Voltage power spectral density measured by the monopole antennas on Voyager/PRA (Warwick et al. 1982) in the microdust rings of Saturn (adapted from Meyer-Vernet et al. 1996), Uranus (adapted from Meyer-Vernet et al. 1986), and Neptune (adapted from Pedersen et al. 1991)

Here $\delta V \equiv \delta V_1$. At frequencies $f = \omega/2\pi \gg 1/2\pi\tau_r \gg 1/2\pi\tau_d$, (11) yields $V_f^2 \propto f^{-4}$, as consistently observed at high frequencies in the presence of dust impacts, whereas the slope decreases at lower frequencies (Figs. 4, 5, 8, and 12).

Equation (11) must be integrated over the mass distribution of the dust impact rate, which yields

$$V_f^2 \simeq \int \mathrm{d}m\, V^2_{(\mathrm{pulse})}\, S(m)\, \mathrm{d}F/\mathrm{d}m \tag{12}$$

where $F(m)$ is the *cumulative* dust flux and $S(m)$ the surface involved, which for this mechanism is the spacecraft area subjected to impacts.

For fast nanodust, we have seen that Q is of the order of magnitude of 10^{-12} C or smaller, so that with the typical spacecraft[9] capacitance $C_{sc} \simeq 200$ pF and $\Gamma \simeq 0.5$, the pulse amplitude is $\delta V_1 \sim \Gamma Q/C_{sc} \simeq 2.5$ mV. Such a small signal requires a sensitive TDS receiver to be detected individually. On the other hand, the impacts can be observed with a frequency receiver if the impact rate is very high; we will show an application for nanodust detection near Jupiter with Cassini/RPWS in Sect. 4. Hence, other mechanisms may dominate the signal.

Perturbation of the Current Balance of the Spacecraft or Antennas

As we already noted, the floating potential of a surface in the interplanetary medium is mainly determined by balance between the photoelectron and ambient electron

[9]These typical values of C_{sc} and Γ hold for both STEREO/WAVES (Bale et al. 2008; Zaslavsky et al. 2011, 2012) and Cassini/RPWS (Gurnett et al. 2004; Meyer-Vernet et al. 2009b).

currents. Let I_{ph0} and I_{e0} be respectively the photo and ambient electron currents per unit surface at zero potential, and $T_{\text{ph(eV)}}$ and $T_{\text{e(eV)}}$ the corresponding electron temperatures in eV. A surface at equilibrium potential $\Phi > 0$ attracts the ejected photoelectrons so that only a fraction $I_{\text{ph}}/I_{\text{ph0}} \simeq (1+\Phi/T_{\text{ph(eV)}})^{\alpha} \exp(-\Phi/T_{\text{ph(eV)}})$ does not return, whereas the attracted ambient electron current becomes $I_{\text{e}} \simeq I_{\text{e0}}(1 + \Phi/T_{\text{e(eV)}})^{\alpha}$, with $\alpha = (n-1)/2$ in n-dimensional geometry[10] (Meyer-Vernet 2007). Hence, since in the solar wind $I_{\text{ph0}} \gg I_{\text{e0}}$ and $T_{\text{e}} \gg T_{\text{ph}}$, the equilibrium potential is $\Phi \sim T_{\text{ph(eV)}} \ln(I_{\text{ph0}}/I_{\text{e0}})$, and any significant perturbation will change this potential by a value of the order of magnitude of $T_{\text{ph(eV)}}$.

Depending on the impact charge, on the photoelectron current, and on the surfaces involved, this may affect the spacecraft and/or the electric antennas. A large grain producing an impact plasma cloud that engulfs the spacecraft may perturb its current balance, which can mainly be observed in monopole mode. In contrast, a fast nanodust impact producing a smaller impact plasma cloud may disturb the equilibrium potential of an antenna affected by this tiny cloud, which can be observed in both dipole and monopole mode.

In particular, if a perturbation disturbs the photoelectrons ejected by an antenna so that a significant fraction of them do not return, the antenna potential will increase by $\delta V \sim T_{\text{ph(eV)}}$ if the perturbation holds during a large enough time. Such an effect was studied by Oberc et al. (1990) to interpret large voltage pulses observed by the Vega two plasma wave (APV-N) instrument in the dust coma of comet Halley, due to impacts of small particles yielding a charge $Q < 2 \times 10^{-12}$ C and producing voltage pulses much higher than the value (10) expected from electron recollection.

For a fast dust impact on a spacecraft, where a fraction x/L of an antenna is affected, the voltage pulse amplitude at the receiver ports is thus expected to be

$$\delta V_x \simeq \Gamma T_{\text{ph(eV)}} x/L, \tag{13}$$

if two conditions are met. First the electric potential perturbation produced by the impact cloud must prevent most of the photoelectrons ejected by the antenna to return to it. One might think naively that this requires the cloud's potential to be of the order of magnitude of the photoelectron energy (in eV). As shown by Pantellini et al. (2012b), this is not so because most of the photoelectrons returning to the antenna do so on elliptic trajectories of high eccentricity, and due to angular momentum conservation, the kinetic energy corresponding to their azimuthal velocity decreases as the square of the radial distance to the antenna axis. Since in the solar wind the photoelectron Debye length is much greater than the antenna radius a, most of the photoelectrons whose trajectory returns to the antenna lie far from its surface and thus have a very small azimuthal kinetic energy. Hence,

[10] For a spacecraft in the solar wind, of size generally greater (respectively, smaller) than the photo (respectively, plasma) electron Debye length, we have $\alpha = 0$ for I_{ph}, and $\alpha = 1$ for I_{e}. For a cylindrical antenna of radius (respectively, length) smaller (respectively, greater) than the Debye lengths, we have $\alpha = 1/2$.

a very small cloud's potential should be sufficient to increase significantly their azimuthal velocity and put them on less eccentric orbits that no longer cross the antenna. The second condition for (13) to hold is that the photoelectron current ejected by the antenna length x affected by the plasma cloud, $I_{ph0} \times 2ax$, must be able to eject the charge $C_A \delta V_x / \Gamma$ during the time available τ (which cannot exceed the lifetime (7) of the impact plasma cloud).[11] Here $C_A \simeq 2\pi\epsilon_0 L / \ln(L_{Da}/a)$ is the low-frequency capacitance of a monopole antenna of radius a and length L in the ambient plasma of Debye length L_{Da} (Meyer-Vernet and Perche 1989). From (13) this requires

$$I_{ph0} \geq \pi\epsilon_0 T_{ph(eV)} / [a\tau \ln(L_{Da}/a)]. \tag{14}$$

We shall see that this condition holds for STEREO at 1 AU (Sect. 4), but not for Cassini at 5 AU (Sect. 5) where the photoelectron current I_{ph0} is smaller by a factor of 25.

Integrating (13) over the probability of impacts at a given distance from an antenna on a spacecraft face of size $R_{sc} > R_{max}$ with a simplified model, one finds the average voltage pulse produced by a plasma cloud of maximum radius R_{max} (Zaslavsky et al. 2012) $\delta V_2 = \langle \delta V_x \rangle \simeq 2\Gamma T_{ph(eV)} R_{max}^3 / [3 L R_{sc}^2]$. Substituting the value of R_{max} given in (6), this yields

$$\delta V_2 \simeq \frac{\Gamma T_{ph(eV)} Q}{2\pi e n_a L R_{sc}^2}. \tag{15}$$

For typical spacecraft and solar wind properties, (10) and (15) yield $\delta V_2 \gg \delta V_1$ at heliocentric distances >0.3 AU, so that upon a nanodust impact on the spacecraft producing Q, the monopole whose photoelectrons are affected by the cloudlet should record a pulse of average amplitude $\delta V_2 + \delta V_1 \simeq \delta V_2$ given by (15), whereas the other monopoles should record simultaneously a pulse δV_1 given by (10) of much smaller amplitude, due to electron recollection by the spacecraft. An example measured on STEREO will be displayed in Sect. 3 (Fig. 9).

Charge Separation in the Cloud

From Sect. 2.3.3 and recent kinetic simulations with a N-body scheme (Pantellini et al. 2012a), the electrostatic field produced by charge separation in the impact plasma cloud is expected to be very small for nanodust impacts, so that this mechanism is not expected to be relevant for nanodust wave detection.

---

[11]From (6) and (2) and a cloud's expanding speed $v_E \sim 20$ km/s, t_{max} is of order 30 μs for a 10-nm grain impact at 300 km/s producing $Q \simeq 10^{-12}$ C.

Electromagnetic Radiation

Plasma oscillations in the impact plasma cloud have been suggested to produce electromagnetic interference at high frequencies (Foschini 1998), and to generate transient magnetic fields (Stamper et al. 1971; Bird et al. 1973). For nanodust in the solar wind, Langmuir waves in the impact plasma cloud do not contribute to the observed spectral density since the frequencies involved are smaller than the plasma frequency in the cloud. The fast motion of the charges might also produce acoustic waves. However, the shape of the observed pulses (Fig. 9 in Sect. 3), as well as that of the observed spectrum (Fig. 8 in Sect. 3), which is close to the spectrum predicted by (11), suggest that this mechanism is not observed in our data.

3 Measurement of Interplanetary Nanodust with STEREO

The STEREO mission consists of two spacecraft in orbit at about 1 AU from the Sun, which, respectively, lead (STEREO A) and trail (STEREO B) the Earth, at heliolongitudes increasing by 22°/year. The WAVES instrument measures electric voltages as described in Sect. 2.1 with three orthogonal antennas of length $L = 6\,\mathrm{m}$ (Bale et al. 2008).

Even though this mission was primarily designed to study the solar and inner heliospheric plasmas in three dimensions, it turned out to be a serendipitous dust detector over a wide range of sizes. Dust grains above several microns are detected by the Heliospheric Imagers in two ways: impacts on the spacecraft blankets produce trails of debris that are imaged by the cameras (St. Cyr et al. 2009), whereas direct impacts on the HI instruments themselves produce pointing offsets (Davis et al. 2012). These impacts are detected simultaneously by the WAVES instrument since they produce large voltage pulses that saturate TDS on the three antennas (St. Cyr et al. 2009). Indeed, a 1-μm radius grain of mass density $2.5 \times 10^3\,\mathrm{kg\,m^{-3}}$ impacting at 20 km/s should produce from (2) and (10) a pulse due to charge recollection $\delta V_1 \sim 0.3\,\mathrm{V}$ on the three antennas, which exceeds the instrument saturation level.

Concerning smaller sizes, the TDS instrument detects beta-meteoroids accelerated by the solar radiation pressure[12] to many tens of km/s (Mann et al. 2010) and interstellar dust of a few tenths micron impacting the spacecraft at several tens km/s (Zaslavsky et al. 2012), which all produce simultaneous pulses on the three antennas via direct charge recollection (10).

[12]The importance of radiation pressure is characterised by the ratio of this force to the solar gravitational force—called β, which has led to the name of the so-called beta-particles which are ejected by radiation pressure. Radiation pressure is negligible for nanodust since their size is much smaller than the radiation wavelength.

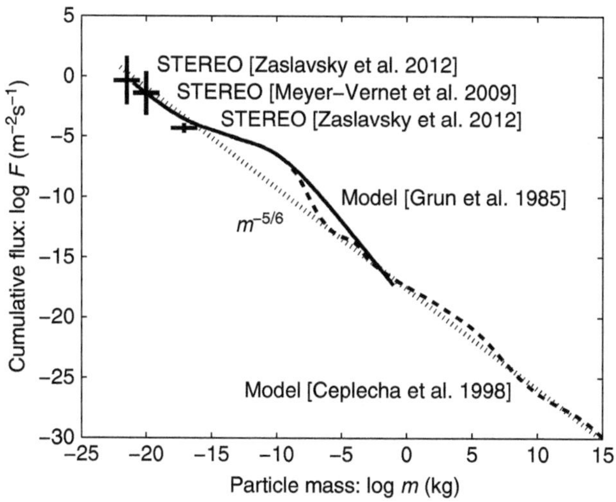

Fig. 6 Cumulative flux of interplanetary dust and bodies at 1 AU. The superposition of models for dust from Grün et al. (1985) (*continuous line*), small bodies from Ceplecha et al. (1998) (*dashed*), and collisional equilibrium $\propto m^{-5/6}$ (*dotted*), is reproduced from Meyer-Vernet (2007), with measurements of nanodust by Meyer-Vernet et al. (2009a) and of both nanodust and beta meteoroids by Zaslavsky et al. (2012) superimposed

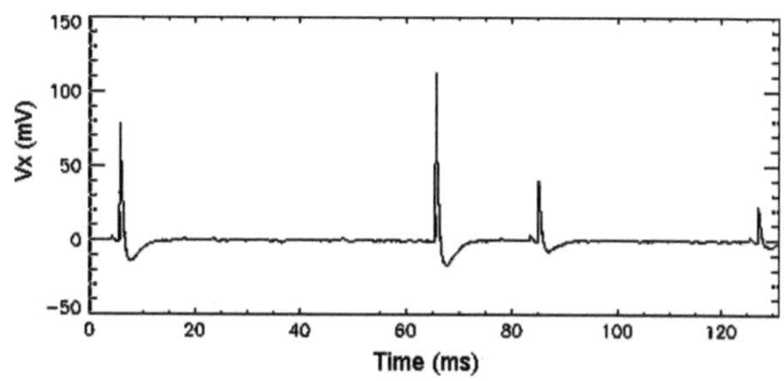

Fig. 7 Sample of voltage on the X monopole of STEREO A during 131 ms, acquired by TDS on the same day as the spectrum of Fig. 8 (3 min before). The recoil is produced by instrumental filtering

However, as predicted by the increase in flux with decreasing mass of the interplanetary dust model (Fig. 6), by far the most frequent observed events are those we interpret as impacts of fast interplanetary nanodust on the spacecraft. As we have seen in Sect. 2.3.4, they produce a very small voltage on the three antennas via direct charge recollection, so that they are mainly detected on the antenna that is directly affected by the impact plasma cloud, via perturbation of the current balance (15), producing large pulses on TDS (Fig. 7) and characteristic f^{-4} spectra on LFR (Fig. 8).

The conspicuous nature of these events, which often dominate the spectrograms (see Fig. 2), enabled to discover these particles and to derive from the measured LFR power spectral density a preliminary order of magnitude estimate of the cumulative flux $\sim 4 \times 10^{-2}\,\mathrm{m^{-2}s^{-1}}$ for $m \sim 10^{-20}$ kg, i.e., radius ∼10 nm (Meyer-Vernet et al. 2009a), reproduced in Fig. 6. This measurement is compatible with

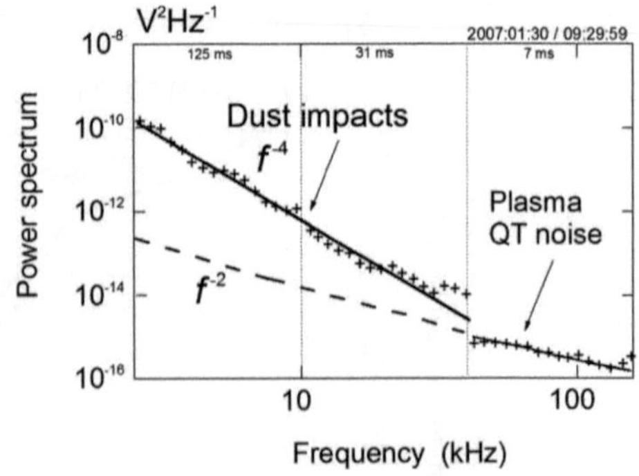

Fig. 8 Example of power spectrum measured with the X–Y dipole of STEREO/WAVES in the three low-frequency bands (*crosses*, acquisition times at *the top*). The two lower bands show dust impacts. The higher band (of much smaller acquisition time) detects only the plasma quasi-thermal noise

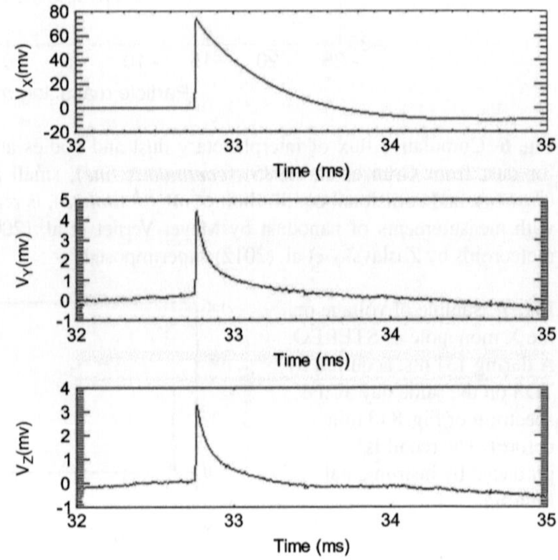

Fig. 9 Example of TDS pulse on the three monopoles upon a nanodust impact on STEREO A. The X antenna records a pulse (15) due to disruption of the antenna current balance (*top panel*), with a recoil due to instrumental filtering. The other antennas record a much smaller pulse (10) due to direct recollection of the cloud's electrons by the spacecraft (*middle* and *bottom panels*). The decay times are close to the timescales of charge equilibrium restoration, for, respectively, an antenna and the spacecraft

the interplanetary dust model (Grün et al. 1985) and with the $m^{-5/6}$ collisional equilibrium (Dohnanyi 1969) curve which rawly approximates the flux at 1 AU for dust (except for the bump around 10^{-10} kg) and small bodies over more than 35 orders of magnitude in mass (Meyer-Vernet 2007).

Let us now check that the shape and amplitude of the voltage pulses observed with the three antennas agree in detail with our interpretation. Figure 9 shows an example of TDS data recorded on the three STEREO monopole antennas. The pulses are simultaneous (to the accuracy of the measurement $\sim$ a few μs), with $\delta V_y \sim \delta V_z \ll \delta V_x$, and the average observed ratio $\delta V_x/\delta V_y \simeq \delta V_x/\delta V_z \simeq 20$ (Zaslavsky et al. 2012). From the calculation of Sect. 2.3.4, one expects the antenna close to the impact location to record a pulse of large amplitude δV_2 given by (15) (top panel of Fig. 9) produced by disruption of the antenna current balance by the impact cloud, whereas the two other antennas record a pulse of much smaller amplitude given by (10), produced by recollection of the cloud's electrons by the spacecraft. On STEREO, we have $L \simeq 6$ m, $R_{sc} \sim 1$ m, $n_a \sim 5\,\text{cm}^{-3}$, $C_{sc} \simeq 200$ pF, $\Gamma \simeq 0.5$, so that (10) and (15) yield

$$\delta V_2 \simeq [1.6 \times 10^{10} T_{\text{ph(eV)}}]Q, \tag{16}$$

$$\delta V_2/\delta V_1 \simeq \frac{T_{\text{ph(eV)}} C_{sc}}{2\pi e n_a L R_{sc}^2} \simeq 6.6\, T_{\text{ph(eV)}}. \tag{17}$$

From the observed value $\delta V_2/\delta V_1 \simeq 20$, we deduce $T_{ph} \simeq 3\,\text{eV}$, which is indeed the typical photoelectron temperature. Furthermore, a detailed study of the pulses detected on the three antennas during 4 years shows a striking agreement with the theoretical voltages expected from this interpretation (Zaslavsky et al. 2012).

Let us now examine whether the observed timescales also agree with this interpretation. First of all, let us verify that the observed rise time enables the antenna affected by the cloud to eject enough photoelectrons for producing the observed pulse amplitude. With a photoelectron current density $I_{ph0} \simeq 50\,\mu\text{A/m}^2$ (Thiebault et al. 2006) and a projected surface area[13] $\simeq 2ax \simeq 0.032x$ m^2 for the affected antenna length x, the charge accumulated during the observed rise time $\tau_r \sim 10$–$20\,\mu\text{s}$ is $Q_{ph} \simeq I_{ph0} \times 2ax \times \tau_r \simeq [1.6 - 3.2]x \times 10^{-11}$ C. With the low-frequency monopole capacitance estimated in Sect. 2.3.4 yielding $C_A \simeq 64$ pF, this produces a pulse amplitude $\delta V_x \sim \Gamma Q_{ph}/C_A \sim [0.25 - 0.5]\Gamma x$ V, which is indeed close to the value given by (13).

The decay of the pulses is governed by restoration of equilibrium by collection of solar wind electrons, which takes a much longer time ~1 ms (Fig. 9, top panel), of the order of magnitude of the theoretical equilibrium time of a STEREO antenna boom in the solar wind (Henri et al. 2011). Note that the pulses shown in the middle and bottom panels of Fig. 9, which are due to variation in spacecraft potential, decay faster since the equilibrium time of the spacecraft is shorter than that of an antenna by roughly the inverse ratio of their surface areas.

The power spectrum produced by averaging the impacts (Fig. 8) agrees with the shape given by (11). As we previously noted, it varies roughly as f^{-4} at frequencies greater than the inverse of the rise time, whereas the plasma noise due to impacts of ambient electrons varies roughly as f^{-2} for $f < f_p$ since the rise time is in that case $\sim 1/\omega_p$. Note that the STEREO antennas, of length $L = 6$ m, are too short compared to the solar wind Debye length $L_D \sim 10$ m to be able to detect a thermal noise peak at f_p, which explains why the plasma line does not appear in Figs. 2 and 8.

The evaluation based on the observed LFR spectra is affected by a large uncertainty because the relation (12) between the observed spectrum and the dust flux involves the rise time of the pulses, whose estimation is difficult. The TDS

[13]Since the antenna radius close to the antenna base is 1.6 cm (Bale et al. 2008).

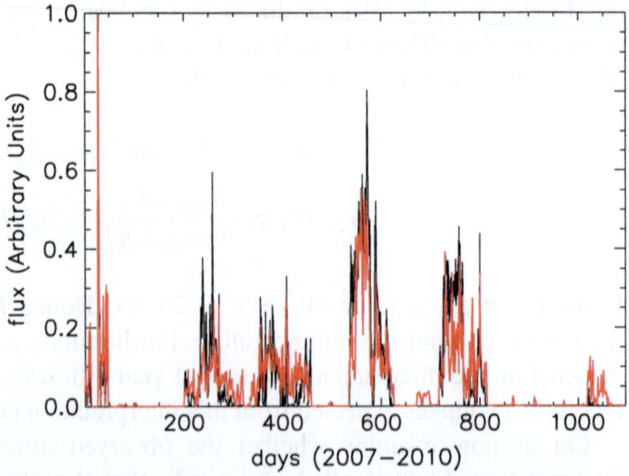

Fig. 10 Nanodust fluxes (day-averages) in 2007–2010 from the TDS (TDS, in *red*) and the LFR (LFR band A, in *black*) on STEREO A. TDS data points are based on direct counting of impacts, whereas LFR data points are based on averages over the receiver acquisition time weighted according to the dust mass and speed (Sect. 2.3.4). Adapted from Zaslavsky et al. (2012)

instrument, which measures directly the individual pulses, reveals more information and has enabled us to refine the above estimate (Zaslavsky et al. 2012).

For a simplified illustration, let us use Fig. 7, which shows an example of TDS data recorded during a period of high dust impact rate, to evaluate the order of magnitude of the flux at this time. The sample exhibits four pulses during 130 ms, i.e., a cumulative flux $\sim 30\,\mathrm{s}^{-1}$, for voltage amplitudes >20 mV. Converting voltages into charges with (16), and charges into masses with (2), assuming a nanodust speed $v \simeq 300$ km/s at 1 AU (Czechowski and Mann 2010, 2012), this yields $m > 3\times10^{-21}$ kg, for which the spacecraft surface for impacts affecting the antenna is from (6) $R^2_{max} \sim 1\,\mathrm{m}^2$. A correct calculation involving integration over the mass distribution as well as statistical averages (Zaslavsky et al. 2012) yields a similar order of magnitude for typical periods of high nanodust flux. However, the measured flux is extremely variable, both on small timescales—as illustrated in Fig. 2 which displays LFR data during 1 day, and on large timescales—as illustrated in Fig. 10 which shows the flux detected during 4 years with both TDS (in red) and LFR (in black).[14]

From the telemetered voltage samples acquired over 4 years (2007–2010) on STEREO A, we find a cumulative flux $F \sim 40\,\mathrm{m}^{-2}\mathrm{s}^{-1}$ for the smaller mass $m \sim 3\times10^{-22}$ kg (Zaslavsky et al. 2012). This flux is an overestimate since the TDS data selection is biased towards greater voltage amplitudes. The actual flux is smaller by an amount which depends on the statistics of the impacts and on the data selection process, and is at most the fraction of total telemetered time per day, i.e., a factor of 10^4. This yields a cumulative flux for mass 3×10^{-22} kg (radius $\simeq 3$ nm) in the range $0.4\times[10^{-2}\text{–}10^{2}]\,\mathrm{m}^{-2}\mathrm{s}^{-1}$ (Zaslavsky et al. 2012).[15] These extreme values are

[14]Note that even in periods of very high flux, the number of impacts during the decay time $\tau_d \sim 1$ ms on the spacecraft surface affected by the impacts ($\sim R^2_{max} \sim 1\,\mathrm{m}^2$) never exceeds unity.

[15]The smaller value is deduced from the number of impacts detected during each day; it is a minimum value since the impacts are counted from about 50 samples (as the one shown in Fig. 7)

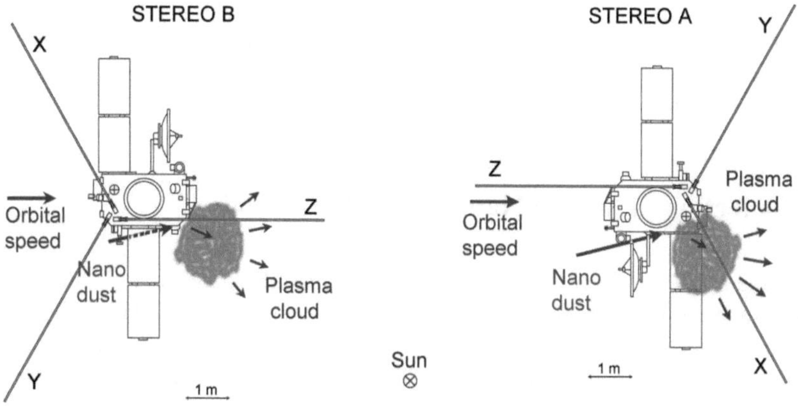

Fig. 11 The spacecraft STEREO A and B orbit the sun respectively ahead (A) and behind (B) the Earth. The views are from an observer looking towards the Sun. They show the three antenna booms X, Y, Z, with a tentative sketch of expanding plasma clouds produced by impacts of nanodust coming from the inner solar system and moving in the prograde sense

displayed in Fig. 6, together with dust models, and to the STEREO measurement of beta-particles, whose size is larger by two orders of magnitude.

As we already noted, the observed flux is highly variable on both short (seconds) and long (months) timescales. Impacts appear clearly in short bursts, but the observed long-term variability is different on STEREO A and B, as well as the antenna affected by the impact cloud (X on A, Z on B as labelled in Fig. 11), and the average impact rate, which is about twice lower on STEREO B than on STEREO A (Zaslavsky et al. 2012).

This large difference in long-term variability, with long periods of very low impact rate on STEREO A which are not observed on STEREO B, whereas the average fluxes on both spacecraft only differ by a factor of two and the short-term variability appears similar on both spacecraft, strongly suggests that a large part of the variability is due to changes in velocity direction of the particles, since both spacecraft have different attitudes, symmetrical with respect to the orbital speed, whereas the appendages such as solar panels—that prevent some spacecraft parts to be impacted from some directions—are positioned differently, as well as the antennas. Such changes in velocity direction may be associated to variations in solar wind properties that change the acceleration of the particles, whereas magnetic field variations at 0.1–0.2 AU from the Sun are expected to detrap nanodust from bound orbits, and enable them to be further accelerated by the solar wind (Czechowski and Mann 2012). Figure 11 illustrates the geometry, and sketches plasma clouds

---

of $\simeq$120 ms each day, i.e., only 0.01% of each day, and the probability that no impact occurs during the rest of the day is extremely small. The higher value is deduced from the number of impacts detected during the telemetered time; it is a maximum value since the telemetered samples are not typical because the onboard data selection is biased towards greater voltage amplitudes.

produced by impacts of prograde nanodust coming from the inner solar system as predicted by dynamics (Czechowski and Mann 2010, 2012). This question is under study, as well as the time periodicities of the fluxes, the statistics of occurrence, and the dust mass distribution.

4 Detection of Jovian Nanodust with Cassini

After the STEREO/WAVES discovery of nanodust, the data from Cassini/RPWS were examined for possible nanodust signatures, even though the latter instrument is less adapted to this goal, for several reasons. First, the electric antennas are longer ($L = 10\,\mathrm{m}$) on Cassini than on STEREO, and they extend farther from the spacecraft, whose shape is quasi-cylindrical, which reduces the probability that the antenna current balance be affected by the impact clouds, in contrast to the quasi-flat surfaces of STEREO. Second, since the primary objective of Cassini was the study of Saturn's environment, it has furnished very few data at smaller heliocentric distances. Third, the waveforms used to detect individual dust impacts are either acquired with the antennas in dipole mode, or via the Langmuir probe which has a high noise level (Kurth et al. 2006).

Finally, the onboard algorithm used to select the waveforms to be transmitted on ground has a high effective threshold; for example, the sensitivity for detection of individual dust impacts in Saturn's E ring corresponds to grains of minimum radius $2.4\,\mu\mathrm{m}$ impacting at 8 km/s (Kurth et al. 2006). From (2), this is equivalent to grains of minimum radius 30 nm impacting at 300 km/s. Interplanetary dust of this size coming from the inner heliosphere are not expected to reach such a large speed (Czechowski and Mann 2012); furthermore, the average size of Jovian and Saturnian stream particles is smaller (Hsu et al. 2012). This suggests that the wide-band receiver on Cassini is not adapted to detect individual nanodust impacts.

In contrast, the high sensitivity of the hf receiver may enable it to detect fast nanodust if the impact rate is sufficiently high for (12) to yield a detectable power by integration over many impacts. An opportunity was offered by the Jovian flyby in December 2000–January 2001, when the Cosmic Dust Analysers on Cassini and Galileo made joint measurements of Jovian nanodust streams (Hsu et al. 2012) and revealed properties of the nanodust impacting Cassini (Graps et al. 2001), whereas the hf receiver of RPWS furnished data in both monopole and dipole mode. Figure 12 shows examples of power spectra acquired at this occasion in the jovian outer magnetosheath, which, from Sect. 2.3.4, correspond respectively to what is expected for nanodust impacts (left) and plasma quasi-thermal noise (right).

We can see in Fig. 13 that the large power corresponding to dust impacts is observed in monopole mode, with similar amplitudes on the different monopoles, the small difference being due to differences in antenna capacitances and receiver gains. This indicates that the signal is produced by recollection by the spacecraft of the impact-generated electrons, and not by perturbation of the current balance on an antenna. This is not surprising, because the photoelectron current at 5 AU is not large enough to produce a large voltage pulse when perturbed. Indeed, with

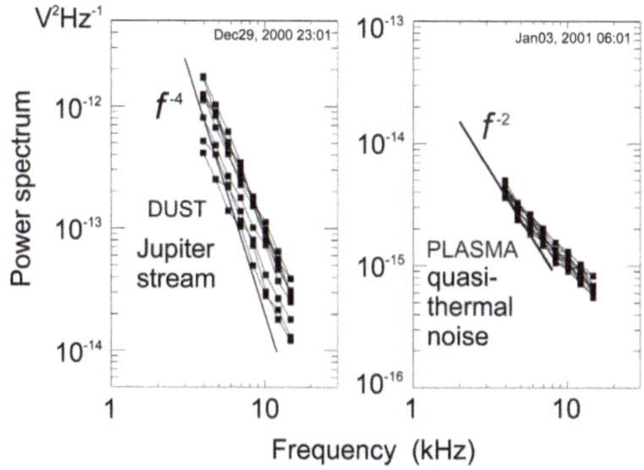

Fig. 12 Typical power spectra measured with the hf receiver on Cassini/RPWS near Jupiter. The *left* and *right spectra* are produced respectively by nanodust impacts and by the plasma quasi-thermal noise (with antennas smaller than the plasma Debye length). Adapted from Meyer-Vernet et al. (2009b)

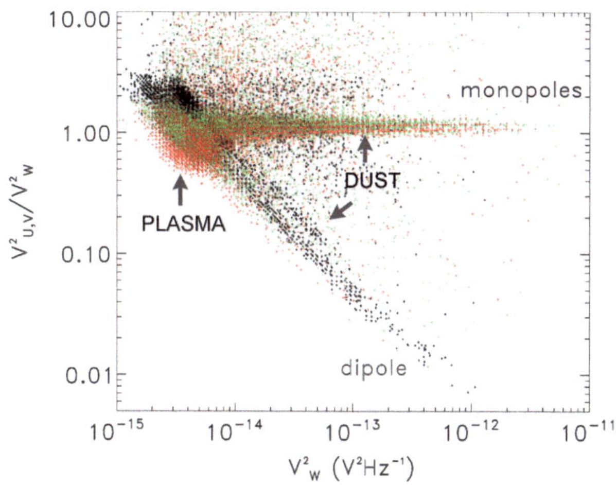

Fig. 13 Cassini/RPWS hf receiver data near Jupiter. Ratios of the power on two monopoles (*red* and *green*) to that on the other one as a function of the latter, and ratio of the power on the dipole to that on the monopole (*black*). The dust yields similar signals on the monopoles, whereas the dipole records mainly the plasma thermal noise (of smaller amplitude). From Meyer-Vernet et al. (2009b)

$a = 1.43$ cm (Gurnett et al. 2004), $T_{\text{ph}} \simeq 3$ eV, $I_{\text{ph}0} \simeq 2 \times 10^{-6}$ μA m^{-2}s^{-1} at 5 AU and τ given by the pulse rise time $\tau_{\text{r}} \simeq 40$ μs deduced from the observed spectral shape, condition (14) does not hold.

Estimating Q from (2), with the particles properties $m \simeq 10^{-21}$ kg, and $v \simeq 450$ km/s deduced from the Cassini/Galileo joint measurements (Graps et al. 2001), we use (10)–(12) to deduce the relation between the power spectral density on the Cassini RPWS monopoles and the dust flux F (Meyer-Vernet et al. 2009b)

$$V^2 \simeq 1.5 \times 10^{-11} F / f^4_{(\text{kHz})} \text{ V}^2\text{Hz}^{-1} \tag{18}$$

From the measured power spectral density (see for example Fig. 12, left) we find the maximum flux $F \sim 10\,\mathrm{m}^{-2}\,\mathrm{s}^{-1}$ in the streams on December 30, 2000, with a time variation similar to that measured by the CDA. This flux is also in the range of values reported by long-term studies of jovian dust streams with Galileo, both in amplitude and variability (Krüger et al. 2005).

5 Conclusions and Perspectives

Contrary to the nanodust wave detection performed with Cassini near Jupiter, which was a mere confirmation of results obtained with classical detectors, the nanodust seen by STEREO at 1 AU had never been observed before. Even though these observations had been predicted theoretically and are compatible with the interplanetary dust model, it is therefore especially important to check their validity. First of all, the possibility that the signals are generated or affected by internal noise or electromagnetic interferences has been eliminated. Numerous tests had been made before and after launch to make sure that the spacecraft was extremely clean and that the instrument was not significantly perturbed by other instruments nor by spacecraft subsystems. Possible remaining perturbations have been carefully eliminated from the data, as well as voltages generated by other radio or plasma wave phenomena or impacts of energetic particles. Second, the observations are made with two independent subsystems of the instrument (LFR and TDS), which yield compatible results (Meyer-Vernet et al. 2009a; Zaslavsky et al. 2012). Third, the instrument also detects other kinds of dust, yielding results that agree with previous measurements by classical dust detectors (Zaslavsky et al. 2012). Finally, the interpretation of the data in terms of voltage pulses produced by impact ionisation explains the amplitude and shape of the observed pulses and the different responses of the antennas, using only known concepts.

However, since nanodust detections with wave experiments have been performed serendipitously, with instruments that were not designed for this purpose, the error bars are presently large, and since these measurements took place very recently, the mechanism of detection and the calibration for dust measurements are still under study. More specifically, the following items are still being investigated. First, a major problem is the use of calibrations such as the charge production described in (2). Not only are they not specific to these instruments, but they rely on laboratory experiments performed with dust grains that are both larger and slower. Specific calibrations should be performed on future instruments. Second, the mechanism of electric field production must be investigated in more detail. This requires a more detailed study of the observed shape of the pulses, as well as a complex simulation taking into account the geometry and environment of the spacecraft and antennas. Finally, the flux variability is being investigated from short ($\sim$1 s) to large (years) timescales, in comparison to the solar wind state, as well as the mass distribution, especially near the low end of the mass range. This latter question is of great importance, since no lower size cut-off has presently been detected in the interplanetary dust distribution, and the properties of these particles lying at the frontier between molecules and bulk matter are badly known, as well as their interactions with the solar wind.

Other data are under study. A number of events suggesting nanodust impacts detected on Cassini/RPWS at heliocentric distances between 1 and 3 AU are being investigated. Few similar events have been detected at farther heliocentric distances and in Saturn environment, and they are much weaker. This is not surprising for several reasons. We have seen that the RPWS waveform mode is not adapted for detecting nanodust impacts, whereas the hf receiver requires a large impact rate to yield a detectable power. Furthermore, the particles produced at Saturn are smaller and slower than those produced at Jupiter (Hsu et al. 2012), and are therefore expected to yield a smaller signal, and furthermore the very small photoelectron current at Jupiter and Saturn distances is not expected to perturb sufficiently the antenna current balance.

Wave data from other interplanetary probes are also being investigated. As a rule, they are not adapted for nanodust detection because they were designed for radio observations, so that they use long antennas extending far from the spacecraft and operated in dipole mode. The length of the antennas and their very small radius imply that the plasma clouds produced by impacts on the spacecraft are not expected to disturb significantly the photoelectrons that control the current balance on antennas and therefore to produce a pulse detectable in dipole mode; for example, with the 2×45 m electric antennas on the spacecraft WIND, (15) yields a pulse amplitude smaller than the value on STEREO by nearly one order of magnitude, and anyway, the antenna radius $a \simeq 0.2$ mm is too small for condition (14) to hold. Recollection of impact electrons by the spacecraft cannot be detected either since the antennas are operated in dipole mode.

The question arises as to whether the radio experiment on Ulysses detected nanodust, especially when the Cosmic Dust Analyser did so near Jupiter in 1992 (Krüger et al. 2006). The radio experiment on Ulysses (Stone et al. 1992) does not have a TDS, but it has an antenna operated in monopole mode, and some low frequency signals have indeed been detected at the times when the CDA detected nanodust impacts near Jupiter in 1992. However, the signals are difficult to analyse because this LFR of old generation is swept through the different frequency channels, doing so in about 2 min. Since STEREO results show that the dust impact rate may vary considerably during this time, this means that a characteristic dust spectrum cannot be observed, since each frequency channel acquired at a different time observes a different impact rate. Furthermore, there is no TDS mode in order to measure the shape of the pulses.

Another major point is the question of the origin of the interplanetary nanodust discovered by STEREO. The main source is expected to be the inner solar system, as proposed by Czechowski and Mann (2010, 2012). A Jovian origin of the particles is clearly excluded, except occasionally, since the average flux measured by the Cosmic Dust Analyser as a function of Jovicentric distance (Krüger et al. 2006) would yield a flux at 4 AU from Jupiter much smaller than the average flux observed by STEREO during 4 years.

On the other hand, a contribution from the Earth environment is not presently excluded. Estimating this contribution requires a calculation of the ejection speed (if any) from the Earth, from which the dust would be accelerated by the solar wind. This calculation is not simple because of the complex Earth's environment in which the dust charge varies during its motion. The fact that the charge is expected to be negative in most of the plasmasphere, i.e., inward of the synchronous orbit, and generally positive outward because of the smaller plasma density, except in shadow (Horányi et al. 1988) might enable outward acceleration by the corotation electric field as in outer planets, albeit to smaller speeds since the magnetic field, rotation rate, and radius are smaller. Assuming that the main forces on nanodust are the corotation electric force and gravitation, a necessary condition for ejection of a charge q of mass m from a Keplerian orbit of radius r_0 (assumed significantly smaller than the magnetospheric size) is that the corotation potential $\Phi \simeq q\Omega_{\oplus}B_{\oplus}R_{\oplus}^3/r_0$ (Burns et al. 2001) exceeds the energy $mM_{\oplus}G/2r_0$. Here $\Omega_{\oplus}$, $B_{\oplus}$,$R_{\oplus}$ are, respectively, the Earth's angular rotation frequency, equatorial magnetic field, and radius. Substituting the Earth properties, this condition yields $q/m > 3 \times 10^{-6}e/m_{\mathrm{p}}$. With a grain's electrostatic potential $\phi \sim 10$ V and mass density $\simeq 2.5 \times 10^3\,\mathrm{kg\,m^{-3}}$, this translates into a grain radius $r < 20\,\mathrm{nm}$. This necessary condition is generally not sufficient since too small grains, of gyroradius smaller than the scale of magnetic variation, are expected to be confined along magnetic field lines. However, the Earth's magnetosphere undergoes frequent perturbations which disrupt the magnetic lines and may enable ejection of these grains, especially along the magnetic tail. Even though a first examination of the data did not show any simple correlation between the observed impact rate and geomagnetic perturbations, which is not surprising since most of the dust acceleration is expected to take place in the solar wind and may confine the grains in narrow streams, a more detailed study is in progress.

Finally, these observations enable one to derive some consequences for the design of wave experiments for measuring dust on future instruments such as RPW on Solar Orbiter and FIELDS on Solar Probe Plus. Solar Orbiter will explore the heliosphere on an elliptic orbit with a perihelion as low as 0.28 AU and increasing inclination up to more than 30° with respect to the solar equator. Solar Probe Plus will explore the solar corona as close as 9.5 solar radii.

In order to optimise the measurements, instrumental modes specifically dedicated to dust detection are currently under development. This may include a specific frequency receiver without AGC, and a TDS mode performing a systematic detection and recording of the nanodust impact main properties, thereby greatly reducing the uncertainties due to the biased event selection of the existing TDS. The realisation of ground-based experiments in dust accelerators is also under study. The aim is to measure the charge Q generated by an impact on the actual spacecraft surface materials, and to perform a ground calibration of the radio detection technique, in order to better understand the processes at the origin of the observed electric signals.

References

Aubier, M., Meyer-Vernet, N., Pedersen, B.M.: 1983, *Geophys. Res. Lett.* **10**, 5.
Auer, S.: 2001, In: Grün, E. et al. (eds.) *Interplanetary dust*, Springer, 385.
Bale, S.D., Ullrich, R., Goetz, K., Alster, N., Cecconi, B., Dekkali, M. *et al.*: 2008, *Space Sci. Rev.* **136**, 487.
Bird, R.S., McKee., L.L., Schwirzke., F., Cooper, A.W.: 1973, *Phys. Rev. A* **7**, 1328.
Bougeret, J.-L., Goetz, K., Kaiser, M.L., Bale, S.D., Kellogg, P.J., Maksimovic, M. et al.: 2008, *Space Sci. Rev.* **136**, 529.
Burns, J.A., D.P. Hamilton, M.R. Showalter: 2001, In: Grün, E. et al. (eds.) *Interplanetary dust*, Springer, 641.
Ceplecha, Z., Borovicka, J., Elford, W.G., Revelle, D.O., Hawkes, R.L., Porubcan, V. et al.: 1998, *Space Sci. Rev.* **84**, 327.
Czechowski, A., Mann, I., : 2010, *Astrophys. J.* **714**, 89.
Czechowski, A., Mann, I., : 2012, this volume.
Davis, C.J., Davies, J.A., St Cyr, O.C., Campbell-Brown, M., Skelt, A., Kaiser, M. et al.: 2012, *Mon. Not. Roy. Astron. Soc.*, **420**, 1355
Dietzel, H., Eichhorn, G., Fechtig, H., Grün, E., Hoffmann, H.-J., Kissel, J.: 1973, *J. Phys. E. Scientif. Instr.* **6**, 209.
Dohnanyi, J.S.: 1969, *J. Geophys. Res.* **74**, 2531.
Drapatz, S. and Michel, K.W.: 1974, *Z. Naturforsh.* **29a**, 870.
Driesman, A., Hynes, S., Cancro, G.: 2008, *Space Sci. Rev.* **136**, 17.
Foschini, L.: 1998, *Europhys. Lett.* **43**, 226.
Göller, J.R., Grün, E.: 1989, *Planet. Space Sci.* **37**, 1197.
Graps, A.L., E. Grün, H. Krüger, M. Horányi, and H. Svedhem: 2001, In: Warmbein, B. (ed.) *Proceed. Meteor. 2001 Conf.*, ESA **SP-495**, ESTEC, Nordwijk, 601–608.
Grün, E., Zook, H.A., Fechtig, H., Giese, R.H.: 1985, *Icarus* **62**, 244.
Grün, E. et al.: 1992, *Science* **257**, 1550.
Grün, E., Baguhl, M., Svedhem, H., Zook, H.A.: 2001, In: Grün, E. et al. (eds.) *Interplanetary dust*, Springer, 295.
Gurnett, D.A., Grün, E., Gallagher, D., Kurth, W.S., Scarf, F.L.: 1983, *Icarus* **53**, 236.
Gurnett, D.A., Kurth, W.S., Kirchner, D.L., Hospodarsky, G.B., Averkamp, T.F., et al.: 2004, *Space Sci. Rev.*, 114, 395–463.
Henri, P. et al.: 2011, *Physics of Plasmas*, **18**, 082308.
Horányi, M., H.L.F. Houpis, D.A. Mendis: 1998, *Astrophys. Space Sci.*, **144**, 215.
Horányi, M., E. Grün, and A. Heck: 1997, *Geophys. Res. Lett.*, 24, 2175–2178.
Hornung, K., Kissel, J.: 1994, *Astron. Astrophys.* **291**, 1.
Hornung, K., Malama, Y., G., Kestenboim, K.S.: 2000, *Astrophys. Space Sci.* **274**, 355.
Hsu, W.H., Krüger, H., Postberg, F.: 2012, this volume.
Johnson, T. V., G. Morfill, and E. Grün: 1980, *Geophys. Res. Lett.*, 7, 305–308.
Kissel, J., Krüger, F.R.: 1987, *Appl. Phys. A* **42**, 69.
Krüger, F.R.: 1996, *Adv. Space Res.* **17**, 1271.
Krüger, H., Linkert, G., Linkert, D., Moissl, R., and Grün, E.: 2005, *Planet. Space Sci.*, **53**, 1109.
Krüger, H., Graps, A.L., Hamilton, D.P., Flandes, A., Forsyth, R.J., Horanyi, M., Grün, E.: 2006, *Planet. Space Sci.*, **54**, 919.
Kurth, W.S., T.F. Averkamp, D.A. Gurnett, and Z. Wang: 2006, *Planet. Space Sci.*, **54**, 988.
Mann, I., Murad, E., Czechowski, A.: 2007, *Planet. Space Sci.* **55**, 1000.
Mann, I., Czechowski, A., Meyer-Vernet, N., Zaslavsky, A., Lamy, H.: 2010, *Plasma Physics and Controlled Fusion* **52**, 124012.
Mann, I., Pellinen-Wannberg, A., Murad, E., Popova, O. et al.: 2011, *Space Sci. Rev.*, 36.
Manning, R.: 1998, In: Pfaff, R. et al. (eds.) *Measurement techniques in Space Plasmas: Fields, Geophys. Monograph Ser.* **103**, AGU, Washington DC., 181.
McBride, N., McDonnell, J.A.M.: 1999, *Planet. Space Sci.* **47**, 1005.

Meuris, P. et al.: 1996, *J. Geophys. Res.* **101**, 24471.
Meyer-Vernet, N.: 1979, *J. Geophys. Res.* **84**, 5373.
Meyer-Vernet, N.: 1985, *Adv. Space Res.* **5**, 37.
Meyer-Vernet, N.: 2001, In: Harris, R. A. (ed.), *Proc. 7th Spacecraft Charging Technology Conf.*, ESA **SP-476**, Nordwijk: ESTEC, 635.
Meyer-Vernet, N.: 2007, *Basics of the Solar Wind*, Cambridge University Press, 339, 351–355.
Meyer-Vernet, N., Perche, C.: 1989, *J. Geophys. Res.* **94**, 2405.
Meyer-Vernet, N., Aubier, M.G., Pedersen, B.M.: 1986, *Geophys. Res. Lett.* **13**, 617.
Meyer-Vernet, N., A. Lecacheux, and B.M. Pedersen: 1996, *Icarus*, 123, 113.
Meyer-Vernet, N., S. Hoang, K. Issautier, M. Maksimovic, R. Manning, M. Moncuquet, R. Stone: 1998, In: Pfaff, R. et al. (eds.) *Measurement techniques in Space Plasmas: Fields, Geophys. Monograph Ser.* **103**, AGU, Washington DC., 205.
Meyer-Vernet, N., Maksimovic, M., Czechowski, A., Mann, I., Zouganelis, Z., Goetz, K. et al.: 2009a, *Solar Phys.* **256**, 463.
Meyer-Vernet, N., Lecacheux, A., Kaiser, M.L., Gurnett, D.A.: 2009b, *Geophys. Res. Lett.* **36**, L03103.
Oberc, P.: 1994, *Icarus* **111**, 211.
Oberc, P.: 1996, *Adv. Space Res.* **17**, (12)105.
Oberc, P., Parzydlo, W., Vaisberg, O.L.: 1990, *Icarus* **86**, 314.
Pantellini, F., Landi, S., Zaslavsky, A., Meyer-Vernet, N. : 2012a, *Plasma Physics and Controlled Fusion*, in press.
Pantellini, F. et al.: 2012b, Astrophys. Space Sci., **54**, 045005
Pedersen, B. et al.: 1991, *J. Geophys. Res.* **96**, 19187.
Scarf, F.L., D.A. Gurnett, W.S. Kurth, and R.L. Poynter: 1982, *Science*, **215**, 587.
Shanbing, Y., S. Gengchen, T. Quingming: 1994, *Int. J. Impact Engng*, **15**, 67.
Stamper, J.A., Papadopoulos, K., Sudan, R.N., Dean, S.O., Mclean, E.A.: 1971, *Phys. Rev. Lett.* **26**, 1012.
Stone, R.G., Pedersen, B.M., Harvey, C.C., Canu, P., Cornilleau-Wehrlin, N., Desch, M.D. et al.:1992, *Science*, **257**, 1524.
St. Cyr, O.C., Kaiser, M.L., Meyer-Vernet, N., Howard, R.A., Harrison, R.A., Bale, S.D. et al.: 2009, *Solar Phys.*, **256**, 475.
Thiebault, B.,Hilgers, A., Masson, A., Escoubet, C.P., Laakso, H.: 2006, *IEEE Trans. Plasma Sci.* **34**, 2078.
Warwick, J. W. et al.: 1982, *Science*, 215, 582–586.
Zaslavsky, A., Meyer-Vernet, N., Hoang, S., Maksimovic, M., Bale, S. D.: 2011, *Radio Sci.*, **46 (2)**, doi: 10.1029/2010RS004464.
Zaslavsky, A., Meyer-Vernet, N., Mann, I., Czechowski, A., Issautier, K., Le Chat, G. et al.: 2012, *J. Geophys. Res.*, in press.
Zook, H. A., Grün, E., Baguhl, M., Hamilton, D.P., Linkert, G., Liou: 1996, *Science* **274**, 1501.

Erosion Processes Affecting Interplanetary Dust Grains

Peter Wurz

Abstract The lifetime of grains in interplanetary space is limited by erosion processes. For grains which are smaller than 1,000 nm the lifetimes against erosion are severely limited, in particular in the inner solar system, mostly by ion-induced sputtering by solar wind ions. Thus, to maintain a stable population of sub-micron grains inside Mercury's orbit, the loss of these grains has to be balanced by a supply of new grains. These grains may have their origin in Sun-grazing comets, decay of larger grains by mutual collisions, and grains released from Mercury's surface by micro-meteorite impacts.

1 Introduction

Grains in interplanetary space are subject to several erosion processes that limit their lifetime. These erosion processes are mainly ion-induced sputtering, photon stimulated desorption and sublimation, with the latter two processes relevant for icy grains and other volatile material contained in the grain. Since the erosion processes act on the surface of a grain and the ratio between surface and volume of a grain becomes larger for smaller grains, the erosion processes become increasingly important for smaller grains, which will lead to a preferential depletion of the smallest grains from the population of grains if this loss is not compensated by a source of such grains.

Here we will review the three major erosion processes and estimate the typical lifetimes of grains. For simplicity we will assume here spherical grains in our estimates on the lifetimes of grains controlled by these erosion processes. A variation of grain size distribution resulting from mutual collisions is not discussed here.

P. Wurz (✉)
Physics Institute, University of Bern, Sidlerstrasse 5, 3012 Bern, Switzerland
e-mail: peter.wurz@space.unibe.ch

I. Mann et al. (eds.), *Nanodust in the Solar System: Discoveries and Interpretations*, Astrophysics and Space Science Library 385, DOI 10.1007/978-3-642-27543-2_8,

For the erosion estimates, we assume that the grains stay in a fixed orbital distance R to the Sun. Note that we consider loss of grains in terms of the three mentioned erosion processes, but we do not consider the loss by destruction, nor the loss of a grain by changing its orbit as a result of external forces.

Actual trajectories in the interplanetary space can be varied, and their calculation is mathematically complex (e.g., Mukai and Yamamoto 1982). In addition to the gravitational forces non-gravitational forces influence the trajectories of grains. The major non-gravitational forces are the solar radiation force, which can completely compensate the gravitational force for grains below the μm size, and the Poynting–Robertson drag. Solar wind pressure will also act against the gravitational pull and becomes increasingly important for the smallest grains. Earlier, the orbital lifetime of obsidian grains in Earth orbit was calculated by Mukai and Yamamoto (1982) considering all these forces. These orbital lifetimes are in the range of 30–1,000 years for grains of 10–1,000-nm size, respectively.

In the following discussion we will distinguish between mineral grains and icy grains. As we will see in later sections, the volatile fraction of a grain will be lost very fast for small grains in the inner solar system and most of the grains will be reduced to their mineral content. Little is known about the mineral composition of grains in interplanetary space. Different compositions are expected depending on the origin of the grain, e.g., from comets or from the asteroid belt. Silicate grains have been observed in the interstellar media in infrared spectra. However, spectral features indicate that other minerals should be present as well. Also, if the grain's origin is in the asteroid belt a chondritic composition is likely.

For the estimates of sputter erosion we use the mineralogical composition of the regolith of Mercury (Wurz et al. 2010), in the absence of detailed information of the dust composition. This mix of minerals consists of 27% feldspar (mostly anorthite), 32% pyroxenes (mostly enstatite), 39% olivines (mostly forsterite), and about 3% of several other minerals (Wurz et al. 2010). Although of different mineralogical composition, the total sputter yield of the lunar regolith is very similar to Mercury's (Wurz et al. 2007), thus we expect that the sputter yields for mineral grains will be about the same. Also sputtering of magnetite and obsidian grains has been investigated before (Mukai and Schwehm 1981; Mukai and Yamamoto 1982).

Water–ice grains have a much larger sputter yield for solar wind ions and are also strongly affected by sublimation.

2 Dust Erosion by Sputtering

Particle sputtering is the release of atoms and molecules from the surface of a solid (e.g., a grain) upon impact of energetic ions or atoms on the surface. The sputter yield is the average number of atoms or molecules removed from the solid per incident particle. Sputtering is a well-studied phenomenon in material science (Behrisch and Eckstein 2007). Particle sputtering will release all species from the surface into space reproducing more or less the local surface composition on an atomic level in the sputtered flux. Preferential sputtering of the different elements of a compound will lead to a surface enrichment of those elements with low sputter yields in the top-most atomic layers. However, the steady-state composition of the flux of sputtered atoms, thus the erosion of a grain, will reflect the average bulk composition (Betz and Wehner 1983). Thus, particle sputtering, when operative, will homogeneously erode the entire gain.

The main population of ions relevant for the survival of interplanetary dust grains against sputtering is the solar wind. The solar wind plasma is a mixture of electrons, protons, alpha particles, and heavier multiply charged ions (Wurz 2005 and references therein). The ions have an average number density of 95% protons and 5% alpha particles, respectively (McComas et al. 2000). Heavy solar wind ions are not important for solar wind induced sputtering because their total abundance is about 0.1% in the solar wind, as was already concluded earlier (Mukai and Schwehm 1981; Wurz et al. 2007). For regular solar wind conditions the solar wind velocities are in the range from $v_{SW} = 300$ to $800\,\mathrm{km\,s^{-1}}$, which covers the typical variation of solar wind conditions. In addition, there are coronal mass ejections (CMEs), which are massive and episodic release of solar matter into interstellar space. CMEs have a larger range in plasma parameters (Wurz et al. 2003; Wimmer et al. 2006). The frequency of CMEs correlates with the solar activity cycle (Yashiro et al. 2004; Riley et al. 2006). At higher particle energies there are a range of transient particle populations with smaller intensities at higher particle energies. Figure 1 summarises the energy spectra for particle populations observed in interplanetary space.

Atoms sputtered from a grain may become ionised via charge exchange with the solar wind ions, mostly by the protons. This newly born ions are accelerated by the electromagnetic fields of the solar wind plasma to become a distinct ion population in the solar wind plasma. Ions introduced into the solar wind plasma flow are called pickup ions (PUI). PUIs originating from sputtering of dust are referred to as "inner source PUI", and have been detected in the solar wind plasma (Gloeckler et al. 2000). Alternatively, it has been proposed that these inner source PUIs are heavy solar wind ions that pass through a dust particle and become neutralised or singly charged by passing through the grain (Wimmer-Schweingruber and Bochsler 2003). This process will only work for grains smaller than the penetration depth of solar wind ions, i.e., for dust particles smaller than approximately 300 nm. Also a fraction of the neutral solar wind arises from the interaction of the solar wind ions with small dust particles from which the dust column density between the Sun and Earth has been estimated (Collier et al. 2003).

2.1 *Sputtering of Mineral Grains*

The sputter yields for the difference species of a grain are obtained using the TRIM.SP calculation (Biersack and Eckstein 1984; Ziegler et al. 1984; Ziegler 2004) and the recent review on computer simulation of sputtering by Eckstein and Urbassek (2007). TRIM, like many other simulation programmes for sputtering,

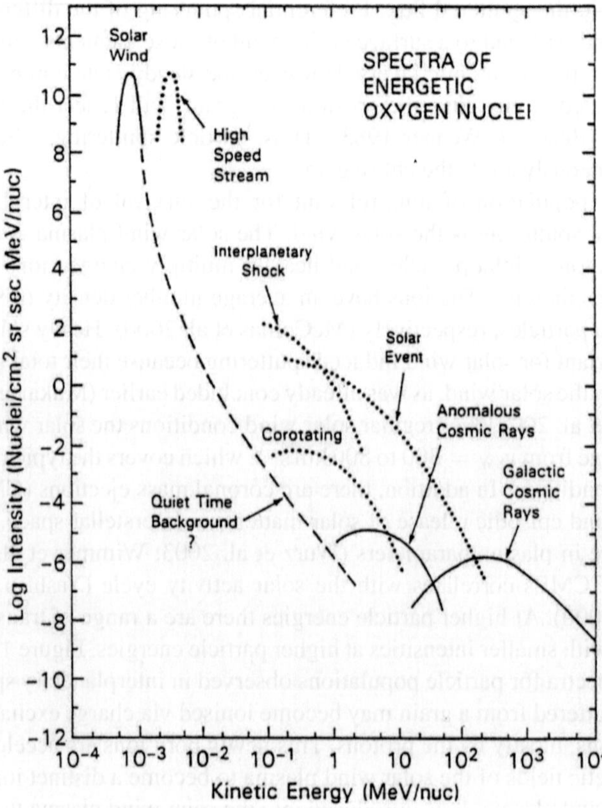

Fig. 1 Typical oxygen differential spectra for different particle populations of interplanetary plasma (Gloeckler and Wenzel 2001). Proton intensities are about a factor of 5,000 higher

assumes that the collisions between atoms can be approximated by elastic binary collisions described by an interaction potential. The energy loss to electrons is handled separately as an inelastic energy loss.

For typical regolith surface compositions, the total sputter yield, i.e., all species sputtered from the surface taken together, is derived as $Y_{\mathrm{tot}} \approx 0.12$ atoms per incoming solar wind ion at $400\,\mathrm{km\,s^{-1}}$, considering protons and alpha particles only. This sputter yield is the integral over all emission angles and all energies of sputtered particles. For details on the energy and angular distribution of sputtered particles from regolith see Wurz et al. (2007). The 5% alpha particles in the solar wind contribute about 30% to the sputter yield. Earlier, Mukai and Schwehm (1981) estimated the solar wind sputter yield for obsidian and magnetite as 0.03 and 0.04, respectively. CMEs can cause increased sputtering of the grains because their ion density can be much larger than that of the regular solar wind. In addition, alpha particles are often more abundant in the CME plasma, which increases the sputter yield even more.

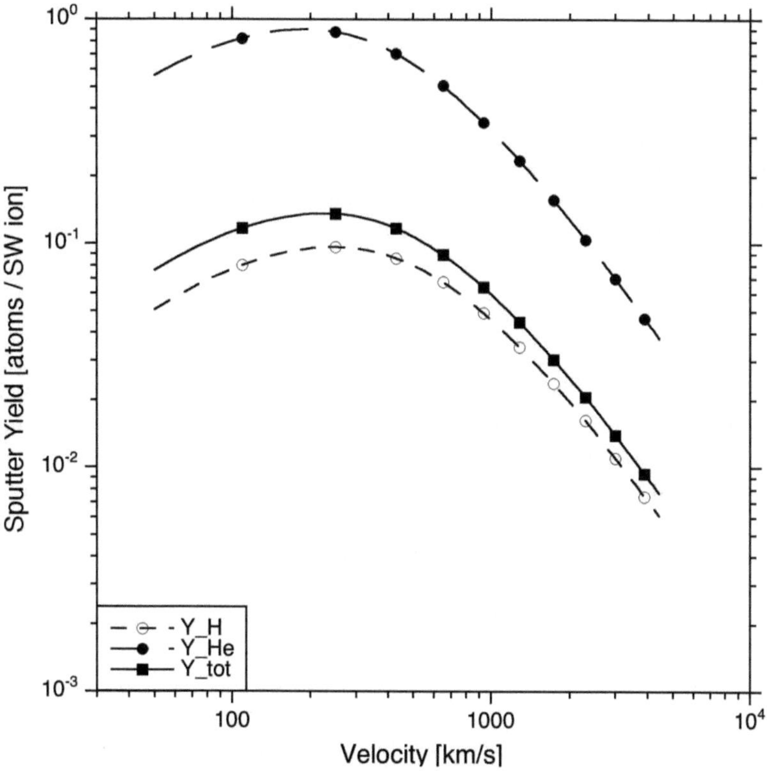

Fig. 2 Estimates of the sputter yield of mineral grains for solar wind sputtering based on Betz and Wien (1994). Traces show the individual proton and alpha particle sputter yield, Y_H and Y_{He}, respectively, as well as the composite yield, Y_{tot}, assuming a composition of 95% protons and 5% alpha particles

The calculated sputter yield is based on the mineralogical composition regolith of the Mercury (Wurz et al. 2010), which we use here since the mineral composition of the grains is not known sufficiently well, and will be varied. This sputter yield calculation assumes a solid grain; if there were porosity in the grain, it will reduce the sputter yield accordingly (Cassidy et al. 2005).

By coincidence the yield for solar wind sputtering of mineral grains peaks for the typical solar wind ion energies, actually around a specific energy of about 1 keV nuc^{-1}, with a rather flat dependence within the typical the solar wind velocity range (Wurz et al. 2007). Draine (1989) has found a similar energy dependence of the sputter yield for interstellar grains. Figure 2 shows an estimation of the sputter yield for solar wind plasma for a large velocity range based on the theoretical formulation by Betz and Wien (1994). More precise sputter yields can be derived from the TRIM simulation programme. For much higher ion energies the sputter yield becomes very small (Behrisch and Eckstein 2007). Energetic particles, which occur episodically in the interplanetary plasma as a result of explosive events on the solar surface and which have much lower intensities than the solar wind ions (see Fig. 1), cause negligible sputtering of mineral grains.

The sputtering mechanism causes particle release only from the uppermost atomic layers of the grains, i.e., from the top-most surface. At solar wind ion energies the impacting particles penetrate typically to a range of 100 nm from the surface of the solid saturating the grains with hydrogen and helium to this depth.

The solar wind flux, $f_{SW}(R)$, scales inversely with the distance to the Sun, R, as (Slavin and Holzer 1981; Russell et al. 1988)

$$f_{SW}(R) = f_{SW}(R_0)\left(\frac{R}{R_0}\right)^2, \tag{1}$$

where R_0 is the astronomical unit. The solar wind flux is $f_{SW}(R_0) = v_{SW}(R_0) \cdot n_{SW}(R_0)$ with the solar wind number density n_{SW}, which at Earth orbit amounts to $f_{SW}(R_0) \approx 8 \times 10^6\,\mathrm{m}^{-3} \times 400 \times 10^3\,\mathrm{m\,s}^{-1} = 3.2 \times 10^{12}$ ions $\mathrm{m}^{-2}\,\mathrm{s}^{-1}$, for typical solar wind conditions in the ecliptic plane (Schwenn 1990; Wurz 2005). At larger solar latitudes, about 30° away from the ecliptic plane, the solar wind speed increases to about $800\,\mathrm{km\,s}^{-1}$ and the density decreases to about $2\,\mathrm{cm}^{-3}$, during solar minimum conditions (McComas et al. 2000). This solar wind organisation with latitude becomes much more chaotic during solar maximum conditions (McComas et al. 2008).

For the solar wind flux of $f_{SW}(R_0) = 3.2 \times 10^{12}$ ions $\mathrm{m}^{-2}\,\mathrm{s}^{-1}$ we get a sputtered flux of $f_{SW}(R_0) \cdot Y_{tot} = 3.2 \times 10^7$ atoms per $\mathrm{cm}^{-2}\,\mathrm{s}^{-1}$ from the surface of a mineral grain. This corresponds to a material removal of about 0.034 nm per year in Earth orbit, which is similar to the erosion of Mercury's and the Moon's surface (Wurz et al. 2010). Earlier estimates for the sputter erosion rate of silicates gave a range of 0.01–0.02 nm per year (Mukai et al. 2001), and for graphite grains in the solar wind the sputter erosion rate was estimated to be 0.005 nm per year assuming a solar wind density of $8\,\mathrm{cm}^{-3}$ (Draine 1989). These earlier estimates are based on a simplified derivation of the sputter yield of solids, which was improved since then (see, e.g., Behrisch and Eckstein 2007). The erosion of a grain by sputtering can be expressed by

$$\frac{\mathrm{d}}{\mathrm{d}r}(n\,V)\frac{\mathrm{d}r}{\mathrm{d}t} = -f_{SW} Y_{tot} A, \tag{2}$$

which gives the erosion rate

$$\frac{\mathrm{d}r}{\mathrm{d}t} = -\frac{f_{SW} Y_{tot}}{4n}, \tag{3}$$

where r is the grain radius, n is the number density of atoms in the grain, V the grain volume, and A the cross section of the grain. Since the grain will be arbitrarily rotating we can assume that sputtering causes homogeneous removal of material from the entire grain surface, although only a hemisphere is bombarded with solar wind ions at the time. Equation (3) can be easily integrated

$$\int_{r_{in}}^{0} \frac{1}{Y_{tot}(r)}\mathrm{d}r = -\int_0^{t_{SP}} \frac{f_{SW}}{4n}\mathrm{d}t \tag{4}$$

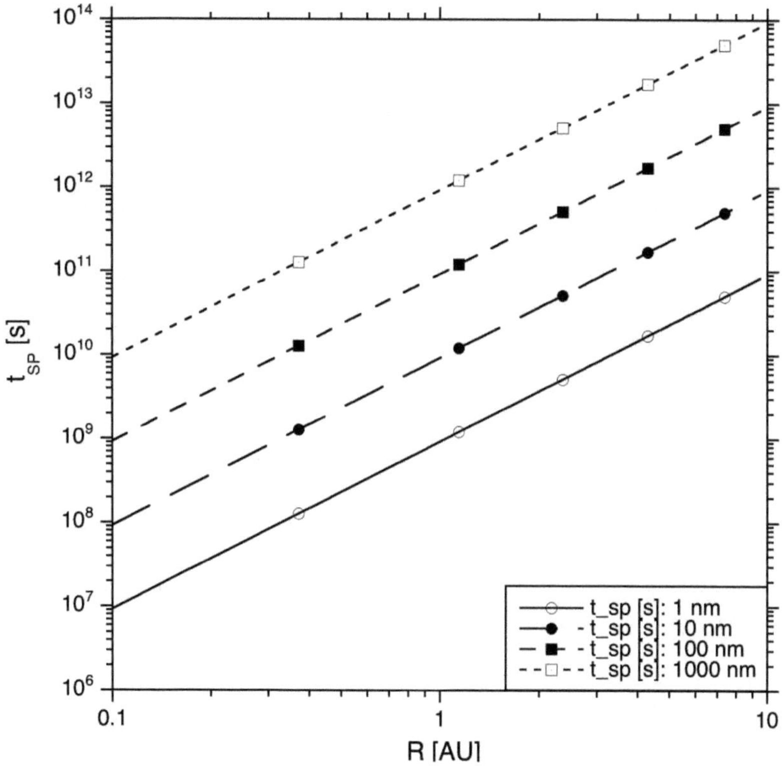

Fig. 3 Lifetime of mineral dust grains against sputtering by solar wind. Calculations are shown for 1,000, 100, 10, and 1-nm grains, *lines* from *top to bottom*

to derive the erosion time, t_{SP}, until a grain of initial size r_{in} is completely consumed by solar wind sputtering

$$t_{SP}(R) = \frac{4\,n\,r_{in}}{f_{SW}(R)\,Y_{tot}} \tag{5}$$

assuming the sputter yield does not depend on the grain radius. The erosion time, t_{SP}, is a function of heliospheric distance, R. The erosion of the grains is a strong function of distance to the Sun, with the erosion time in the range of years to hundreds of years at Earth orbit and becomes rapidly shorter closer to the Sun. Figure 3 shows the results of this calculation for 1, 10, and 100-nm size using the composition of Mercury's regolith as a proxy for the grain composition (Wurz et al. 2010).

In this calculation we assume that the sputter yield is independent of grain size r, even for the smallest grains. There are two scenarios for this size dependence: The sputter yield could be higher for smaller grains, because the energy deposited by the impacting ion cannot be distributed and accommodated so effectively in lattice phonons and thus causes an increased release of atoms. Or, on the contrary, the sputter yield could be lower for smaller grains because the impacting ion is not completely stopped in the solid and thus a lower energy is deposited in the grain resulting in a lower sputter yield. There are no experimental laboratory data of sputter yields for sub-micron-sized grains. However, there are sputter investigations on thin carbon and diamond foils of nm thickness under ion irradiation (Funsten and Shappirio 1997; Liechtenstein et al. 2010). It is observed that under prolonged ion bombardment the energy loss of ions transmitted through the foil becomes smaller, thus the foil thickness is reduced, because of sputtering of the foils. This reduction of energy loss is observed to be linear with the ion fluence indicating that the sputter yield stays constant during the thinning of the foil until it is consumed. In analogy to these foil measurements we made the assumption of a constant sputter yield independent of grain size.

2.2 *Sputtering of Icy Grains*

For solar wind protons the sputter yield of water ice is about a factor of 10 higher than for sputtering of mineral grains (Shi et al. 1995), with a sputter yield of ≈ 0.7 water molecules per proton. Figure 4 shows the sputter yield of water molecules

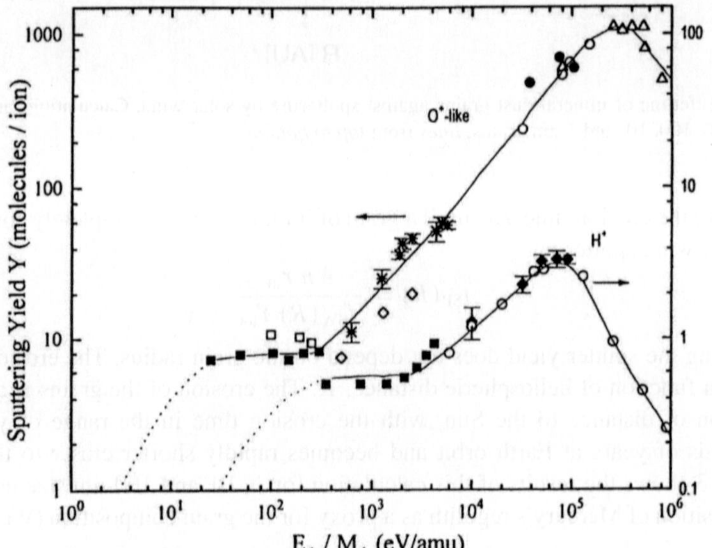

Fig. 4 Compilation of sputter yields of ice for incident protons and oxygen-like ions (from Shi et al. 1995). *Open circles* are for H^+, C^+, and O^+; *solid circles* are for N^+; *open triangle* for F^{+q}; *solid square* for H_2^+, and Ne^+; *open diamonds* for Ne^+; *open squares* for N^+ and Ne^+; *solid diamonds* for H^+; *asterisks* for O^+. *Dashed lines* are extrapolations based on estimations of the nuclear stopping power

from ice for protons and heavy ions for a large energy range. At lower energies the curve has been extrapolated on the basis of the nuclear stopping power. Recently, measurements have been performed in the lower energy range (Famá et al. 2008) and the correlation with nuclear stopping has been demonstrated. Famá et al. (2008) also found a temperature dependence of the sputter yield

$$Y_{H_2O}^{SP} = Y_0 + Y_1 \exp\left(-\frac{E_{SP}}{k_B T}\right), \tag{6}$$

where Y_0 is the sputter yield below 80 K, Y_1 the prefactor for the temperature dependent sputter yield, E_{SP} the activation energy of 0.06 eV, k_B the Boltzmann constant, and T is the absolute temperature.

Compared to sputtering of minerals (see (5)), the erosion time for icy grains becomes smaller by a factor of 10, i.e., of the order of weeks to years for 1, 10, and 100 nm size grains. The high sputter yield for ice will render the grains ice-free on the surface.

For ice grains the sputter yield increases strongly with the energy of the impacting ion (Shi et al. 1995), unlike the sputter yield of mineral grains, which becomes lower at higher ion energies as discussed above (see also Behrisch and Eckstein 2007). Thus, if the ice grains pass through a magnetosphere, or originate inside a magnetosphere, with a substantial population of energetic particles, their erosion is much faster.

The lifetime of a 1,000 nm ice grain in Saturn's magnetosphere has been estimated as 50 years (Jurac et al. 2001; Johnson et al. 2008), which is much shorter than the 10^6 years a mineral grain would survive in the solar wind at Saturn's distance. Other estimates for 1,000 nm size grains inside the Saturnian magnetosphere give a range of lifetimes of 10^2–10^4 years (Burns 1991). The erosion rate due to magnetospheric ions in Saturn's magnetosphere is 1 μm year^{-1} (Burns et al. 1999).

For grains inside the Gossamer rings of Jupiter's magnetosphere, Burns (1991) and Burns et al. (1999) estimated that grains of 1,000 nm size have a lifetime of 10^2–10^4 years because of sputtering by magnetospheric ions, and a lifetime of 10^4–10^6 years for catastrophic fragmentation. The range indicates the uncertainty in the quantitative calculation of the lifetime in the Jupiter system. The corresponding lifetime for sputtering of mineral grains in the solar wind outside Jupiter's magnetospheres for a 1,000 nm mineral grain is about 10^6 years (5), and about a factor 10 shorter for an icy grain in the solar wind.

3 Dust Erosion by Evaporation

For a given temperature every solid compound has an associated vapour pressure as function of temperature. In thermodynamics this reflects the equilibrium condition between material from the solid entering the gas phase (evaporation) and material from the gas phase condensing at the surface. In interplanetary space, even in the vicinity of a grain's surface, the pressure will be effectively zero (i.e., perfect vacuum), thus any material set free from the surface will be lost to space, which is referred to as sublimation. Here we will consider only the sublimation of water from the grains, sublimation of other species has been discussed in the literature as well (e.g., Mukai et al. 2001). Other volatiles, possibly present in a grain, can be treated with the same formalism. For a perfect vacuum the rate of evaporation, f_{sub}, is equal to the rate of molecules hitting the surface, which is given by ideal gas theory as

$$f_{sub} = \frac{1}{4} n \overline{v} = \frac{p}{\sqrt{2\pi m k_B T}} \tag{7}$$

with the unit (part. $(m^2\,s)^{-1}$). In (7) n is the number density of the vapour, $\overline{v}$ is the most probable velocity, p is the vapour pressure in Pa, m is the molecular or atomic mass in kg, k_B is the Boltzmann constant, and T is the absolute temperature of the grain's surface. For water ice the data for vapour pressure, p_{vap}, have been recently reviewed by Grigorieva et al. (2007)

$$p_{vap} = \begin{cases} 3.56 \times 10^{12} \exp\left(-\frac{6{,}141.667}{T}\right) & \text{for } T \geq 170\,\text{K} \\ 7.59 \times 10^{14} \exp\left(-\frac{7{,}043.51}{T}\right) & \text{for } T < 170\,\text{K}, \end{cases} \tag{8}$$

which is given in Pa here. The vapour pressure is a strong function of the temperature. Using the vapour pressure dependence on the temperature (8) we can calculate from the sublimation flux (7) the lifetime of an ice grain against sublimation

$$t_{sub} = \frac{n_{H_2O}\, \eta\, r_{in}}{f_{sub}(T)}, \tag{9}$$

where the surface density of water ice is $n_{H_2O} \approx 9.75 \times 10^{14}\,\text{cm}^{-2}$, η is the fraction of ice on the surface, and r_{in} is the initial grain radius. Sublimation severely affects the ice loss of small grains, since the grain temperature depends on the heliospheric distance. Moving away from the Sun, there is a certain heliospheric distance beyond which the rate of sublimation is negligible compared to other erosion processes or the orbit lifetime, which is often referred to as the "snow line" (Artymowicz 1997). In the heliosphere the distance is approximately at the location of the asteroid belt. Note that there is significant loss of water ice even beyond the snow line because of photon stimulated desorption, as will be discussed in the next chapter, and by solar wind induced sputtering.

For smaller objects, for example a 100-nm ice grain at $T = 150\,\text{K}$ (approximately the temperature at the "snow line") is lost already after about 6μs. At 110 K, about Jupiter's orbit, a 100-nm ice grain will be evaporated after 140 s, and at 90 K, about Saturn's orbit, after 6 years. For grains which are a mixture of icy materials and minerals the timescales for ice loss can become much shorter (Grigorieva et al. 2007). Figure 5 shows the sublimation time derived from (9) using a simple

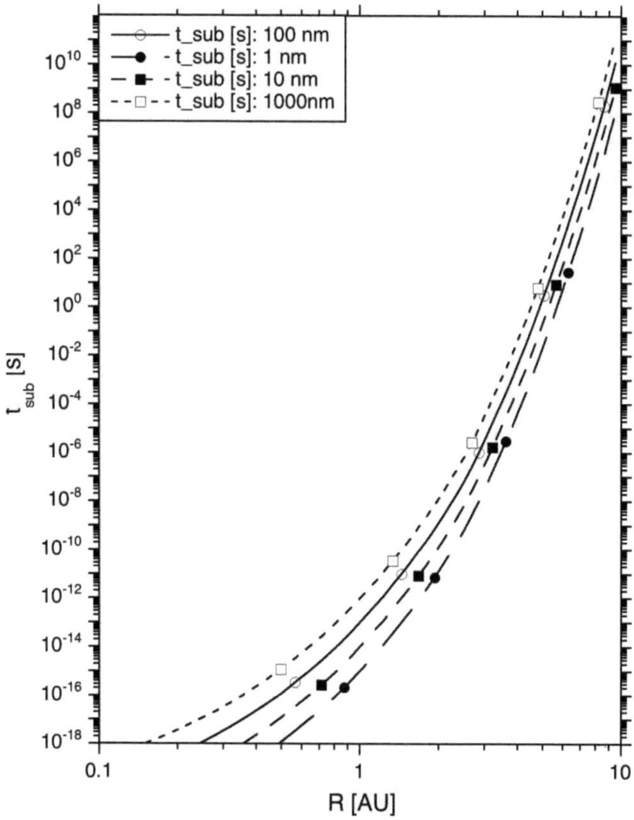

Fig. 5 Lifetime of ice dust grains in seconds against sublimation using (9). Calculations are shown for 1,000, 100, 10, and 1-nm grains, *lines* from *top to bottom*. The temperature of the grains has been approximated by assuming a simple equilibrium between absorption of solar light and thermal emission

approximation for the grain temperature, i.e., the equilibrium between absorption of solar light and thermal emission. The dramatic decrease in lifetime for icy grains for smaller heliocentric distances can clearly be seen.

The temperature of a grain in interplanetary space depends mainly on three parameters (Grigorieva et al. 2007): its size, its mineral composition, and its distance from the Sun. The temperature can be calculated by considering the energy balance between absorbed solar energy and energy loss by re-radiation and sublimation (Mukai and Schwehm 1981). Grigorieva et al. (2007) performed these calculations for a range of grain sizes and a variety ice-mineral mixtures (see also Li 2011). Below a grain size of 100-μm size-dependent heating and cooling becomes important, with grains smaller than about 1,000 nm being significantly hotter or cooler than their larger counterparts. Also the latent heat of sublimation was considered in these calculations. However, Grigorieva et al. (2007) found that at grain temperatures $T \geq 150\,\text{K}$ the effect of the latent heat on the grain temperature becomes negligible. For small grains, in the size range of 1–10 nm, Aannestad (1989) calculated that the absorption of a photon increases the temperature of the grain momentarily, followed by a slow decrease of the temperature with a time constant of the order of 1,000 s. For 5-nm grains in the interstellar medium temperature rises in the range of 20–50 K will occur upon absorption of a photon.

4 Dust Erosion by Photon Stimulated Desorption

Photon-stimulated desorption (PSD), which sometimes is referred to as photon sputtering, is when a photon is absorbed by the surface of a grain and an atom or molecule is released from the surface eventually, via an electronic excitation process at the surface.

We can calculate the flux of PSD ϕ_i^{PSD} of a species i from the surface by

$$\begin{aligned} \phi_i^{\text{PSD}} &= f_i N_S \int \phi_{\text{ph}}(\lambda) Q_i(\lambda) \text{d}\lambda \\ &\approx \tfrac{1}{4} f_i N_S \phi_{\text{ph}} Q_i, \end{aligned} \tag{10}$$

where the factor of 1/4 gives the surface-averaged value. ϕ_{ph} is the solar UV photon flux at the grain's surface, Q_i is the PSD cross section, N_S is the surface atom density, and f_i is the species fraction on the grain surface. The experimentally determined PSD cross section for Na is $Q_{\text{Na}} = (1-3)\times 10^{-20}\,\text{cm}^2$ in the wavelength range of 400–250 nm (Yakshinskiy and Madey 1999) and for K the PSD cross section is $Q_{\text{K}} = (0.19 - 1.4) \times 10^{-20}\,\text{cm}^2$ in the wavelength range of 400–250 nm (Yakshinskiy and Madey 2001).

PSD is highly species selective, and works efficiently for the release of Na and K from mineral surfaces. PSD is considered the major contributor for the Na and K exospheres of Mercury and the Moon (Killen et al. 2007; Wurz et al. 2010). However, the release Na and K from the mineral matrix is not very important for a significant erosion of the dust grain since it will cease once the surface is void of Na and K.

Water is also desorbed via the absorption of photons, thus PSD plays an important role for icy dust grains. PSD of water has been studied in the laboratory (Westley et al. 1995), and the photodesorption yield per incoming photon, $Y_{\text{H}_2\text{O}}^{\text{PSD}}$, is temperature dependent

$$Y_{\text{H}_2\text{O}}^{\text{PSD}} = Y_0 + Y_1 \exp\left(-\frac{E_{\text{PSD}}}{k_{\text{B}} T}\right) \tag{11}$$

with $Y_0 = 0.0035 \pm 0.002$, $Y_1 = 0.13 \pm 0.10$, and $E_{\text{PSD}} = (29 \pm 6) \times 10^{-3}\,\text{eV}$ (Westley et al. 1995). The temperature dependence is very similar to the one for sputtering of ice (see (6)), which was found later. The photodesorption flux is thus

$$\phi_{H_2O}^{PSD} \approx \frac{1}{4}\phi_{ph} f_{H_2O} Q_{H_2O}^{PSD} n_{H_2O} = \frac{1}{4}\phi_{ph} f_{H_2O} Y_{H_2O}^{PSD}. \tag{12}$$

The photodesorption yield given in (11) has been experimentally determined for Lyman-α photons, i.e., with a photon energy of 10.2 eV. Photodesorption yields comparable to Westley et al. (1995) results have been derived by theoretical modelling of the interaction of photons and icy surfaces, however at photon energies in the range of 8–9.5 eV (Andersson et al. 2006). Since the solar UV spectrum, and the interplanetary UV spectrum as well, is dominated by Lyman-α radiation the experimental values from Westley et al. (1995) are a good choice for quantitative estimates for photon stimulated desorption.

Photon sputtering depends on the UV photon being actually absorbed on the surface and not being reflected. Grigorieva et al. (2007) showed that the photon absorption of UV photons is very low at longer wavelengths, and only for photons with energies exceeding about 7.5 eV the absorption reaches unity.

With a surface density of water ice of $n_{H_2O} \approx 9.75 \times 10^{14}\,\mathrm{cm}^{-2}$ we get a cross section $Q_{H_2O} = Y_{H_2O}^{PSD}/N_{H_2O} = 1.8\times10^{-18}\,\mathrm{cm}^2$ at 30 K. For typical solar Lyman-α fluxes of $5 \times 10^{11}\,\mathrm{cm}^{-2}\,\mathrm{s}^{-1}$ this results in a UV erosion rate of water molecules of $4.4 \times 10^8\,\mathrm{cm}^{-2}\,\mathrm{s}^{-1}$. This corresponds to a water–ice removal of about 7.5μm per year in Earth orbit.

Using (12) we can estimate the lifetime of an icy grain at a certain heliospheric distance, R, which is

$$t_{UV}(R) = \frac{4\, n_{H_2O}\, r_{in}}{\phi_{ph}(R)\, Y_{H_2O}^{PSD}}, \tag{13}$$

where the photodesorption yield $Y_{H_2O}^{PSD}$ is a function of temperature (11), which is a function of heliospheric distance R. Figure 6 shows the lifetime against photon stimulated desorption, $t_{UV}(R)$, for ice grains of 100, 10, and 1-nm radius, for traces from top to bottom, respectively, for temperatures below 80 K. These lifetimes have to be considered as upper limit since the grains will be warmer at smaller heliospheric distances. Lifetimes of small icy grains against photon sputtering are relatively short; even at Jupiter's orbit the lifetime is about 1–100 years for 1–100-nm grains, respectively, and decreasing to days to weeks at Earth orbit. Around stars that have a higher UV luminosity than our Sun the effect of photodesorption will be even stronger Artymowicz (1996) and the position of the actual snow line is no longer determined by sublimation alone.

Desorption yields for water for fast protons and electrons impacting on the icy grain are similar to the PSD yields (Brown et al. 1978, 1980; Heide 1984; Shi et al. 1995); however, in interplanetary space their fluxes are usually much lower than the Lyman-α fluxes, unless the grain is inside the magnetosphere of giant planet.

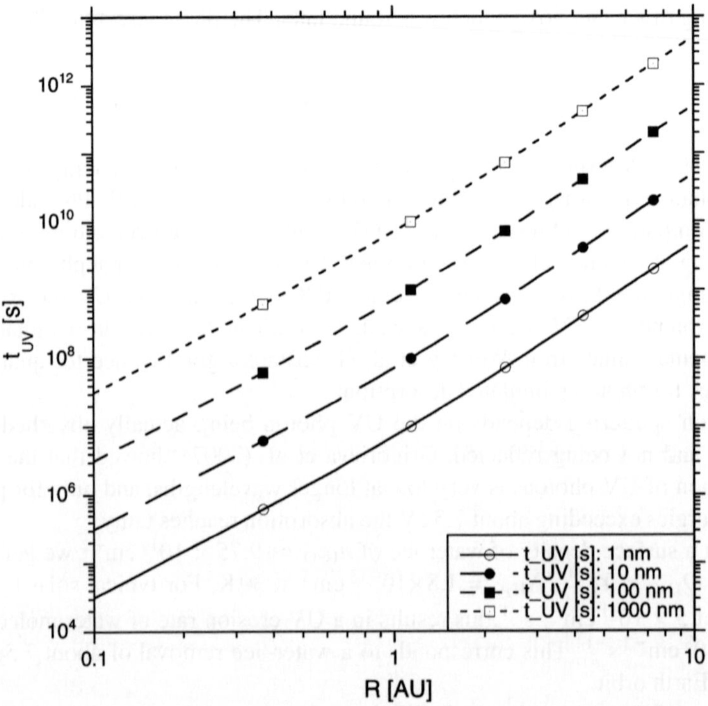

Fig. 6 Lifetime of water–ice grains against photon stimulated desorption by solar Lyman-α photons. The life is calculated using a simple temperature estimate for the grain temperature for the temperature-dependent photodesorption yield (11). Calculations are shown for 1,000, 100, 10, and 1-nm water–ice grains, *lines* from *top to bottom*

5 Conclusions

The lifetime of nanoparticles in the inner solar system is severely affected by various erosion processes, which can also be seen in the summary given in Fig. 7. Ion-induced sputtering of grains by solar wind ions occurs at all heliocentric distances and it becomes increasingly important the closer to the Sun the grain is residing. Inside Mercury's orbit the lifetime ranges from tens of years to fractions of a year, depending on grain size. Large improvements in the quantitative understanding and modelling of ion-induced sputtering have been made during the recent decades, e.g., Behrisch and Eckstein (2007). Unfortunately, these advances were accomplished only for large solids and not for nanograins. Thus, there remains an uncertainty in the sputter yield for nanograins, which still has to be resolved, when the penetration depth of the impacting ions in the matter of the grain becomes comparable or larger than the grain size, although experiments with thin foils indicate that the sputter yields also hold for small bodies.

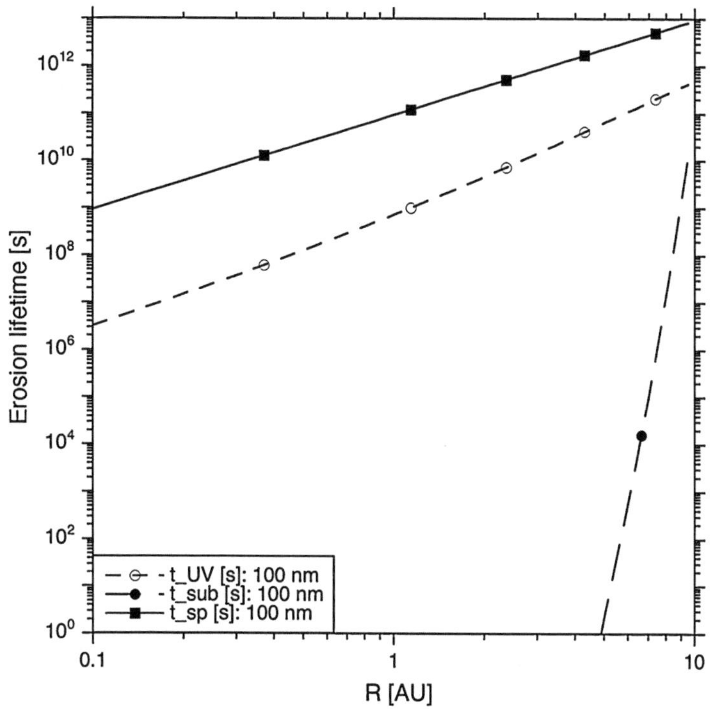

Fig. 7 Comparison of the lifetime of grains against the three erosion processes for 100-nm grains. Data are from the previous figures: sputtering of mineral grains (Fig. 3), sublimation of ice (Fig. 5), and photon stimulated desorption of ice (Fig. 6)

Since the erosion processes affect the smaller grains more than the larger grains, because of the increasing ratio between surface and volume for smaller grains, the grain size distribution will be altered, with a depletion of the smallest grains. If there is a stable population of small grains inside Mercury's orbit the loss of grains by sputtering has to be balanced by a source of grains. This source of grains could be material released sun-grazing comets, destruction of larger grains, and grains released from Mercury by micro-meteorite bombardment. For example, the production of small dust grains has been identified as an important source process to maintain the dust distribution (Ishimoto 2000).

Mercury's mass accretion rate is 10.7–23.0 tons/day, for Mercury's apocentre and pericentre, respectively (Müller et al. 2002). Recently, Borin et al. (2009) reported a meteoritic flux of $2.382 \times 10^{14}\,\mathrm{g\,cm^{-2}\,s^{-1}}$ that corresponds to 1,540 tons/day, which is about a factor of 80 higher than the earlier estimates. The bodies impacting on Mercury will release matter from its surface, atoms, molecules, up to grains of various sizes, and a certain fraction of this material will leave the gravitational field of Mercury. If we take the latter rate of meteoritic flux and assume that 0.1% of the impacting flux causes the release of 100-nm grains from Mercury, the source flux of these grains would be about 5×10^{15} grains $\mathrm{s^{-1}}$. Kameda et al. (2007) found a correlation of Na atoms emitted from Mercury's surface and interplanetary dust.

If the grains are composed partly or fully out of ice they are also subject to photon stimulated desorption and sublimation of water molecules. Sublimation is the dominant process for ice loss inside the "snow line" severely limiting the lifetime of icy grains (see Fig. 7). But even outside the snow line there is erosion of icy grains because PSD is highly effective, at least by a factor of 100 more effective than ion sputtering, in removing the ice from the grain. Thus, we can assume that all grains coming directly from the asteroid belt or from the outer solar system to locations inside Earth's orbit will have lost their ice. For grains being initially a mixture of minerals and ice they might become porous after having lost their water ice. However, it was suggested that packing forces produced by anisotropic sublimation of mantle material of the grain will result in compaction of the grain, i.e., an increase in its density (Mukai and Fechtig 1983). The timescale for such a process was estimated to be 10^4–10^5 years.

Since the lifetime of ice in the inner solar system is very limited, icy grains have to be delivered by a larger object, i.e., a comet, and released in the inner solar system. For example, ice grains have been observed in the coma of comet Hartley 2 (A'Hearn et al. 2011) and have been observed in the plume after the impact of the *Deep Impact* impactor on comet Temple 1 (Schulz et al. 2006).

Acknowledgements This work was initiated and partly carried out with support from the International Space Science Institute (ISSI) in the framework of an International Team entitled "Nano Dust in the Solar System: Formation, Interactions, and Detections".

References

Aannestad, P.A.: Temperature fluctuations and small particles, in *Evolution of interstellar dust and related topics*. Bonetti A., and Greenberg, J.M. (edt.), North-Holland, pp. 121–141, (1989).

A'Hearn, M.F.: A different class of cometary activity. in *Proceedings of the 42nd Lunar and Planetary Science Conference* Abstr. 2516 (2011).

Andersson, S., Al-Halabi, A., Kroes, G.-J., and van Dishoeck, E.F.: Molecular-dynamics study of photodissociation of water in crystalline and amorphous ices. Jou. Chem. Phys. **124**, (2006) doi: 10.1063/1.2162901

Artymowicz, P.: Vega-Type Systems, in *The Role of Dust in the Formation of Stars*, H.U. Käufl and R. Siebenmorgen (edt.), Springer, 137–148, (1996).

Artymowicz, P.: Beta Pictoris: an Early Solar System? Ann. Rev. Earth Plan. Sci. **25**, 175–219 (1997).

Behrisch, R., and Eckstein, W.: Sputtering by particle bombardment: experiments and computer calculations from threshold to MeV energies. Springer, Berlin (2007).

Betz, G., Wehner, G.K.: Sputtering of multicomponent materials, in *Sputtering by Particle Bombardment II*, Behrisch, R. (ed.), Springer-Verlag, New York, pp. 11–90, (1983).

Betz, G., and Wien, K.: Energy and angular distributions of sputtered particles. Int. Jou. Mass Spectr. Ion Proc. **140**, 1–140 (1994).

Biersack, J.P., and Eckstein, W.: Sputtering of solids with the Monte Carlo program TRIM.SP. Appl. Phys. **A 34**, 73–94, (1984).

Borin, P., Cremonese, G., Marzari, F., Bruno, M., and Marchi, S.: Statistical analysis of micrometeoroids flux on Mercury. Astron. Astrophys. **503**, 259–264, (2009).

Brown, W.L., Lanzerotti, L.J., Poate, J.M., and Augustyniak, W.M.: "Sputtering" of ice by MeV light ions. Phys. Rev. Lett. **40**, 1027–1030 (1978)

Brown, W.L., Augustyniak, W.M., Lanzerotti, L.J., Johnson, R.E., and Evatt, R.: Linear and Nonlinear Processes in the Erosion of H_2O Ice by Fast Light Ions. Phys. Rev. Lett. **45**, 1632–1635 (1980)

Burns, J.A.: Physical Processes on Circumplanetary Dust, in *Origin and Evolution of Interplanetary Dust*, Astrophys. Sp. Sci. Lib. **173**, 341–348, (1991).

Burns, J.A., Showalter, M.R., Hamilton, D.P., Nicholson, P.D., de Pater, I., Ockert-Bell, M.E., and Thomas, P.C.: The Formation of Jupiter's Faint Rings. Science **284**, 1146–1149 (1999).

Cassidy, W., and Johnson, R.E.: Monte Carlo model of sputtering and other ejection processes within a regolith. Icarus **176**, 499–507, (2005).

Collier, M.R., Moore, T.E., Ogilvie, K., Chornay, D.J., Keller, J., Fuselier, S., Quinn, J., Wurz, P., Wuest, M., and Hsieh, K.C.: Dust in the wind: The dust geometric cross section at 1 AU based on neutral solar wind observations, in *Solar Wind X*, American Institute Physics, **679**, 790–793, (2003).

Draine, B.T.: Destruction processes for interstellar dust, in *Evolution of interstellar dust and related topics*. Bonetti A., and Greenberg, J.M. (edt.), North-Holland, pp. 103–119, (1989).

Eckstein, W., and Urbassek, H.M.: Computer simulation of the sputtering process, in *Sputtering by particle bombardment: experiments and computer calculations from threshold to MeV energies*, R. Behrisch, and W. Eckstein (edt.). Springer, Berlin, pp. 21–31 (2007).

Funsten, H.O., and Shappirio, M.: Sputtering of thin carbon foils by 20 keV and 40 keV Ar^+ bombardment. Nucl. Instr. Meth. B, **127** 905–909 (1997).

Famá M., Shi, J., and Baragiola, R.A.: Sputtering of ice by low-energy ions. Surf. Sci. **602** 156–161, (2008).

Gloeckler, G., Fisk, L.A., Geiss, J., Schwadron, N.A., and Zurbuchen, T.H.: Elemental composition of the inner source pickup ions. Jou. Geophys. Res. **105**, 7459–7464 (2000).

Gloeckler, G., and Wenzel, K.P.,: Acceleration processes of heliospheric particle populations, in *The Century of Space Science*, Bleeker, J.A.M., Geiss, J., and Huber, M.C.E. (edt.), Kluwer Academic Publishers, pp. 963–1005 (2001).

Grigorieva, A., Thébault, Ph., Artymowicz, P., and Brandeker, A.: Survival of icy grains in debris discs — The role of photosputtering. Astron. Astrophys. **475**, 755–764 (2007)

Heide, H.-G.: Observations on ice layers. Ultramicroscopy **14**, 271–278, (1984)

Ishimoto, H.: Modeling the number density distribution of interplanetary dust on the ecliptic plane within 5AU of the Sun. Astron. Astrophys. **362**, 1158–1173 (2000).

Johnson, R.E., Famá, M., Liu, M., Baragiola, R.A., Sittler, E.C., and Smith, H.T.: Sputtering of ice grains and icy satellites in Saturn's inner magnetosphere. Plan. Sp. Sci. **56**, 1238–1243 (2008).

Jurac, S., Johnson, R.E., and Richardson, J.D.: Saturn's E Ring and Production of the Neutral Torus. Icarus **149**, 384–396 (2001).

Kameda, S., Yoshikawa, I., Ono, J., and Nozawa, H.: Time variation in exospheric sodium density on Mercury. Plan. Sp. Sci. **55(11)**, 1509–1517, (2007).

Killen, R., Cremonese, G., Lammer, H., Orsini, S., Potter, A.E., Sprague, A.L., Wurz, P., Khodachenko, M., Lichtenegger, H.I.M., Milillo, A. and Mura, A.: Processes that Promote and Deplete the Exosphere of Mercury. Space Science Rev. **132**, 433–509, (2007)

Li, A.: Nanodust properties derived from interstellar medium observations and their implications for the solar system. This volume, (2011).

Liechtenstein, V., Jaggi, V., Olshanski, E., Scheer, J.A., Wurz, P., and Ziesler, S.K.: Investigation of sputtering of thin diamond-like carbon (DLC) target foils by low energy light ions. Nucl. Instr. Meth. A **613**, 429–433 (2010).

McComas, D.J., Barraclough, B.L., Funsten, H.O., Gosling, J.T., Santiago-Muñoz, E., Skoug, R.M., Goldstein, B.E., Neugebauer, M., Riley, P., and Balogh, A.: Solar wind observations over Ulysses' first full polar orbit. Jou. Geophys. Res. **105**, 10419–10434 (2000).

McComas, D.J., Ebert, R.W., Elliott, H.A., Goldstein, B.E., Gosling, J.T., Schwadron, N.A. and Skoug, R.M.: Weaker solar wind from the polar coronal holes and the whole Sun, Geophys. Res. Lett. **35**, L18103, doi:10.1029/2008GL034896 (2008).

Mukai, T., and Schwehm, G.: Interaction of grains with the solar energetic particles. Astron. Astrophys. **95** 373–382 (1981).

Mukai, T., and Yamamoto, T.: Solar Wind Pressure on Interplanetary Dust. Astron. Astrophys. **107**, 97–100, (1982).

Mukai, T., and Fechtig, H.: Packing effect of fluffy particles. Plan. Sp. Sci. **31**, 655–658, (1983)

Mukai, T., Blum, J. Nakamura, A.M., Johnson, R.E., and Havnes, O.: Physical Processes on Interplanetary Dust, in *Interplanetary Dust*, E. Grün, B.A.S. Gustafson, S.F. Dermontt, and H. Fechtig (edt.), Springer, pp. 445–507 (2001).

Müller, M., Green, S.F., McBride, N., Koschny, D., Zarnecki, J.C., and Bentley, M.S.: Estimation of the dust flux near Mercury. Planet. Sp. Sci. **50**, 1101–1115, (2002).

Riley, P., Schatzman, C., Cane, H.V., Richardson, I.G., and Gopalswamy, N.: On the Rates of Coronal Mass Ejections: Remote Solar and In Situ Observations. Astrophys. Jou. **647** 648–653 (2006).

Russell, C.T., Baker, D.N., and Slavin, J.A.: The Magnetosphere of Mercury, in *Mercury*, University of Arizona Press, Tucson, AZ, USA, pp. 514–561 (1988).

Schulz, R., Owens, A., Rodriguez-Pascual, P.M., Lumb, D., Erd, C., and Stüwe, J.A.: Detection of water ice grains after the DEEP IMPACT onto Comet 9P/Tempel 1. Astron. Astrophys. **448**, L53–L56, (2006).

Schwenn, R.: Large-Scale Structure of the Interplanetary Medium, in *Physics of the Inner Heliosphere I*, Springer-Verlag, Berlin, pp. 99–181 (1990).

Shi, M., Baragiola, R.A., Grosjean, D.E., Johnson, R.E., Jurac, S., and Schou, J.: Sputtering of water ice surfaces and the production of extended neutral atmospheres. Jou. Geophys. Res. **100(E12)**, 26387–26395, (1995).

Slavin, J.A., and Holzer, R.E.: Solar wind flow about the terrestrial planets. I - Modeling bow shock position and shape. Jou. Geophys. Res. **86**, 11401–11418, (1981).

Westley, M.S., Baragiola, R.A., Johnson, R.E., and Baratta, G.A.: Photodesorption from low-temperature water ice in interstellar and circumstellar grains. Nature **373**, 405–407, (1995)

Wimmer-Schweingruber, R.F., and Bochsler, P.: On the origin of inner-source pickup ions. Geophys. Res. Lett. **30(2)**, 1077, doi:10.1029/2002GL015218, (2003).

Wimmer-Schweingruber, R.F., Crooker, N.U., Balogh, A., Bothmer, V., Forsyth, R.J., Gazis, P., Gosling, J.T., Horbury, T., Kilchenmann, A., Richardson, I.G., Richardson, J.D., Riley, P., Rodriguez, L., von Steiger, R., Wurz, P., and Zurbuchen, T.H.: Understanding Interplanetary Coronal Mass Ejection Signatures. Space Sci. Rev. **123**, 177–216, (2006).

Wurz, P., Wimmer-Schweingruber, R., Bochsler, P., Galvin, A., Paquette, J.A., and Ipavich, F.: Composition of magnetic cloud plasmas during 1997 and 1998. Proceedings of *Solar Wind X*, American Institute Physics, **679**, 685–690, (2003).

Wurz, P.: Solar Wind Composition, in *The Dynamic Sun: Challenges for Theory and Observations*, ESA **SP-600** 5.2, 1–9 (2005).

Wurz, P., Rohner, U., Whitby, J.A., Kolb, C., Lammer, H., Dobnikar, P., and Martín-Fernández, J.A.: The Lunar Exosphere: The Sputtering Contribution. Icarus **191**, 486–496 (2007) DOI:10.1016/j.icarus.2007.04.034.

Wurz, P., Whitby, J.A., Rohner, U., Martín-Fernández, J.A., Lammer, H., and Kolb, C.: The contribution to Mercury's exosphere by sputtering, micrometeorite impact and photon-stimulated desorption. Planet. Space Sci. **58**, 1599–1616 (2010); Corrigendum **58**, 2051 (2010).

Yakshinskiy, B.V., and Madey, T.E.: Photon-stimulated desorption as a substantial source of sodium in the lunar atmosphere. Nature **400**, 642–644, (1999).

Yakshinskiy, B.V. and Madey, T.E.: Electron- and photon-stimulated desorption of K from ice surfaces. Jou. Geophys. Res. **106** 33303–33308 (2001).

Yashiro, S., Gopalswamy, N., Michalek, G., St. Cyr, O.C., Plunkett, S.P., Rich, N.B., and Howard, R.A.: A catalog of white light coronal mass ejections observed by the SOHO spacecraft. Jou. Geophys. Res. **109**, CiteID A07105, DOI: 10.1029/2003JA010282 (2004).

Ziegler, J.F., Biersack, and J.P, Littmark, U.: The Stopping and Range of Ions in Solids, vol. 1 of series Stopping and Ranges of Ions in Matter, Pergamon Press, New York, (1984).

Ziegler, J.F.: SRIM-2003, Nucl. Instr. Meth. **B 219**, 1027–1036, (2004).

Charge-Exchange and X-ray Processes with Nanodust Particles

Vasili Kharchenko and Nicholas Lewkow

Abstract The interaction of nanodust particles with X-ray photons and solar wind ions is considered in detail. Collisions of nanoparticles with solar wind protons and heavy ions are investigated as a semiclassical process. Cross sections of the ion charge-exchange collisions, which reduce ion charges and increase grain charges, are calculated. The production rate of energetic neutral hydrogen atoms, the neutral H wind, and the intensity of the He^+ flux are computed for the inner heliosphere and results are compared with observations. Collisions with highly charged solar wind ions may dramatically modify the electronic structure of nanodust particles and lead to fragmentation of nanograins. The yield of X-ray photons and energetic Auger-electrons produced in collisions of heavy ions and nanograins is calculated. The dependence of the charge-exchange X-ray intensity on the dust size, ion charge, and collision velocity is analyzed. Scattering and fluorescence of X-ray photons by nano-size particles are investigated inside 1 AU. Fluxes of scattered X-ray photons with energies between 350 eV and 1 keV are computed, and the calculated intensities of the diffuse X-ray emission are compared with other mechanisms of production of the heliospheric X-ray background. Sensitivity of the scattered X-ray flux to spatial distributions of nanograins and to solar conditions is investigated.

V. Kharchenko (✉)
Department of Physics, University of Connecticut, 2152 Hillside Road, Storrs, CT 06269, USA

Harvard-Smithsonian Center for Astrophysics, 60 Garden Street, Cambridge, MA02138, USA
e-mail: kharchenko@phys.uconn.edu

N. Lewkow
Department of Physics, University of Connecticut, 2152 Hillside Road, Storrs, CT 06269, USA
lewkow@phys.uconn.edu

I. Mann et al. (eds.), *Nanodust in the Solar System: Discoveries and Interpretations*, Astrophysics and Space Science Library 385, DOI 10.1007/978-3-642-27543-2_9,

1 Introduction

Intrinsic properties of nano-size dust particles and their influence on observable characteristics of various astrophysical environments have been discussed for many years (Draine 2003; Smith 2008; Witt et al. 2002) and, recently, more attention has been paid to the presence of the nano dust in the Solar system (Meyer-Vernet et al. 2009; Czechowski and Mann 2010; Vasilyev et al. 2004). On the one hand, nanodust particles, observed in different astrophysical systems, have different origin, composition, and demonstrate specific spectral properties, but, on the other hand, nanoparticles show a fundamental similarity in their interaction with space plasma and electromagnetic radiation. All physical parameters of nanoparticles are sensitive to their geometric size and shape. This sensitivity of particle properties to geometric parameters reflects a dependence of electronic wave functions on boundary conditions associated with the surface of nano-size grains (Witt et al. 2002). Manifestation of nano-scale properties can be seen in size-dependent shifts of infrared, optical, and Extreme Ultraviolet (EUV) absorption and emission spectra in quantum crystals, dots, atomic and molecular clusters and other nanosystems. The spectral shift may be qualitatively explained by quantization of electronic wave functions: confinement of electrons inside the nanodust grains or clusters of a size a leads to the energy shift $\delta E \sim \hbar^2/2m_* a^2$, where m_* is the effective mass of a conduction band electron or valence band hole in considered nanoobjects (Kharchenko and Rozen 1996). Electronic and optical properties of nanoparticles differ from the bulk (macroscopic scale) material ones, and the size dependence of absorption or emission spectra may have a very complicated characteristic because of individual size dependencies of different electronic states (Rinnen and Mandich 1992; Efros et al. 1996). An increase in geometric dimensions of dust to a micron size transforms the electronic and optical properties of grains to that of bulk material. Nanodust particles with sizes of nanometers or several angströms show an actual molecular behavior. Astrophysical nano-size dust particles can be considered representative of a large family of atomic and molecular nano-clusters. Physical and chemical properties of clusters have been intensively studied for several decades in laboratory conditions. For that period, a valuable information on electronic and optical properties of atomic and molecular clusters have been accumulated using various methods of atomic and molecular physics, chemistry, and condensed matter physics. Physical conditions in laboratory dusty plasmas (Shukla and Eliasson 2009) and cluster beam experiments (Rinnen and Mandich 1992) may adequately simulate parameters of astrophysical environments with nanodust particles and provide reliable data for modeling and interpretation of observational data.

The majority of astrophysical observations of dust particles have been carried out for micron-size grains and only a few observations have been done to detect nano-size particles. Nano-size particles are not efficient in scattering of the infrared and optical radiations and, because of this, cannot be seen well in ground-based and satellite observations. However, nano-size particles should be more efficient in scattering of the X-ray radiation, because their typical size a is comparable with wavelengths of X-ray photons: $a \sim \lambda$. Scattering of X-rays by nanodust is similar to molecular X-ray processes and cannot be accurately represented by modeling of bulk material grains. Physical models of collision interactions between nanodust particles and the Solar Wind (SW) plasma electrons and ions may also differ from the model of the SW plasma interactions with micron grains, which are relatively well described by bulk material parameters of crystal or amorphous materials. In this chapter, we will consider specific features of nanoparticle interactions with X-ray photons and SW ions. Both processes are important for a detection of nanoparticles in the heliosphere and planetary and cometary atmospheres. X-ray observations of nanodust particles in planetary atmospheres may be a formidable task because of a relatively strong fluorescence and scattering by atmospheric atoms and molecules (Dennerl 2002).

2 Nanoparticle Collisions with Plasma Ions: Source of Neutral Wind

The SW ion flux consists of protons H^+ and alpha-particles He^{2+} moving with energies of a few keV per nuclear mass. A small fraction of heavy highly charged ions, such as C^{6+}, N^{7+}, O^{6+}, O^{7+} and others, provide a minor contribution to the total SW plasma flux. Nevertheless, because of the high charges of these ions, their interaction with nanoparticles is important for a production of the charge-exchange X-ray/EUV radiation and fragmentation of nanodust grains. Nanodust collisions with the SW ions involve a significant amount of inelastic and reactive scattering processes in which ions and nanograins can change their intrinsic states and even identity. One of the examples of such inelastic collisions is a fragmentation of the initial dust nano-particle D_N, containing N atoms, into the ensemble of smaller dust particles $D_{N-k}^{m_k+}$ in collisions with the heavy SW ion X^{q+}:

$$X^{q+} + D_N \rightarrow X^{(q-m)+} + \sum_k D_{N-k}^{m_k+}. \tag{1}$$

Nano-size fragment particles produced in these collisions may be charged and, in our example in (1) the sum of their charges is equal to the number of electrons m captured by the heavy SW ion X^{q+} : $m = \sum m_k$.[1] Other examples of inelastic collisions are the Charge-Exchange (CX) collisions of H^+ and He^{2+} ions with nano-size grains:

$$H^+ + D_N \rightarrow H^* + D_N^+, \tag{2}$$

[1] The relation between m and m_k should include a number of free electrons, if they are produced in ion–grain collisions.

$$He^{2+} + D_N \rightarrow He^{*+} + D_N^{+}, \tag{3}$$

$$He^{2+} + D_N \rightarrow He^{*} + D_N^{2+}. \tag{4}$$

The asterisk symbol in (2)–(4) indicates a possibility of the one- or two-electron capture into final excited states of atoms or ions. Populations of different excited electronic states of energetic neutral atoms and ions depend on the collisional velocity. Fast H and He atoms produced in ion collisions with nanodust create the neutral wind, propagating from the Sun to outer regions of the heliosphere.

Collisions between ions and nano-size grains may be formally described as ion collisions with giant molecules. The majority of reactions and particle transformations occur when ions penetrate inside grains. Nevertheless, neutral and charged nanoparticles interact with SW ions on distances r_i which are significantly larger than grain geometric dimensions. The empirical "optical" potential of interaction between ions and nanoparticles includes both the real $V(r_i)$ and complex $W(r_i, E)$ parts, which are responsible respectively for the elastic and reactive collisions:

$$V_{ig}(r_i) = V(r_i) + \mathrm{i}\, W(r_i, E), \tag{5}$$

where r is the radius vector of the ion from the Center of Mass of the grain, and E is the kinetic energy of the grain–ion relative motion. The long-range asymptotic behavior of the interaction potential $V(r_i)$ reflects a multipole nature of the electrostatic interaction between ion and grain: $V(r_i) = \sum_{n=0}^{\infty} C_n / r_i^n$ for $r_i > a$. The real part of the potential $V(r_i)$ can include the Coulomb interaction $(n = 1)$ between the initially charged grain and ion. The expansion coefficients C_n depend on the charge state and intrinsic structure of the nanodust particle and could be computed, if this structure is known. Thus, the long-range interaction can be responsible for momentum and energy transfer without the ions crossing any grain boundaries. The cross sections of such elastic collisions may be significantly larger than the geometric cross sections of nanograins, but the values of transferred momentum and energy are relatively small. Rate of the momentum transfer is important for a consideration of the dust acceleration and dynamical description of nano-particle motion in astrophysical plasmas.

Inelastic and reactive channels, if their cross sections are known, have to be included in the complex part of the "optical" potential $W(r_i, E)$. We focus our attention on the charge-exchange collisions between ions and nanograins and will evaluate the cross sections for these collisions at different geometric and electronic parameters of nanograins and ion charges. Charge-exchange collisions of the SW ions with interstellar gas are a major source of Energetic Neutral Atoms (ENAs) in the outer heliosphere (Heerikhuisen et al. 2007), and the flux of energetic neutrals, mostly hydrogen and helium atoms, may provide a detailed information on the interaction between the SW plasma and heliospheric gas (Möbius et al. 2009). We will consider the charge-exchange collisions in the inner heliosphere, where nanodust particles may contribute to the formation of the ENA flux. A brief introduction in the theory of ion–atom collisions is given in Bransden and McDowell (1992).

Key parameters of charge-exchange collisions, required for modeling the interaction between interstellar gas and SW plasma, were determined in experimental and theoretical investigations and reported in the literature, but little is known about ion collisions with nanodust particles. Charge-exchange collisions between nanodust particles and ions in the range of energies of astrophysical interests, from a few meV to a few keV, can be described using simplified methods of the ion–atom and ion–molecule collision theory. In Sect. 2.1, we will employ the semiclassical overbarrier model (Bransden and McDowell 1992) to calculate charge-exchange cross sections for collisions of the SW ions with nano-size grains. In Sect. 2.2, the next level of complexity, the quantum-mechanical corrections, will be introduced and used in computations of the charge-exchange cross sections for slow ion collisions with carbon and silicon nanograins. In Sect. 2.3, the formation of the neutral wind in charge-exchange collins between the SW ions and nanodust particles will be analyzed.

2.1 *Overbarrier Model for Ion Collisions with Nanodust Particles*

All semi-classical models of collisions are based on the assumption that nuclear and electronic motion can be separated. Such separation is possible because of the large difference in electronic and nuclear masses (Mott and Messey 1995). The separation procedure is valid for an entire set of interparticle distances R, except some isolated areas of nonadiabatic behavior. These nonadiabatic regions are responsible for transitions between different electronic states of the considered system of colliding ion and nanograin. Examples of nonadiabatic transitions in collisions of atomic and molecular particles are Landau–Zener transitions between crossing electronic terms (Landau and Lifshitz 1981). In Fig. 1, a colliding ion and charged dust particle are schematically shown at the interparticle distance R. The potential energy V_e of electrons is represented by the sum of several specific potentials: the Coulomb

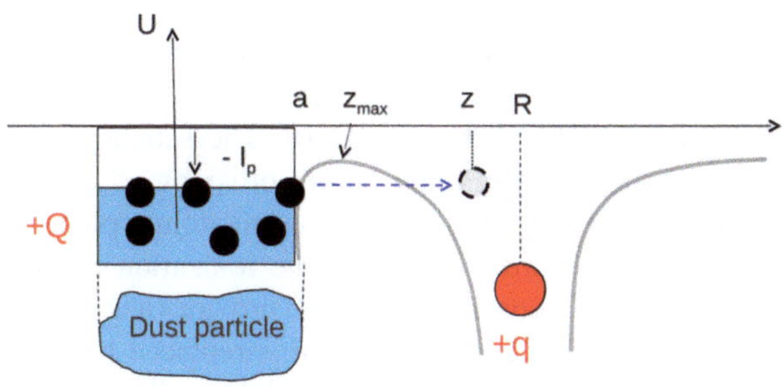

Fig. 1 Electron capture in collisions between the ion $+q$ and nanodust particle. The Q-value is the charge of the dust particle, a is the grain radius and I_P is the nano-particle ionization potential. The electron potential is shown along the z-direction, the direction of the ion–grain axis **R**

potentials of the ion and charged grain, the self-consistent potential V_g produced by nanoparticle atoms, and the long-range "image" potential V_{ind} induced in the ion–grain interaction:

$$V_e(\mathbf{r}_e, \mathbf{R}, q, Q, a) = -\frac{qe^2}{|\mathbf{R} - \mathbf{r}_e|} - \frac{Qe^2}{r_e} + V_{ind}(\mathbf{r}_e, \mathbf{R}, q, Q, a) + V_g. \quad (6)$$

The overbarrier model assumes that charge-exchange transitions occur only, if the Fermi energy of the grain electrons is higher than the maximal value V_e^{max} of the potential barrier separating the grain's electrons from the ion (Cederquist et al. 2000; Thumm 1994). This is a purely classical criterion with a complete neglect of electron tunneling effects. Electron tunneling is fundamentally important at low velocities of collision and will be analyzed as a quantum-mechanical correction to the overbarrier model. If the spherical nanograin consists of N atoms and has the radius $a = a_N$, the electronic coordinate of the barrier saddle point z_{max} can be found from the equation:

$$\frac{\partial V_e(z_e \mathbf{e}_z, R\mathbf{e}_z, q, Q, a)}{\partial z_e} = 0 \rightarrow z_e = z_{max}(R, Q, q, a) \quad (7)$$

and the criterion of classical ionization of the nanoparticle by the ion electric field may be written as the critical value for the ion–grain distance R_p, which separates ionizing and nonionizing collisions:

$$V_e(z_{max}(R, Q, q, a), R, q, a) = -I_p \rightarrow R_p = R_p(Q, q, a, I_p). \quad (8)$$

Charge-transfer occurs at all distances R smaller than R_p and the total cross sections σ_{CX} of electron capture collisions are evaluated as

$$\sigma_{CX} = \frac{1}{2}\pi R_p^2(Q, q, a, I_p). \quad (9)$$

The factor of 1/2 in (9) reflects the equal probabilities for an electron to be captured by an ion or to return back to the initial state inside a nanoparticle. The overbarrier model has been used for analysis of ion collisions with atoms, molecules, and solid state surfaces. For example, collisions of highly charged ions with nano-size carbon molecules C_{60} have been successfully analyzed with this model (Cederquist et al. 2000; Schwartz 2000). The satisfactory results for the comparison of the absolute cross sections for collisions of highly charged ions with C_{60} have been obtained by modeling C_{60} as an atomic point-like object with the ionization potential I_p, corresponding to the ionization potential of the C_{60} molecule (Schwartz 2000). The overbarrier model provides a very simple expression for the critical radius, if the charged nanograin is replaced by an ion-like object with the I_P ionization potential (Bransden and McDowell 1992; Schwartz 2000). In Figs. 2 and 3, we show the results of our calculations of the charge-exchange cross sections for collisions of H^+, He^{2+}, and O^{7+} ions with the carbon and silicon nanoparticles. Hydrogen and helium ions are the major ion constituents of the SW

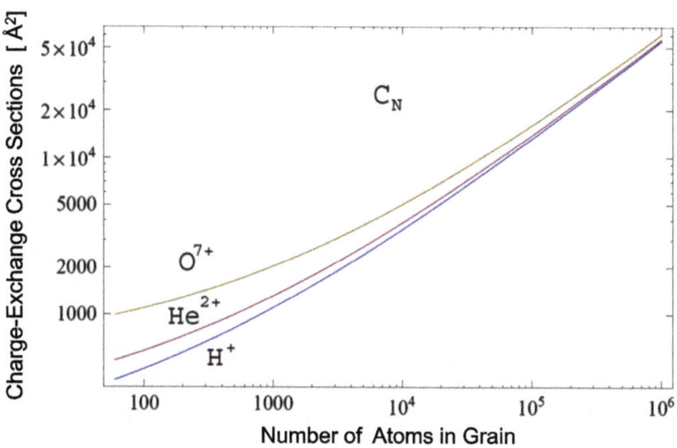

Fig. 2 Charge-exchange cross sections computed with the overbarrier model for collisions of the H^+, He^{2+}, and O^{7+} SW ions with the C_N nanoparticles consisting of N carbon atoms. Cross sections are given in Å^2

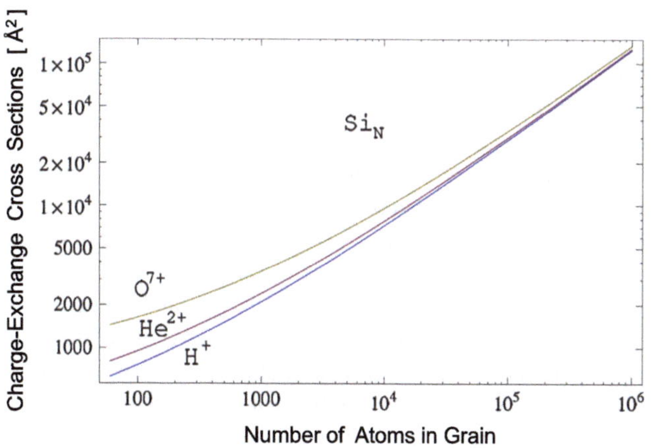

Fig. 3 Charge-exchange cross sections computed with the overbarrier model for collisions of the H^+, He^{2+}, and O^{7+} SW ions with the Si_N nanoparticles consisting of N-silicon atoms. Cross sections are given in Å^2

plasma flux and the cross sections of their charge-exchange collisions with dust particles are important for analysis of available data on the H and He^+ wind in the inner heliosphere. In our calculations, the nanoparticles have been constructed as spherical grains, containing N atoms, and these grains represent the simplest model for interstellar and interplanetary nanodusts. For further simplification, the ionization potentials I_p and an averaged interatomic distance, which is two times larger than the Wigner–Seitz radius, have been respectively selected as the same values as in the C and Si bulk materials. The cross-sections, shown in Figs. 2 and 3, depend on the ion charge and increase with the number of atoms N. For very large grains with $N \sim 10^5 - 10^6$ atoms, charge-exchange cross sections become asymptotically close to geometrical cross sections. This is reflected in Figs. 2 and 3 by the straight line N-asymptotics of the cross sections as function of N. The geometric radii of grains containing about 10^6 atoms are around 36 nm and 56 nm, respectively, for the C and Si dust materials. Collisions of ions with larger grains could be considered rather as ion interaction with the surface of a bulk material particle. Charge-exchange cross sections for collisions between ions and very small grains, with hundreds or less atoms, are visibly larger than their geometric cross sections. For highly charged ions, such as O^{7+} on Figs. 2 and 3, the saddle points of potential barriers are significantly shifted toward the ion. Thus, the O^{7+} charge-exchange cross sections are larger than the cross sections of low-charged ions H^+ and He^{2+} for any fixed value of N. It is important to note that the overbarrier models discussed above do not include any dependence on collision velocity.

2.2 Quantum-Mechanical Corrections to the Overbarrier Model

In the heliospheric environment, the nanodust grains may have different relative velocities with respect to the SW plasma flux. Averaged energies of collisions between the SW ions and nanodust particles can vary between a few eV and several keV. Low energy collisions require a more accurate computation of the dynamics of transferred electrons. Quantum-mechanical behavior of electrons becomes critically important at low collisional velocities v. In slow collisions, a collision time τ_c is significantly larger than a typical time required for the electron tunneling from the grain to the ion. Electrons can be captured by ions without approaching the critical distance R_p. The electron tunneling time τ_t can be calculated using the quasi-classical approximation, if the electron potential barrier in the ion–dust particle system is known. More accurate information on the tunneling time $\tau_t = \hbar / V_{ex}(\rho)$ may be extracted from the value of the electron "molecular" exchange energy $V_{ex}(\rho)$, where ρ is the shortest distance from the ion to the boundary of the nanograin. The electron exchange energy decreases exponentially with the distance ρ and this decreasing is slower for grains with lower values of the ionization potential I_p (Hodgkinson and Briggs 1976). The tunneling time and the time of collision $\tau_c = \rho/v$ become equal at the quantal critical distance ρ_c. The value of ρ_c depends logarithmically on the collision velocity $v = (2E/m_I)^{1/2}$ and increases as the collision velocity decreases. The total cross section of charge-exchange collisions with a consideration of electron tunneling can be written as

$$\sigma_{CX}(a, E, I_p) = \frac{\pi}{2}\left[a + (1 + 2\sqrt{q})\frac{e^2}{I_p} + \left(\frac{\hbar^2}{2m_e I_p}\right)^{\frac{1}{2}}\left(\ln\left(\sqrt{\frac{E_0}{E}}\right) + A_q\right)\right]^2, \tag{10}$$

where E is the energy of ion–grain collision; m_e is the electron mass; $E_0 \sim (M_i/m_e)I_p$ is the energy scaling constant with a value of several keV. The last term in (10) represents the quantal critical distance ρ_c. Equation (10) includes the numerical coefficient A_q, which may be computed in the closed analytic form, if the grain is considered as an atomic point-like object (Hodgkinson and Briggs 1976). At very low energies, the charge transfer cross sections can be several times larger than the geometric cross sections of nanoparticles. This feature resembles ion–atom or ion–molecule charge-exchange collisions, where ratios of the charge exchange to geometric cross sections may reach two or three orders of magnitude, depending on the collision velocity.

2.3 Neutral Wind

Charge-exchange collisions of the SW ions with neutral atoms and molecules lead to the formation of energetic neutral H and He atomic fluxes, the neutral wind. Small fractions of highly charged heavy SW ions, interacting with the interstellar and atmospheric gases, can also produce EUV and X-ray emissions. Nano-size grains can be efficient in neutralization of the SW plasma because of their large charge-exchange cross sections. The contribution of nanoparticles to a formation of the neutral wind depends on their spatial density n_d and may be detectable in cometary and planetary atmospheres, regions of zodiacal dust, and areas of nanodust particle trapping. The optical and infrared detection of nanoparticles and determination of their spatial density in various environments is a formidable task because nanoparticles are very inefficient in scattering of optical or infrared photons, contrary to the micron dust grains. Density of nanodust particles could be inferred from observations of the ENA fluxes, produced through the interaction between the SW ions and dust. The charge-exchange cross sections for H^+ and He^{2+} collisions with nanoparticles, derived in the previous section, are used to compute the neutral H and He wind and the flux of He^+ ions produced by the interacting SW and nanograins in the inner heliosphere. Computation of the neutral He wind requires cross sections of double-electron capture in collisions of He^{2+} with nanograins. Cross sections of multiple-electron capture may be larger than the single electron charge-exchange cross sections, if a multicharged ion collides with many-electron atoms, molecules, or dust particles. The parameters of the velocity, charge, and size distributions of nanodust particles have been analyzed and computed recently for the heliospheric region inside 1 AU (Czechowski and Mann 2010). The nanodust particles with radii smaller than 10 nm may be trapped in the region between 0.1 and 0.2 AU (Czechowski and Mann 2010) and contribute to the neutral wind formation in the inner heliosphere (Collier et al. 2003). Assuming that after charge-exchange collisions with nanograins in the region inside 1 AU energetic neutral H and He atoms continue their propagation in antisunward directions, we may compute the line of site intensities of the energetic fluxes for the both H and He components of the neutral wind. The macroscopic (line of sight) cross section Γ for the neutral wind production by nanodust particles at 1 AU is defined as an averaged value of the inverse length of the SW ion neutralization (Collier et al. 2003): $\Gamma = < n_d * \sigma_{CX} >$. This value has been reported in several observations (Collier et al. 2003) for the neutral H wind and the flux of He^+ ions. For the line of sight observations of the neutral wind in the ecliptic plane, the macroscopic cross section can be computed as an average over the nanodust column inside 1 AU. To evaluate Γ, the computed cross sections of charge-exchange collisions have been integrated with a simplified formula for the spatial and size distribution of nano-particle grains in the region between 0.1 and 1 AU (Czechowski and Mann 2010):

$$\Gamma = \frac{1}{R_0 - r_t} \int_{r_t}^{R_0} \int_{a_{\min}}^{a_{\max}} dr\, da\, \frac{\partial n_d(r,a)}{\partial a} \sigma_{CX}(a), \tag{11}$$

where the grain size distribution $\frac{\partial n}{\partial a}$ is defined between grain radii $a_{\min} = 3$ nm and $a_{\max} = 250$ nm and the spatial distribution is determined between the region of trapping at $r_t = 0.1$ AU and the Earth orbit $R_0 = 1$ AU. To show scaling values for nanodust parameters, a simplified formula for the distribution of grains with radii less than 10 nm ($a \leq 10$ nm) may be presented as

$$\frac{\partial n(r,a)}{\partial a} \simeq 2.5 n_d(R_0) \frac{a_{\min}^{3.5}}{a^{3.5}} \left(\frac{R_0}{r}\right)^2, \tag{12}$$

where $n_d(R_0) = 1.5 \times 10^{-10}$ cm^{-3} is the total density of these small-radius grains (3 nm< a <10 nm) near the Earth orbit $R_0 = 1$ AU. This density is close to the upper limit inferred from the nanodust flux observed by Meyer-Vernet et al. (2009), if we assumed grain speeds to be around a value of 300 km/s, expected from the nanoparticle dynamics (Czechowski and Mann 2010). Our estimation provides an upper limit of the intensity of X-rays induced by nanodust. Theoretical estimations of the Γ values for H and He^+ fluxes induced in the SW interaction with nanodust particles are

$$\Gamma(\mathrm{H}) = < n_d(r,a)\sigma_{CX}^{H}(a) >_{r,a} = 1.2 \times 10^{-21} \mathrm{cm}^{-1}, \tag{13}$$

$$\Gamma(\mathrm{He}^+) = < n_d(r,a)\sigma_{CX}^{He^+}(a) >_{r,a} = 1.7 \times 10^{-21} \mathrm{cm}^{-1}. \tag{14}$$

The theoretical value of $\Gamma(He^+)$, computed without any adjustable parameters, is in a good agreement with the result of *in situ* satellite observations: 4.6×10^{-21} cm^{-1} (Collier et al. 2003). The computed value of Γ(H) for the neutral hydrogen wind, which is produced in collisions with the same nanodust particles as He^+ ions, yields a significantly lower value, which was indicated in Collier et al. (2003) as an upper limit of Γ(H): $\Gamma(\mathrm{H}) < 6 \times 10^{-19}$ cm^{-1}. Although, there is no contradiction between the theoretical Γ(H) value and the upper limit from Collier et al. (2003), more detailed investigations would be important to clarify the role played by other mechanisms of production of the neutral hydrogen wind. In the inner heliosphere, where the density of neutral interstellar atoms is very small, dust particles could be a major source of the neutral wind.

3 X-ray Emission of Heliospheric Nanodust

Investigations of the interaction between nano-particles and electromagnetic (EM) radiation cannot be considered as a new area of physics. Laser beam experiments with free atomic and molecular clusters in molecular beam physics, scattering of the EM waves by aerosols in atmospheric science, luminescence and fluorescence of nanocrystals, and nano-size quantum dots in condensed matter physics, X-ray scattering, and fluorescence of interstellar nano-size dust particles in astrophysics and many other examples illustrate general features of the interaction between nano-size objects and electromagnetic radiation. The interaction of a nano-size particle with radiation is governed by the properties of the electronic ground and excited states in a close analogy with atoms and molecules. Computation of the electronic structure and rate of radiative transitions for nano-size grains is a challenging task because of the large number of degrees of freedom and a large uncertainty of the atomic configuration of dust particles. Simplified models, based on the many body physics results, have been developed for a description of absorption and emission spectra of metallic atomic clusters containing from a few to a few million atoms. Laboratory analysis and computations of the electronic structure of free carbon or silicon nanoparticles have been carried out mostly for condensed matter systems. Absorption and emission spectra of free nanoparticles are different from the bulk material spectra of both crystal and amorphous materials. An example of a precise experimental analysis of absorption spectra for small free Si_N particles with N changed from 18 to 41 was given by Rinnen and Mandich (1992). Significant progress has been achieved in the self-consistent computations of properties of Si_N nanocrystals for optical and infrared emissions (Pennycook et al. 2010; Fischer et al. 2011; Idrobo et al. 2006). Less information is available on the X-ray scattering and fluorescence of nanodust particles, although X-ray scattering is the major method of investigation for interstellar nanograins (Draine and Allaf-Akbari 2006; Draine 2003; Smith 2008) and their composition (Lee et al. 2009).

We will consider two mechanisms of X-ray emission, which involve heliospheric nano-particles:

(a) Scattering and fluorescence of the solar X-rays by nanodust particles in the inner part of the heliosphere.
(b) X-ray emission induced in charge-exchange collisions between highly charged SW ions and nanograins.

We will evaluate the contribution of each of these mechanisms into the heliospheric diffuse X-ray background.

3.1 X-rays Produced in Collisions of Heavy SW Ions with Nanodust

The charge-exchange mechanism of X-ray production in astrophysical environments has been investigated for more than a decade, after the surprising discovery of X-ray emission from comets (Lisse et al. 1996). The charge-exchange collisions of the highly charged SW ions, such as C^{q+} ($q = 5, 6$), N^{q+} ($q = 5, 6, 7$), O^{q+} ($q = 5, 6, 7, 8$) and others, had been identified as a major mechanism of the X-ray emission from comets (Cravens 1997; Krasnopolsky 1997):

$$X^{q+} + A \rightarrow X^{*(q-1)+} + A^{+} \rightarrow X^{(q-1)+} + h\nu + A^{+}, \tag{15}$$

where X^{q+} is a heavy SW ion, A is a cometary atom or molecule, and $\hbar\nu$ is X-ray photon induced by the excited ion $X^{*(q-1)+}$, which is produced in the charge-transfer collisions between SW heavy ions and cometary gas. Now more than 20 comets have been observed as a source of the charge-exchange X-ray emission. Also, charge-exchange X-rays have been observed from regions of interaction of the SW plasma with planetary atmospheres and the interstellar gas (Bhardwaj et al. 2007). Highly charged SW ions, colliding with nano grains, may capture electrons in highly excited states and radiate X-ray photons. The nano-size grains behave as large molecules in these collisions. For micron-size grains, the probability of multi-electron capture in a single charge-exchange collision increases significantly with grain size. Doubly- or multiple-excited ions, produced in such collisions, quickly decay with emission of Auger electrons. Auger processes reduce the probability of X-ray emissions by orders of magnitude. Thus, the charge-exchange collisions of highly charged SW ions with nanoparticles have to be considered as an additional source of the charge-exchange X-ray emission, but collisions with micron-size grains suppress X-ray emission and yield energetic Auger electrons. The rate of production of X-ray photons could be calculated with the same method as it was done for the cometary X-rays (Kharchenko and Dalgarno 2001), if the density and size distribution of dust particles are known. The Line of Sight (LOS) intensities of the X-ray emission induced in collisions between the SW heavy ions and nano-size dust are given by formula (Pepino et al. 2004):

$$I_{\mathrm{CX}} = \frac{1}{4\pi} \int_{R_0}^{\infty} \mathrm{d}x \int_{a_{\min}}^{a_{\max}} \mathrm{d}a \frac{\partial n_{\mathrm{d}}(\rho + \mathbf{x}, a)}{\partial a} n_i(\rho + \mathbf{x}) v_i(\rho + \mathbf{x}) \sigma_{\mathrm{CX}}^{(h)}(a), \tag{16}$$

where the vector $\mathbf{x}$ describes the variable of integration along line of sight of observation of the X-ray emission, the vector ρ indicates a point of the closest approach of the LOS towards Sun The X-ray detector is located near the Earth's orbit at $R_0 = 1$ AU. In (16), the density n_i and velocity v_i of heavy SW ions are integrated along the LOS and the charge-exchange cross section $\sigma_{\mathrm{CX}}^{(h)}(a)$ includes only collisions of heavy SW ions. Numerical estimations of the LOS intensity I_{CX} of X-ray emission induced in the charge-exchange collisions between SW ions and nanodust grains yield the value of $I_{CX} \sim 0.03$ photons/cm^2 s ster. The estimated value of the diffuse X-ray flux is smaller by one or two orders of magnitude than the X-ray background emission induced by charge-exchange with heliospheric gas and by cosmic sources. The charge-exchange X-ray flux associated with nano-size dust I_{CX} may increase by more than an order of magnitude to 0.3–1 photons/cm^2 s ster when considering the interaction of nano-size particles with the Coronal Mass Ejections (CMEs), since fast CMEs have large velocities and larger fractions of heavy ions such as O^{q+} and Fe^{q+}. It is important to note that the density of neutral interstellar atoms is extremely small in the region of the large concentration of nanodust particles (0.1–0.2 AU), and the charge-exchange X-ray mechanism for the neutral gas—CMEs interaction can be neglected in this region.

3.2 *Scattering of Solar X-rays by Nanodust Particles*

Scattering of X-rays by nano-size grains in the astrophysical environment has been investigated for the interstellar nano-size dust particles (Predehl and Schmitt 1995; Draine 2003; Smith 2008). The analysis of the X-ray scattered emission from remote cosmic objects requires an accurate description of a small angle scattering, but an analysis of dust-scattered X-rays in the inner heliosphere has to consider also large scattering angles because the local objects appear extended. In such conditions, even simplified models should take into account the strong anisotropy of X-ray scattering. Accurate calculations of the X-ray scattering and fluorescence parameters for nano-particles with sizes smaller than 10 nm require a knowledge of electronic wave functions of ground and involved excited states in analogy with scattering and fluorescence by atoms and molecules (Hopersky and Yavna 2010). Semi-empirical models have been used for a description of X-ray scattering by nanoparticles to avoid a complicated quantum-mechanical analysis. We have employed the spherical homogeneous Mie model to estimate the efficiency of nanodust particles in the scattering of solar X-rays. The size and spatial distributions of the nanoparticles with radii between 1 and 10 nm are given in (12). This nanodust distribution function $\partial n(\mathbf{r}, a)/\partial a$ was used in the previous section to calculate the intensity of the neutral wind. We can evaluate the total intensity $I_s(\varepsilon)$ [photons/cm^2 s] of solar X-ray photons of energy $\varepsilon = h\nu$, scattered by the nano-size dust particles located inside a region with the heliocentric radius r boundaries $r_t \le r \le R_0$ ($r_t = 0.1$ AU and $R_0 = 1$ AU). The ratio of $I_s(\varepsilon)$ to the solar X-ray intensity $I_{R_0}(\varepsilon)$ at 1 AU is given by the formula:

$$\frac{I_s(\varepsilon)}{I_{R_0}(\varepsilon)} = \int_{r_t}^{R_0} \int_0^{\pi} \int_0^{2\pi} \mathrm{d}r \, \sin\theta \mathrm{d}\theta \mathrm{d}\phi \int_{a_{\min}}^{a_{\max}} \mathrm{d}a \frac{\partial n(\mathbf{r}, a)}{\partial a} \frac{|f(\Theta_{sc}(\theta), a, h\nu)|^2}{1 + \frac{r^2}{R_0^2} - 2\frac{r}{R_0}\cos\theta},$$

$$\Theta_{sc}(\theta) = \theta + \mathrm{arctn}\frac{r \sin\theta}{R_0 - r\cos\theta}, \qquad (17)$$

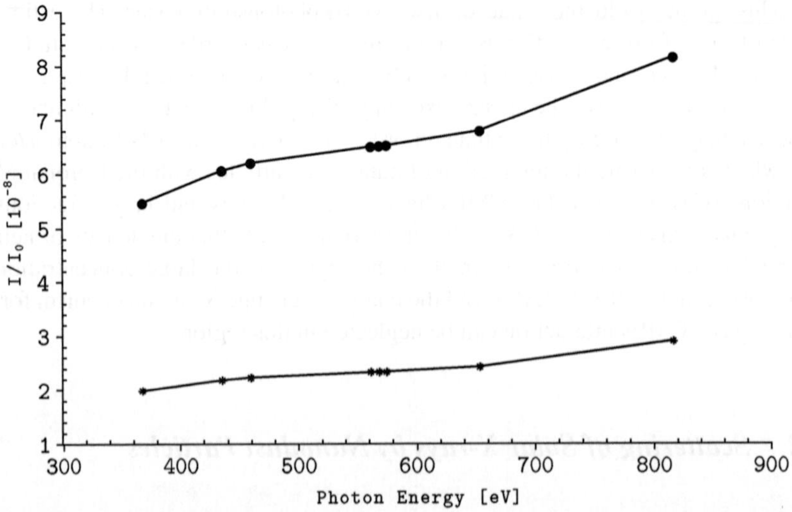

Fig. 4 Scattering efficiency of solar X-rays by nano-size carbon dust grains. (*asterisk*) indicate the results computed for the uniform distribution shell model, (*black circle*) indicate the results for the realistic dust distribution given by (12)

where $\Theta_{sc}(\theta)$ and $f(\Theta_{sc}(\theta), a, h\nu)$ are the Mie scattering angle and amplitude for the grain of size a and X-ray photon energy $\varepsilon = h\nu$. The detector of the X-ray emission is located near Earth at $r = R_0$, and the angles θ and ϕ are the heliocentric polar and azimuthal angles with respect to the axis Sun-detector. The major contribution to the scattering flux is expected from the ecliptic plane, which contains more dust particles than regions of higher heliospheric latitudes.

We have computed the total fraction of scattered X-ray photons with the distribution function nano-size dust grains given by (12) and with a simplified assumption that this distribution is valid for all latitudes. The results of our calculations are shown in Fig. 4 for X-ray photons with energies from 350 to 850 keV. The calculations have been performed for the carbon dust grains and the dependence of the real and imaginary parts of the X-ray refractive index has been taken from Henke et al. (1993). The total intensity of X-ray photons, scattered by nanoparticles can be evaluated for the known flux of the solar X-ray photons $I_{R_0}(h\nu)$. For the interval of photon energies ε between 290 and 530 eV the photon flux is about $1.6 \times 10^8 \mathrm{ph/cm^2\,s}$ and $0.2{\times}10^8 \mathrm{ph/cm^2\,s}$ for photons $\varepsilon \geq 530\,\mathrm{eV}$ at the solar minimum (Krasnopolsky 1997). The X-ray flux can increase at the solar flare condition from 10 to 100 times. The total intensities of scattered X-rays, given by (17), can be considered as the intensity of the LOS X-ray signal integrated over all directions of observations. The total intensity of scattered X-ray photons in the energy interval between 350 eV and 1 keV can be estimated at the solar minimum I_{sc}^{min} and during the solar flare at solar maximum I_{sc}^{f} (Krasnopolsky 1997), using the results of our calculations presented in Fig. 4:

$$I_{sc}^{min} \sim 8\frac{\text{ph}}{\text{cm}^2\,\text{s}} \quad \text{and} \quad I_{sc}^{f} \sim 350\frac{\text{ph}}{\text{cm}^2\,\text{s}}.$$

During solar minimum conditions, the intensity of X-ray emission induced by nanodust scattering is slightly smaller than the diffuse X-ray background, but during strong X-rays flares the nano-particles could be seen in scattered X-rays for short duration of the flare. Scattered X-ray flare signals have been recently detected in the first time from the Jupiter atmosphere (Bhardwaj et al. 2007) and could be observed from the inner sources of nanodust particles.

To clarify a role of the spatial dust distribution in the formation of the X-ray scattering flux, the same total amount of grains with the identical a-size distribution has been uniformly distributed near the trapping point, inside the thin spherical shell $r_{min} \leq r \leq r_{max}$ (r_{min}= 0.08 AU and r_{max} =0.1 AU). The flux of X-ray photons produced by this distribution decreases by factor of 3 as it is shown in Fig. 4. This effect clearly indicates that the flux of scattered X-rays is sensitive to the spatial distribution of nanoparticles and characteristics of the nanodust distribution in the inner heliosphere could be measured by investigating scattered solar X-rays.

4 Summary

We have investigated two major processes that govern the physical properties of environments containing nanoparticles, plasmas, and neutral gas: the charge-transfer collisions of nanodust with the solar/stellar wind ions and scattering of the solar/stellar X-ray radiation by nanodust. Analysis of both processes indicates that nanodust particles behave as large molecules. The cross sections of their charge-exchange collisions are larger than geometric cross sections of nanograins. This difference may reach an order of magnitude for dust particles constructed from dozens of atoms, if collision velocities are low. Charge-exchange collisions with nano-size dust in the inner heliosphere can explain a formation of the anti-sunward flux of energetic neutral H atoms (the neutral wind) and He^{+} ions. These fluxes are produced in the regions of the inner heliosphere, where the interstellar gas is almost ionized. Charge-exchange collisions of nanograins with highly charged solar wind ions, O^{q+}, C^{q+}, Fe^{q+} and others, produce X-ray photons, but intensity of this flux is by one or two orders of magnitude below the observed diffuse X-ray background. Solar X-ray scattering and fluorescence, induced by nano-size grains in the inner heliosphere, may provide the X-ray flux, which is comparable with the diffuse X-ray background at the solar minimum. Solar flares increase the scattered flux of X-rays by one or two orders of magnitude and make nano-size dust "visible" with respect to the diffuse X-ray background. The total X-ray flux produced by nanoparticles is sensitive to the spatial distribution of nanodust in the inner heliosphere.

Acknowledgments We are grateful to A. Czechowski and I. Mann for the data on the density and size distributions of nanosize grains in the inner heliosphere.

References

Bhardwaj, A. et al.: 2007, *Planetary and Space Science* **55** , 1135.

Bransden, H., McDowell, M.: 1992, *Charge-exchange and the theory of ion-atom collisions*, Oxford University Press, USA, New York, ISBN-10: 0198520204.

Cederquist, H. et al.: 2000, *Phys. Rev. A* **61**, 22712.

Collier M. R. et al.: 2003, in: Velli,M., Bruno,R., Malara, F. (Eds.), *SolarWind 10 Proceedings* **679** AIP, Melville, New York, 790.

Cravens, T. E.: 1997, *Geophys Res. Lett.* **24**, 105.

Czechowski, A., Mann, I.: 2010, *ApJ.* **714,** 89.

Dennerl, K.: 2002, *A&A* **394**, 1119.

Draine, B.T.: 2003, *Annu. Rev. Astron. Astrophys.* **41**, 241.

Draine, B. T., Allaf-Akbari, K.: 2006, *ApJ.* **652**, 1318.

Efros, Al.L., Kharchenko, V.A., Rosen, M.: 1996, *Solid State Communications* **93**, 281.

Fischer, S.A., Isborn, C. M., Prezhdo, O. V.: 2011, *Chemical Science* **2**, 400.

Heerikhuisen, J., Pogorelov, N. V., Zank, G. P., Florinsky, V.: 2007, *ApJ.* **655**, L53.

Henke, B.L., Gullikson, E.M., Davis, J.C.: 1993, *Atomic Data and Nuclear Data Tables* **54**, 181–342.

Hodgkinson, D.P., Briggs, J.S.: 1976, *J. Phys. B* **9**, 255.

Hopersky, A. N., Yavna,V. A.: 2010, *Scattering of Photons by Many-Electron Systems,* Springer Series on Atomic, Optical and Plasma Physics, New York, Springer.

Idrobo, J. C. et al.: 2006, *Phys. Rev. B* **74**, 153410.

Kharchenko, V., Rozen, M.: 1996, *Journal of Luminescence* **70**, 158.

Kharchenko, V., Dalgarno, A.: 2001, *Astrophys. J. Lett.* **554**, L99.

Krasnopolsky, V. A.: 1997, *Icarus* **128**, 368.

Landau, L. D., Lifshitz, E. M.: 1981, *Quantum Mechanics,* Butterworth-Heinemann (3 edition).

Lee, J.C. et al.: 2009, *ApJ.* **702**, 970.

Lisse, C.M. et al.: 1996, *Science* **274**, 205.

Meyer-Vernet, N. et al.: 2009, *Solar Phys.* **256**, 463.

Mott, N.F., Massey, H.S.W.: 1995, *Theory of Atomic Collisions,* World Scientific, ISBN-10:9810222521.

Möbius, E., Bochler, P. et al.: 2009, *Science* **326**, 969.

Pennycook, T. J. et al.: 2010, *Phys. Rev. B* **82**, 125310.

Pepino, R., Kharchenko, V., Dalgarno, A., Lallement R.: 2004, *ApJ.* **617**, 1347.

Predehl, P., Schmitt, J.H.M.M.: 1995, *A&A* **293**, 889.

Rinnen, K.-D., Mandich, M.L.: 1992, *Phys. Rev. Lett.* **69**, 1823.

Schwartz, S.H.: 2000, *Phys. Rev. A* **63**, 013201.

Shukla, P.K., Eliasson, B.: 2009, *Rev.Mod. Phys.* **81**, 25.

Smith, T.L., Witt, A.N.: 2002, *ApJ.* **565**, 304.

Smith, R. K.: 2008, *ApJ.* **681**, 343.

Thumm, U.: 1994, *J. Phys. B* **27**, 3515.

Vasilyev, V.P., Kalinichenko, A.I., Vasylyev, S.V.: 2004, *A&A.* **415,** 781.

Causes and Consequences of the Existence of Nanodust in Interplanetary Space

Ingrid Mann and Andrzej Czechowski

Abstract Nanodust is observed in the solar system in situ from spacecraft when the particles impact onto the detectors with high speed. Nanodust is now observed with the STEREO spacecraft in interplanetary space near 1 AU for more than 3 years and it is plausible to assume that this nanodust is a component of the interplanetary dust cloud of the solar system. The total mass flux observed in the nanodust is a small fraction of the mass that is destroyed by mutual collisions inside 1 AU, although the collision models have large uncertainties. Further studies are needed in order to understand the exact processes by which the nanodust forms in interplanetary space. Measuring its composition and mass distribution will help to understand the dust cloud inside 1 AU and the formation of nanodust during hypervelocity collisions.

1 Introduction

While the existence of nanodust in the interstellar medium is inferred from astronomical observations, there is no such evidence for the nanodust in the solar system (Li and Mann 2012). The interplanetary medium that fills the space between the Sun and the planets consists of the hot solar wind plasma, energetic particles (mainly atomic nuclei), neutrals, and cosmic dust. An obvious reason

I. Mann (✉)
Belgium Institute for Space Aeronomie, 3 Avenue Circulaire, 1180 Brussels, Belgium

Laboratoire d'Etudes Spatiales et d'Instrumentation en Astrophysique, Observatoire de Paris, Universite Pierre et Marie Curie, Universite de Paris Diderot, CNRS, Meudon, France

EISCAT Scientific Association, P. O. Box 812, SE-981 28 Kiruna, Sweden
e-mail: ingrid.mann@eiscat.se

A. Czechowski
Space Research Centre, Polish Academy of Sciences, Bartycka 18A, 00-716 Warsaw, Poland
e-mail: ace@cbk.waw.pl

I. Mann et al. (eds.), *Nanodust in the Solar System: Discoveries and Interpretations*, Astrophysics and Space Science Library 385, DOI 10.1007/978-3-642-27543-2_10,

that astronomical observations do not reveal the existence of nanodust in the interplanetary medium is that according to the size distribution of dust measured near 1 AU the maximum contribution to the total dust geometric cross section per volume in space comes from dust in the size range of micrometers. If the measured size distribution can be extrapolated to smaller sizes with roughly the same power law, then the nanodust barely contributes to the total cross section. Astronomically, dust particles can be observed in the Zodiacal light, which, above the Earth atmosphere, is the major unresolved brightness at the nightsky at wavelengths 0.3–100 μm (Leinert et al. 1998). It originates from thermal emission and sunlight scattered by the dust with sizes 1–100 μm that is located near the ecliptic plane of the solar system in majority within 3 AU from the Sun. Meteor observations confirm the existence of larger particles (i.e., meteoroids). In situ instruments on spacecraft observe smaller particles and until recently the interplanetary dust was measured to sizes as small as several ten nanometers.

Smaller particles, nanodust, are observed in certain regions in interplanetary space: there are scattered observations of nanodust near comet Halley and streams of nanodust ejected from the magnetospheres of Jupiter and Saturn were quite intensively studied during several space missions. Hsu et al. (2012) discuss the detection of stream particles in this issue. Nanodust is also observed in the upper Earth atmosphere with instruments that are similar to those on spacecraft.

Observations by the STEREO mission show for the first time that nanodust also exists widely distributed in interplanetary space (Meyer-Vernet et al. 2009). These measurements are discussed in detail by Meyer-Vernet and Zaslavsky (2012) in this issue. Contrary to the larger dust particles, the nanodust is characterized by high velocities and highly time-variable fluxes. But average flux rates are close to the extrapolation to smaller masses of the interplanetary flux model (IFM),[1] that describes observational data near Earth orbit. The nanodust forms most likely during collisions of larger dust in the inner solar system, is accelerated in the solar wind, as described by Czechowski and Mann (2012) in this issue, and observed near Earth orbit when moving outward. The goal of this chapter is to shed some light on this formation path.

We describe the different observations of nanodust with space instruments in Sect. 2. We then describe the interplanetary dust cloud (Sect. 3) and consider the dust production by fragmentation and the collisional evolution of the dust cloud inside 1 AU that possibly generates the nanodust (Sect. 4). We discuss some of the implications in Sect. 5 and in Sect. 6 we give a summary and a number of questions for future observations.

[1]This is the model of dust flux vs. mass at 1 AU suggested by Grün et al. (1985) which will be further discussed in Sect. 3.3 and Sect. 4.

2 Observations of Nanodust

2.1 Detection of Nanodust Near Comet Halley

Nanodust was probably observed in situ for the first time during the space missions to comet Halley, when the two Vega and the Giotto spacecraft crossed the path of the comet with flyby velocities of 78 km/s and 70 km/s, respectively.

The three spacecraft carried similar dust instruments that measured the number and mass of ions that are generated by the impacting dust particles (*impact ionization detectors*). Aside from measuring mass spectra of dust particles $\geq 10^{-18}$ kg for which the instruments were designed, they observed a large number of ion signals that were accumulated without the trigger signal that is usually generated due to dust entry into the instruments. Analyzing the data provided the following results published by Utterback and Kissel (1990, 1995): The events are caused by impacts of dust with masses of the order 10^{-21} kg, at that time often called attogram particles or very small dust grains (VSG). The nanodust is observed up to large distance from the nucleus (730,000 km), which implies that it survived for days after leaving the nucleus. The derived slope of number density with distance from the nucleus is flatter than $1/r^2$ and a distinct break occurs in the slope measured from VEGA-1 (later denoted as glitch). But this glitch was identified as instrumental.

A flux of dust with constant speed outward from the nucleus would generate a $1/r^2$ decrease in number density and the observed flatter profile suggests that a fraction of the nanodust forms at larger distance from the nucleus. Its formation by fragmentation of larger cometary dust seems plausible. Typical dust velocities relative to the comets are only ~0.1–1 km/s, but vary with size, so that collisions possibly cause fragmentation of fragile particles. Fragmentation can also occur when the volatiles that are contained within the porosities of large heterogeneous particles vaporize. Considering the heating and sublimation of water ice imbedded in a structure of silicates Minato and Mann (2006) have shown that it is plausible that such fragmentation events occur when the comet is at distance 0.89 AU from the Sun (i.e., the solar distance of comet Halley around the time of the flybys of the three spacecraft). Evidence for fragmentation events is also found in measurements of the larger dust at comet Halley (Simpson et al. 1987) and during the Stardust mission to comet Wild 2 (Tuzzolino et al. 2004).

The observations of nanodust near Halley motivated several theoretical studies, none of them showing a clear agreement of the theory with the observations: Fomenkova and Mendis (1992), for instance, suggested that nanodust forms by electrostatic disruption of larger grains. Mendis (2001) unfortunately quotes exactly the glitch in the VEGA-1 observations that was found to be instrumental (see above) as the most compelling evidence for the electrostatic disruption of cosmic dust. Ip and Chow (1997) proposed that the charged nanodust would be trapped in the coma and by collisions generate soft X-ray emission (the major source of the cometary X-ray emission are charge exchange reactions though, cf. Dennerl

2010). Ip and Jorda (1998) discuss the possible contribution of nanodust collision vaporization to the sodium tail of comets. Since the Halley observations also led to speculations about PAHs in comets, it is worthwhile to point once again to the experimenters (Utterback and Kissel 1995) who emphasized that it is not possible to derive composition information of the nanodust from these measurements with the dust instruments.

We finally note that also during the Giotto mission, Kirsch et al. (1991) observed unexpected signals in measurements by the energetic particle analyser (EPA) and speculated that these signals may have been caused by the impacts of heavy ions and charged dust particles. The dust masses that they derived based on assuming typical dust surface charges are 10^{-23} kg $\leq m \leq 10^{-20}$ kg.

2.2 Detection of Nanodust in Interplanetary Space

So-called stream particles of nanometer-size are observed with the dust instruments onboard Ulysses, Galileo and Cassini. They are discussed in detail by Hsu et al. (2012) in this issue and we refer the reader to discussion and references given there. The stream particles originate from the vicinity of Jupiter and Saturn and are accelerated in the magnetospheres and in the solar wind. The majority of Jovian stream particles most likely form by condensation from Io plume gas. The Saturnian stream particles are remnants of the sputtering of larger dust in the rings of the planet. The Jovian stream particles have sizes ~6–24 nm (Zook et al. 1996), the Saturnian stream particles ~2–9 nm (Hsu et al. 2012). The flux of the stream particles extends into interplanetary space but decreases with distance from the planets and is small near Earth orbit, e.g., for Jovian stream particles it is $\leq 2 \times 10^{-4}$ m^{-2} s^{-1} (Krüger et al. 2006),

Nanodust that is widely distributed in interplanetary space is first observed with the wave instruments onboard the two STEREO spacecraft near 1 AU (Meyer-Vernet et al. 2009, 2010). Both spacecraft move away from Earth by 22 degrees per year. The measured flux variations do not reveal any regular pattern, nor do they suggest that the nanodust originates from Earth or any other local source. The most recent analysis of the ongoing observations indicates that the mass range of the dust is 3×10^{-22}–2×10^{-20} kg (Zaslavsky et al. 2012). The measurements are discussed in detail by Meyer-Vernet and Zaslavsky (2012) in this issue.

These observations pose the question, why this nanodust was not observed before. The Ulysses space mission explored interplanetary space during more than a solar cycle at distances between 1 and 5.4 AU from the Sun, near the ecliptic and at higher latitudes. The dust instrument, apart from the Jovian stream particles, observed dust down to the mass 10^{-20} kg (corresponding to the radius of 10 nm for the typical silicate bulk material with density of 2,500 kg m^{-3}), but the experimenters note that possibly not all of the dust particles in the smallest mass interval were observed (Grün et al. 1995). The detector opening was for most of the

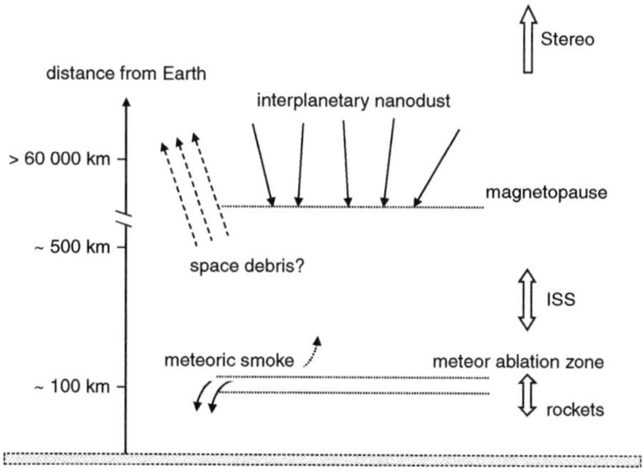

Fig. 1 Nanodust contained in the solar wind can possibly enter the magnetosphere of the Earth but its number is small compared to the nanometric meteoric smoke particles that form in the meteor ablation zone. Although the nanodust is accelerated in the solar wind, it has near 1 AU aside from its radial velocity also a large azimuthal velocity component. The nanoscale craters observed on ISS are either from impacts of the interplanetary nanodust or from accelerated space debris. Distances from Earth not drawn to scale

orbit pointing away from the Sun so that the detection probability for dust particles that come from the inner solar system was generally low (Wehry and Mann 1999).

2.3 *Detection of Nanodust in Space Near Earth*

Interplanetary dust enters the near Earth space and the accelerated nanodust can probably enter the Earths magnetosphere (Fig. 1). The upper Earth atmosphere contains dust and meteoroids[2] originating from the solar system dust cloud, meteoric dust that forms from recondensation of the meteoroid material that vaporizes during entry into the Earth atmosphere and finally the anthropogenic particles, at high altitude mainly space debris. Solid objects (when roughly $\geq 50\,\mu$m) are heated to melting and evaporation temperature when they enter the Earth atmosphere. A column of partly ionized meteoroid and atmospheric gas forms along their trajectory and generates the observed meteors. The entering solid objects lose most of their mass at altitudes 80–120 km, i.e., the meteoroid ablation zone. Rosinski and Snow (1961) first suggested that condensation from the meteor gas phase forms nanometric dust ("meteoric smoke"). We are not aware of a detailed study of the initial dust condensation process in meteors. Starting from an initial condensate different studies consider coagulation, diffusion, and

[2]Meteoroids are larger solid particles that cause meteor phenomena when entering the atmosphere.

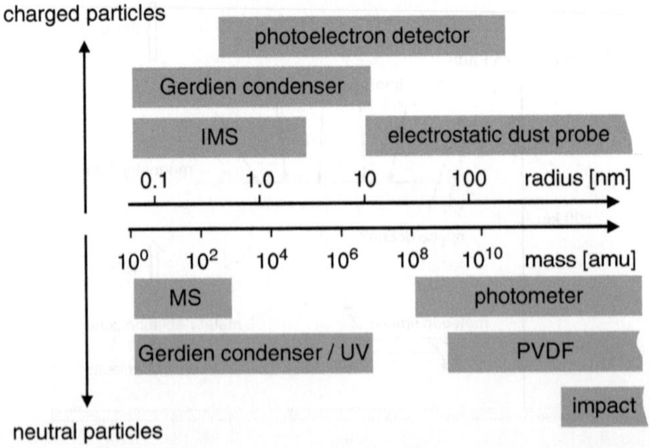

Fig. 2 The detection range of particle detectors and dust instruments that observe neutral and electrically charged dust from rockets at ∼70–100 km altitude. The figure is adapted from Rapp et al. (2003) with the addition of the detection range of the newly developed photoelectron detectors (Rapp et al. 2010). Except for the impact ionization detectors (marked "impact") all instruments detect fluxes of particles rather than single impact events. Further information is given in the text

sedimentation in the atmosphere (Hunten et al. 1980; Kalashnikova et al. 2000; Megner et al. 2008) and predict altitude profiles of the meteoric smoke with sizes ≤100 nm.

Dust measurements are made from sounding rockets in the meteor ablation zone and Rapp et al. (2003) compare the different detection methods that are illustrated in Fig. 2 and briefly summarized here. Nanodust was first found with conventional particle measurements by neutral (MS) and ions mass spectrometers (IMS). Dedicated in situ measurements often base on the detection of the surface charges of dust particles hitting a charge-collecting surface (i.e., Faraday cup). Such measurements are made by electrostatic dust probes. Photoelectron detectors are in addition equipped with a UV flash lamp to enhance the dust charge before it is measured; Gerdien condensers utilize fields to charge-select the particles before they are detected. Some early measurements were made with impact ionization detectors (cf. section 2.1) and some with PVDF detectors that measure a change of capacitance when dust particles penetrate a target foil. The measurements from sounding rockets do not detect single particles, and their interpretation requires assuming a certain surface charge. An additional complication is the fact that the measured charges are a combination of dust primary surface charges and secondary charges generated by the dust impact (see, e.g., Havnes et al. 1996). Measurements are further hampered by aerodynamics that prevents a fraction of the smallest nanodust particles from entering the detector. When number densities of nanodust are sufficiently high, it can be optically detected by the scattered sunlight. In this case, the smallest observed mass is limited by the light scattering efficiency of the dust particles at the wavelength of observation and measurements using UV photometers on rockets so far allowed to quantify results for dust sizes ≥40–50 nm (Gumbel et al. 2001). We are not aware of a measurement indicating the smallest size limit of meteoric smoke particles. The nanodust densities observed in the meteoroid ablation zone are ~1–1,000 cm^{-3} (Megner et al. 2008), which exceeds by far the densities of interplanetary dust (both nanodust and larger dust) in interplanetary space near Earth.

Nanodust was also observed at 350-km altitude with foils that were exposed to space for about 3 years onboard the international space station (ISS) (Carpenter et al. 2007). The authors analyze the craters on the foil and find evidence for a population of nanodust with diameters ≤10 nm. The flux derived from the ISS crater analysis lies on the IFM curve. However, Carpenter et al. (2007) point out that the craters could also result from impacts of nanometer-sized space debris. Similar to the nanodust in planetary magnetospheres space debris is possibly accelerated in the Earth magnetosphere (Horanyi et al. 1988).

2.4 Possible Detection of Nanodust in the Brightness Observations

The nanodust in the solar system is not observed in the astronomical observations and we are only aware of one work that suggested such an observation of nanodust. During the 2001 June 21 solar eclipse, Habbal et al. (2003) observed an emission in the spectral interval around 1,074.7-nm. The wavelength interval of this observation covers the emission of the coronal Fe XIII line, but the observed emission falls off less steeply than the coronal brightness that is observed at other wavelengths or locations in the corona. Polarization of the coronal brightness is commonly described with the component in the direction radial from the center of the Sun and the tangential component perpendicular to that. Radial polarization is characteristic of scattering at coronal electrons, which is the major source of brightness in the inner corona, Since in addition to its different radial profile the brightness in the 1,074.7 nm spectral interval was tangentially polarized the observers suspected that it is not produced by electrons, neither by ions, but generated by dust. They suggested photoluminescence of silicon nanoparticles being the source.

Photoluminescence is one of the emission mechanisms that are considered to apply for the nanodust in the interstellar medium. It arises as a result of quantum confinement in nanoparticles and its wavelength is characteristic of the material as well as of the size. Silicon nanodust of size 3.5 nm shows photoluminescence in a narrow spectral interval around the wavelength of the coronal observations. Mann and Murad (2005) objected that this requires that the nanodust forms in a very narrow size interval and showed that from thermal consideration silicon would be unlikely to survive at the temperatures of several hundred Kelvin that are typical for the dust in the inner solar system. Moreover laboratory experiments (Fig. 3) show near infrared photoluminescence only at temperatures ≥150 K (Fujii et al. 1998). Other coronal observations do not substantiate the silicon nanodust hypothesis either (Kuhn et al. 1996; Singh et al. 2004).

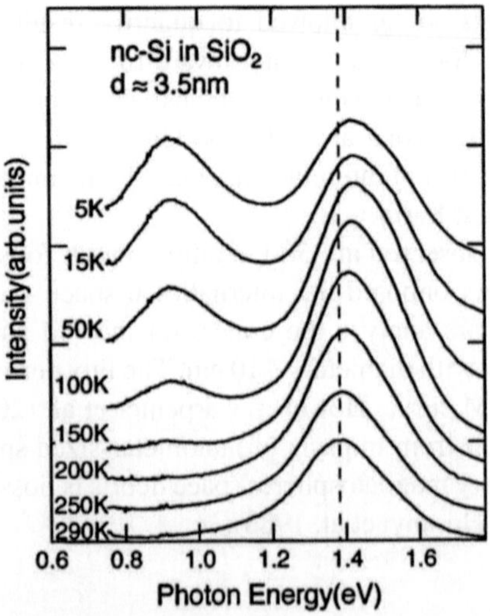

Fig. 3 Photoluminescence as function of temperature measured at 3.5 nm nanocrystals imbedded in SiO_2 illuminated at wavelength 457.9 nm. The ∼1.4-eV peak is in the range of emissions observed in the interstellar medium. The ∼0.9-eV peak that was suggested to explain near IR coronal emission disappears at temperatures ≥150 K. Figure reproduced from Fujii et al. (1998)

2.5 *Discussion of Observations*

Table 1 summarizes the discussed observations of nanodust in interplanetary space, the given mass or size intervals are those given in the works that we reference in the previous sections. In situ detection of single particles was made for nanodust with a high-impact velocity. This is the case for the fast flybys at comet Halley (Sect. 2.1), for stream particles ejected from Jupiter and Saturn and for nanodust accelerated in the solar wind (Sect. 2.2). Velocities are low in the Earth atmosphere, but the nanodust is present with large number density and accumulated fluxes are measured (Sect. 2.3). Brightness observations of nanodust near the Sun are not confirmed (Sect. 2.4). As far as the formation of the nanodust is concerned, the Jovian stream particles and the meteoric smoke particles condense from the gas phase and the Saturnian stream particles form by sputtering of larger single grains. The nanodust observed near Halley and the nanodust in the interplanetary dust cloud form by fragmentation. The dust fragmentation near comets may result from mutual collisions and/or from the vaporization of volatiles that are embedded in the larger dust. High-velocity collisions within the interplanetary dust cloud most likely produce the nanodust that is observed near 1 AU. This formation process is quite different from those of the other nanodust populations that are observed in the interplanetary space and we shall now discuss this collisional production in more detail.

Table 1 The observations of nanodust in the solar system discussed in the text

Observation	Mass	Size	Formation model
Comet Halley			Fragmentation/ break-up
Giotto/PIA and VEGA/PUMA	$\sim 10^{-21}$ kg		
Giotto/EPA	10^{-23}–10^{-20} kg		
Jovian Streams			Condensation
Cassini/CDA	–	$\sim$6–24 nm	
Saturnian Streams			Sputtering
Cassini/CDA	–	$\sim$2–9 nm	
Solar Wind			High-velocity collisions
STEREO/WAVES	3×10^{-22}–2×10^{-20} kg		
Near Earth $\sim$350 km			Undetermined
ISS crater analysis		$\leq$5 nm	
Near Earth $\sim$70–100 km			Condensation
Rocket experiments	$> 3 \times 10^{-24}$ kg		

3 Dust Cloud Overview

We give an overview of the interplanetary dust cloud by first discussing the forces that act on the dust particles (Sect.3.1) and that distinguish the three populations: large dust particles, β-meteoroids, and nanodust. The large dust particles have a disk-like spatial distribution (Sect.3.2) and generate smaller dust by collisions. The mass distribution is best described near 1 AU (Sect.3.3).

3.1 Acting Forces and Dust Populations

The major forces acting on the dust in interplanetary space are (solar) gravitational force, solar radiation pressure force, and electromagnetic (Lorentz) force. The forces vary with dust mass and one may roughly distinguish three dust populations in interplanetary space near 1 AU (Mann et al. 2010):

- *Large dust particles* (typically, mass $m \geq 10^{-14}$ kg or size $s \geq 1\,\mu$m) are mainly influenced by gravity. They move in roughly Keplerian orbits, with an additional migration inward caused by the nonradial component of the radiation pressure force (i.e., the Poynting–Robertson effect see Burns et al. 1979). For a typical

Poynting–Robertson lifetime of, for example, 10^4 years the additional radial velocity is $\sim$0.5 m/s.

- The *β-meteoroids* cover the approximate mass interval 10^{-19} kg $\leq m \leq 10^{-14}$ kg. Direct radiation pressure is comparable to gravitation for them and when released from a larger parent body they move outward in hyperbolic orbits and near 1 AU have velocities $\leq$80 km/s (cf. Mann et al. 2010).
- The *nanodust particles* with masses $m \leq 10^{-19}$ kg are mainly influenced by electromagnetic forces and are deflected in a way similar to the pickup ions in the solar wind (cf. Mann et al. 2007). They move away from the Sun and near 1 AU can reach velocities of 300 km/s and higher.

3.2 Overall Structure

The majority of dust in interplanetary space consists of collision-generated fragments of the larger solid objects (dust particles and meteoroids) that originate from asteroids and comets. Similar to these parent bodies, the majority of large dust particles move in prograde orbits with small orbital eccentricity and inclinations. They form a disk-like distribution near the ecliptic. Flux observations as well as Zodiacal light observations confirm the existence of this ecliptic-concentrated dust cloud (denoted as "disk component" in Sect. 4.2). Dust particles that originate from long-period comets may also move in high-inclination and retrograde orbits (denoted as halo in Sect. 4.2).[3] They make up a smaller amount of the dust cloud near 1 AU that is not well quantified (Mann et al. 2004). The dust particles of the halo component cross the disk with high relative velocities. Assuming the existence of the halo component is plausible for two reasons: (1) the detailed study by Ishimoto (2000) has shown the importance of (presumably cometary) dust sources inside 1 AU and (2) the sungrazing comets that are observed with SOHO for now almost 20 years proof the existence of objects in retrograde high inclination orbits in the inner solar system. Collision rates and collision velocities of the dust particles are determined by the dust number density distribution and the distribution of the orbital elements in the dust cloud (Fig. 4). The relative velocities change with distance $\sim r^{-1/2}$ and number densities $\sim r^{-1}$ or $\sim r^{-1.3}$ so that collision rates increase toward the Sun. This is where the majority of nanodust and β-meteoroids form. Aside from collision rates, relative velocities, and sizes of the colliding particles, the numbers and sizes of the forming fragments depend on the physics of the collision process.

[3] An additional component is the interstellar dust entering the solar system (cf. Mann 2010), which we do not consider here.

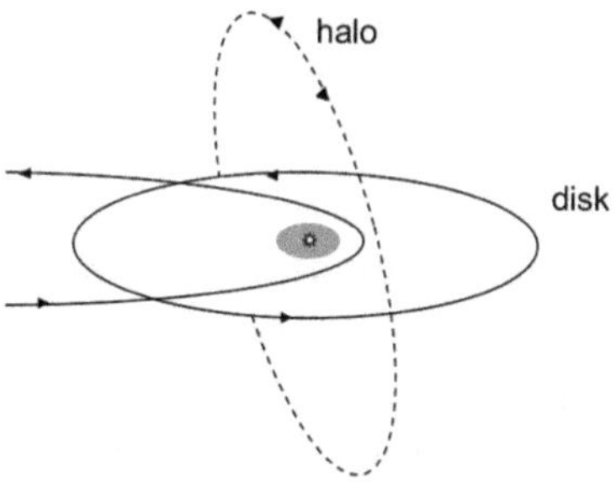

Fig. 4 The majority of interplanetary dust particles move in prograde orbits near the ecliptic plane and the nanodust forms as collision fragments in this circumsolar disk. We show in the disk a circular orbit and a highly elliptic orbit for a dust particles with large radiation pressure-to-gravity ratio. Dust particles in high inclination orbits (halo) induce high collision rates and collision velocities

3.3 Dust Distribution Near 1 AU

The dust fluxes for different masses in interplanetary space are derived from meteor observations, measurements from spacecraft, and analysis of the craters found on lunar samples. Number densities are derived from brightness observations. Most of these observations are restricted to the region near 1 AU. Conversions between fluxes and number densities are often made assuming that dust velocities and material compositions do not vary with size. This is an approximation, since lunar sample analyses suggest that smaller grains have larger velocity (Fechtig et al. 1974) and also a larger bulk density (Le Sergeant D'Hendecourt and Lamy 1980).

Dohnanyi (1969) initially developed a model to describe the evolution of asteroids and their dust debris particles as a result of mutual collisions and fragmentation. Subsequently different researchers studied the collisional evolution in the interplanetary dust cloud using similar approaches (Gruen and Zook 1980; Le Sergeant D'Hendecourt and Lamy 1980, 1981; Leinert et al. 1983).

Grün et al. (1985) reviewed the distribution models derived from observations and based on that suggested the so-called IFM for the mass interval $10^{-21}\,\mathrm{kg} \leq m \leq 10^{-1}\,\mathrm{kg}$. For the case of large grains, the IFM represents the result of balance between the collisional production, Poynting–Robertson drift and collisional destruction of the grains at 1 AU. The small mass part of the IFM is not derived from observations, but from using the large mass part of the IFM to calculate the flux of the collisional fragments that form within 1 AU and move outward (see Sect. 4.2).

4 Nanodust Formation in the Interplanetary Dust Cloud

We discuss collisional fragmentation and the sizes of forming fragments in Sect. 4.1; combination of the fragmentation models with dust collision models allows to estimate the fluxes of nanodust (Sect. 4.2); and, finally, in Sect. 4.3 we discuss to what extent the model assumptions are applicable.

4.1 Collisional Fragmentation and Nanodust Sizes

There are no experimental results that are directly relevant to the case of the nanodust production in high-velocity collisions between the dust grains. The results obtained in the context of studying the collisional evolution of asteroids are discussed by Nakamura and Michel (2009). Measurements are typically performed at small impact velocities and using large projectiles accelerated by compressed gas in so-called gas guns. The measured parameters are, for instance, crater volume, size of largest fragments, and size distribution of fragments. The studies of dust collisions in the interplanetary dust cloud often assume the empirical distribution of fragment sizes that Fujiwara et al. (1977) derived from laboratory measurements with projectile sizes of the order of 1 mm and impact velocities of 3 km/s. Laboratory studies at mechanically compressed meteorite samples suggest that the meteorite composition and optical properties change during collisions (Morlok et al. 2010).

Experimental results concerning larger impact velocities are rare. Electrically charged particles of 10–100 μm size can be accelerated in electrostatic accelerators. This method is limited to certain projectile materials and velocities of several kilometer per second in rare cases up to about 100 km/s. It permits measuring the impact-generated charges (e.g., McBride and McDonnell 1999). Hypervelocity impacts are also studied by spectroscopy of vapor clouds generated by laser pulses simulating the impact (Sugita et al. 1998) and by means of numerical simulations (Hornung and Kissel 1994; Hornung et al. 2000). To our knowledge, none of these studies constrains the size of the smallest forming fragment.

The shock wave that the impact of a projectile generates in the target and in the projectile material can be studied using the theory of solids (cf. Zel'dovich and Raizer 1967). The compression and subsequent expansion of the material may cause vaporization, melting, transition to different type of solid state (i.e., high pressure phase), and mechanical fragmentation (shattering).

Tielens et al. (1987, 1994) and Jones et al. (1996) developed a semi-empirical model to describe the dust collisions that we will apply below to the collisions in the inner solar system. According to their model, in a collision between the projectile with mass m_{P} and the target with mass m_{T} the fragmented mass of the target is proportional to the projectile mass and the proportionality factor depends on grain material and collision velocity. The upper and lower fragment mass limits (m_+ and m_-, respectively) depend on the collision parameters.

Figure 5 shows a calculation that we base on their model for silicate target and projectile material for impacts on a target with mass $m_{\mathrm{T}} = 10^{-9}$ kg. Shown are total vaporized mass, m_{VAP}, total fragmented mass, m_{FRAG}, minimum fragment mass, m_-, and maximum fragment mass, m_+. The impact speed $20/(0.2)^{1/2}$ km/s

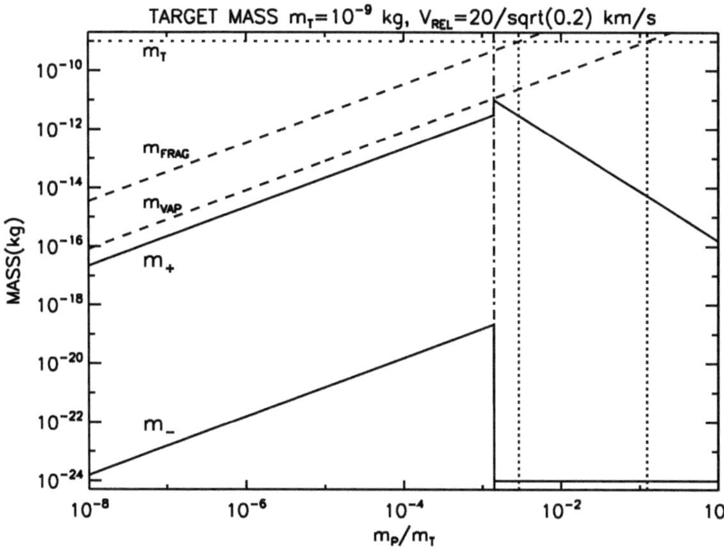

Fig. 5 Calculated fragment masses for impacts of silicate particles with speed ~45 km/s on a target with mass $m_T = 10^{-9}$ kg. Shown are as functions of projectile/target mass fraction m_P/m_T with *solid lines* the minimum (m_-) and maximum (m_+) mass of fragments, with *dashed lines* the total vaporized (m_{VAP}) and fragmented (m_{FRAG}) mass. The values of m_P/m_T at which all of the target becomes fragmented or vaporized are indicated by *vertical dotted lines*. The *vertical dashed–dotted line* shows the boundary between the cratering and the catastrophic collision case (the catastrophic region is to the right of the boundary)

~45 km/s is the average collision velocity within the interplanetary dust disk component at 0.2 AU (see Sect. 4.2). For cratering ("erosive") collisions, the fragment sizes are derived from considering the generated stresses and the shear strength of the material. The lower limit of the fragment size depends on projectile and target size and composition and on the impact velocity. For large velocities, the collisions become destructive ("catastrophic") and fragmentation also occurs due to stresses produced by the reflected wave created when the shock breaks out through the back of the target. The vertical dashed–dotted line in Fig. 5 shows the boundary between the cratering and the catastrophic collision case (the catastrophic region is to the right of the boundary). The formula that determines the largest fragment is different from that for the cratering collision and the smallest mass, m_-, is not specified. Jones et al. (1996) point out the limitations of the model in describing smaller fragments and apply a fixed lower limit $m_{min} = 10^{-24}$ kg, which corresponds to a spherical compact grain with radius 0.5 nm for bulk density 2,500 kg m^{-3}. The values of m_P/m_T at which all of the target becomes fragmented or vaporized are indicated by vertical dotted lines.

4.2 Dust Mass Distribution and Estimated Fluxes

In the report of the first observations of nanodust in the interplanetary dust cloud at 1 AU, Meyer-Vernet et al. (2009) point out that their flux estimate is close to the values obtained by extrapolating the empirically derived flux vs. mass curve near Earth (IFM) to smaller masses. This is confirmed by the ongoing measurements (Zaslavsky et al. 2012). We therefore consider the conditions that lead to this flux curve. Grün et al. (1985) assume that collision fragments generated inside from 1 AU by the intermediate and high mass part of the IFM all move outward across 1 AU and generate the flux of small dust ($m \leq 10^{-15}$ kg) there. To calculate the number and sizes of fragments, they apply the empirical fragmentation law based on laboratory experiments with low collision velocities (see Grün et al. (1985) for details). They assume that the average collision velocity within the dust cloud is 20 km/s at 1 AU increasing to the Sun as $(r/(1\,\mathrm{AU}))^{-1/2}$ and the number density vs. mass following their IFM with a radial increase toward the Sun by a factor $(r/(1\,\mathrm{AU}))^{-1.3}$. This radial increase describes the variation of Zodiacal light brightness with distance from the Sun measured onboard Helios between 0.3 and 1 AU. They multiply the calculated flux by a factor of 0.6 and in this way obtain a smooth transition to the measured fluxes at larger masses. We extrapolate the curve to smaller masses (from 10^{-21} kg given in the original work to 10^{-23} kg) by assuming the same parameters used in the original work for describing the small mass part of the distribution.

For another estimate, we use the fragmentation model from Tielens and Jones that is described in Sect. 4.1 and the following dust distribution model: The parent dust grains form a "disk" ($i = 1$) and a "halo" ($i = 2$) component of the circumsolar cloud with the number densities $n_1(m,r)$ and $n_2(m,r)$. We assume that $n_1(m,r) \sim 1/r$ and $n_2(m,r) \sim 1/r^2$ with the ratio $n_1/n_2 = 9$ at 1 AU. In some calculations, this power-law behavior is taken to change near the Sun, where the number densities are assumed to become flat within approximately 0.04 AU from the Sun center (Czechowski and Mann 2010). The mass dependence is the same as in the IFM. The average relative velocities between the grains in the same component v_{ii} and different components v_{ij} are specified ($v_{11} = 20$ km/s, $v_{12} = 40$ km/s, $v_{22} = 30$ km/s at 1 AU) and assumed to vary as $r^{-1/2}$ with distance from the Sun. The collision velocities near the Sun are comparable to those discussed for the collisional evolution in the interstellar medium (Jones et al. 1996). For a detailed discussion and comparison of other assumptions for the dust distributions and collisions, we refer to Mann and Czechowski (2005).

In Czechowski and Mann (2010), we have used the model with a fixed lower mass limit and for the case of destructive collisions to calculate the nanodust production rate as a function of distance from the Sun (as shown in Fig. 8 of that previous paper or with the solid line in Fig. 6). In a further step, the shown production rate together with the dust trajectories was used to estimate the flux of the nanodust at 1 AU. In this step, an arithmetic error occurred in the flux estimate given (Czechowski

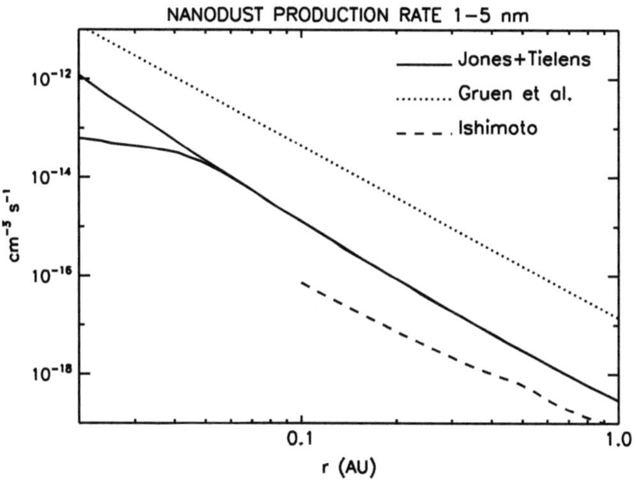

Fig. 6 Effect of different spatial distributions of the dust on nanodust production. Calculated production rate (in the disk region) of nanodust in the 1–5 nm radius range is plotted as a function of distance from the Sun. The models illustrated are: the fragmentation model based on Tielens et al. (1994) and Jones et al. (1996) (*solid line*) with the two-component (disk + halo) spatial distribution (both for the pure power law in r for all distances and for the case of flat distribution near the Sun); the same fragmentation model but with the spatial distribution calculated by Ishimoto (2000) (*dashed line*); the fragmentation model and the spatial distribution of Grün et al. (1985) extrapolated to nm size (*dotted line*)

and Mann 2010). The corrected flux is $\sim 1\,\mathrm{m}^{-2}\mathrm{s}^{-1}$.[4] The IFM, on the other hand, gives $\sim 10\,\mathrm{m}^{-2}\mathrm{s}^{-1}$ and both values are within the uncertainty limits of the STEREO data. The trapping of the nanodust produced near the Sun found in Czechowski and Mann (2010) can reduce the fraction of particles reaching 1 AU by a factor of order 10 compared to these numbers.

The difference between the two models used to calculate the nanodust production rates shown in Fig. 6 can be traced to widely different assumptions about the impact energy needed to shatter the whole of the target. For 20-km/s collision velocity and target mass 10^{-9} kg ($\sim$ the target mass within the IFM with maximum collision rate), the fragmented mass is 8×10^4 times the projectile mass in the Grün model. According to Jones et al. (1996) (assuming the case of silicate grains) the same collision velocity fragments only $\sim$100 times the projectile mass.

It is also worthwhile to note that for both fragmentation models the nanodust is generated from particles in the mass interval $10^{-13}\,\mathrm{kg} \leq m \leq 10^{-3}\,\mathrm{kg}$. This is the range of particles that generate the peak of the mass flux and also the peak of the geometric cross sections in the mass distribution that was assumed in both cases.

[4]The calculated flux in the 1–5 nm size range (radii) for the particular set of parameters used is $0.9\,\mathrm{m}^{-2}\mathrm{s}^{-1}$ and not $150\,\mathrm{m}^{-2}\mathrm{s}^{-1}$ as given in Czechowski and Mann (2010).

The curve is different for a calculation that we made for a different mass distribution that will be discussed below.

4.3 Critique of the Collision Models

The flux estimates from the previous section are roughly comparable with the observations, even though the calculated fluxes would be lower by roughly one order of magnitude, if we assume that a fraction of the produced nanodust is trapped near the Sun and therefore does not reach 1 AU.

The production rate of nanodust is strongly affected by the spatial and the velocity distributions of the parent bodies. This is an uncertainty for both models, since they are based on the assumption that the shape of the dust flux vs. mass distribution is the same over the entire considered region in space. This hypothesis is supported by the Zodiacal light observations that were carried out from spacecraft at distances 0.3–1 AU from the Sun (Leinert et al. 1998). The interpretation of these brightness observations is possible assuming the same size distribution as near 1 AU also at smaller distances from the Sun. This suggests that the shape of the distribution does not change (significantly) in the approximate mass range 10^{-14} kg $\leq m \leq 10^{-8}$ kg. There is no direct information about the larger particles inside 1 AU and we are not aware of a collision model that really shows that the size distribution is sustainable within 1 AU.

The first observations of β-meteoroids have caused already debates of the collisional evolution in the interplanetary dust cloud. Based on conservation of mass flux Le Sergeant and Lamy (1978) argued that the observed β-meteoroids cannot be explained as collision fragments of the larger dust. In a more detailed study Le Sergeant D'Hendecourt and Lamy (1981) suggested that the β-meteoroids are an independent dust population, not generated by the larger dust particles. Leinert et al. (1983) took up this problem and based on their calculations suggested that the cloud be replenished by larger objects in the size range of radar and photographic meteors. Later Ishimoto (2000) included the flux of interstellar dust into the solar system to study the collisional evolution. He also found that the dust distribution near 1 AU with masses $\leq 10^{-2}$ kg cannot sustain the dust cloud inside 1 AU and suggested that comets are a significant source of the dust inside 1 AU. The number density of dust with masses 10^{-9}–10^{-8} kg and larger according to his model is suppressed compared to the IFM inward from 1 AU and this reduces nanodust production. Since this latter model was only calculated for distance ≥ 0.1 AU from the Sun and dust masses $\geq 10^{-16}$ kg, we cannot directly compare the results to the other calculated fluxes. We compare instead the nanodust production rates as function of distance from the Sun. Figure 6 shows the rates for the previously discussed models in comparison to the rates that we calculate for one of the distributions found by Ishimoto (we use the model shown in his Fig. 7b). It suggests a comparatively low production rate, since there is a lack of parent bodies. Ishimoto, on the other hand, suggests dust productions from larger parent objects and in this context also suggests the possible occurrence of dust avalanches, i.e., enhanced dust collision rates and dust fluxes within the dust cloud during limited time after disintegration of a larger single object. Since the nanodust is observed near 1 AU within days after being formed, it is quite possible that nanodust production varies with time, while until today no such variations were observed in the cloud of larger dust particles near 1 AU.

5 Discussion

5.1 *Size and Composition of Nanodust*

Our model assumption of the formation of nanodust by fragmentation is in contrast to the observations of experimentalists who in laboratory experiments failed to generate nanodust by fragmentation (Kimura, personal communication) and rather apply condensation experiments to generate the nanodust. The laboratory experiments are discussed by Kimura (2012) in this issue.

The fragmentation law used by Grün et al. (1985) is solely based on laboratory data from experiments at large particles with low speed. The fragmentation model used by Tielens and Jones was established based on considering shock wave propagation in the solid. It should be applicable to the large collision velocities, but the authors point out the limitations to describe small fragment sizes (Jones et al. 1996). Following the presentation by Kimura, the silicate particles in the size range 3–12 nm (radius, i.e., the size of nanodust observed with STEREO) consist between roughly 40% and 10% of surface atoms and the properties of the bulk solid cannot form on these scales. Consequently the considerations on the basis of the properties of the solid do not apply and different distributions of the fragments may form at small size end of the distribution.

Alternatively large dust particles may form nanodust when they disintegrate into subunits that are already present within the material. The collected interplanetary dust particles are typically heterogeneous in material composition and structure and are often porous; the major common materials are corundum (Al_2O_3), kamacite (FeNi), forsterite (Mg_2SiO_4), enstatite ($MgSiO_3$), troilite (FeS), and organic refractory compounds with unknown chemical appearance (see, e.g., Rietmeijer 2002). The average size of subunits is of the order of 100 nm, but the size distribution of subunits is broad and includes some nanometer-sized constituents. Laboratory analyses also show the presence of nanodiamonds of about nanometer size. These nanodiamonds were either formed before the formation of the solar system (e.g., presolar grains, see Li and Mann (2012) in this issue for further references); or they are formed in the solid of the larger particles as a result of heating or the impact of energetic particles (Kamitsuji et al. 2004). Also individual iron–nickel particles of sizes 10 nm and smaller were found in interplanetary dust (Bernatowicz et al. 1999).

When considering the impact process (Sect. 4.1), it seems also possible that the nanodust forms either from a phase of the solid that was shocked under high pressure or that the nanodust forms from the gas phase. In the latter case, the nanodust possibly forms in a similar way as the meteoric smoke particles in the Earth atmosphere. Condensation experiments of gases with meteoroid composition show the formation of nanometer-sized refractive particles that form the meteoric smoke (Rietmeijer 2000) and similar condensation processes also occur in astrophysical environments (Nuth et al. 2002). Refractory compounds form first during condensations and this would possibly explain the existence of nanodust at small distances from the Sun. This process of nanodust formation would critically depend on the initial parameters of the impact-generated vapor cloudlets and on their expansion in the solar wind. Given the large plasma temperatures in solar corona and solar wind $\geq 10^5$ K, condensation is not likely to occur there.

Although we are not aware of any better models, the extrapolation of dust fragmentation laws to the sizes of few nanometers should be reconsidered and the measured nanodust fluxes near 1 AU provide an opportunity to reconsider these fragmentation laws. It is particularly interesting to find out whether the size distribution from the larger dust to the nanodust is continuous and what its slope is.

5.2 Dust Pile-Up

Our current knowledge of dust in the inner solar system does not provide any evidence for a significant pile up of dust near the Sun.

Wimmer-Schweingruber and Bochsler (2003) previously suggested that nanodust piles up near the Sun, since it is not influenced by the Poyting–Robertson effect (cf. Sect. 3.1). Although we do not know the exact optical properties of the nanodust in the solar system dust cloud, it is indeed likely that their radiation pressure to gravity ratio "β" is smaller than that of the dust in the adjacent size interval from 100 nm to several μm.[5] The Poynting–Robertson effect is not the limiting effect, however, since (1) the nanodust does not move in bound Kepler orbits and (2) it is destroyed by sublimation.

Trajectory calculations (Czechowski and Mann 2012, in this issue) show that nanodust that is not ejected follows complex bound orbits for which the perihelia are close to the Sun. This trapping zone forms at distance $\leq$0.2 AU from the Sun. Since the perihelia of trapped trajectories are typically $\leq$0.05 AU it is safe to assume that the majority of trapped particles are destroyed by sublimation. Nanodust that is not destroyed by sublimation is destroyed by sputtering. The destruction probability during one orbit assuming the sputtering rates of obsidian under average solar wind conditions are between 0.002 and 0.04 for a 10-nm

[5]As a caveat to the considerations of Poynting–Robertson lifetimes, we point out that those are not only influenced by the radiation pressure force, but also by a drag force that is caused by the momentum transfer from the solar wind particles that hit the dust and this also applies to the nanodust (Minato et al. 2004).

particle Czechowski and Mann (2010). We point the reader to the discussion of dust destruction processes by Wurz (2012) in this issue.

Since observations of the solar corona in near infrared (IR) have shown several humps in the slope of the brightness with distance from the Sun, the formation of a dust ring near the Sun was considered and we mention it here for completeness. This pile-up of some of the large dust particles bases on a different dynamical effect than that of the nanodust. Mukai et al. (1974) have shown that the dynamics of sublimating dust for which the radiation pressure changes with the changing size can enhance the dust number density in the region of the sublimation zones. Large particles, with the masses above the value corresponding to the maximum of radiation pressure force, are exposed to a larger radiation pressure force when their size is reduced. Hence their orbits become more elliptic and they stay at the given distance from the sun longer than the dust in circular orbit. This can occur for particle sizes $\sim 1\,\mu\mathrm{m}$ and it depends on dust composition and structure (Kimura et al. 1997). This enhanced dust number density occurs only for the highly absorbing fraction of the interplanetary dust, the enhancement is comparatively small and restricted to a small region (Mann et al. 2004).

Coming back to the nanodust, the dynamics that leads to the trapping may also cause flux variations.

5.3 *The Variation of the Nanodust Flux*

Flux variations of the nanodust are possibly connected to the trapping that the trajectory calculations reveal (Czechowski and Mann 2010). The trapping disappears for nanodust with high radiation pressure to gravity ratio and it varies with the solar magnetic field. Since nanodust production is largest at small distance from the Sun, small variations of the size of the trapping zone can cause flux variations that are possibly observed near 1 AU. Flux variations possibly also result from other effects, as the acceleration of the dust, for instance, depends on solar wind velocity and takes place already near the Sun. The nanodust crosses 1 AU within $\sim$ days after its formation, while the β-meteoroids pass Earth orbit after weeks and dust in bound orbits may stay in the solar system for thousands of years. As a result, stochastic variations of the dust collision rates might be visible in the nanodust fluxes, although they are not observed for larger dust.

5.4 *Nanodust as Source of Pickup Ions and Neutral Solar Wind*

Dust particles also interact with the solar wind by different processes (Mann et al. 2011): Solar wind ions impinging on the dust particles are decelerated within a distance of several 10 nm. The ions remain in the dust particles or, when their path through the grain is shorter than the penetration depth, leave the particle after charge exchange. Solar wind ions may also recombine on the surface of dust particles and finally dust destruction by sublimation and collisions generates new ions. Some traces of these interactions are possibly measured, but based on our current knowledge none of these measurements clearly confirm or require the presence of nanodust. The contribution of dust particles is often discussed in the context of the formation of pick up ions and neutral solar wind.

Neutral solar wind forms by charge exchange reaction of solar wind protons with neutrals, typically hydrogen, but possibly also with dust particles. Collier et al. (2003) use neutral solar wind observations to place an upper limit on the dust geometric cross section in the sunward direction. Kharchenko and Lewkow (2012) present a detailed study of the possible formation of neutrals by charge exchange with nanodust in this issue and show that their cross section for charge exchange exceeds the geometric cross section by about one order of magnitude, nonetheless the cross section stays below the limit derived by Collier et al. (2003).

Pickup ions are minor species in the solar wind that do not originate from the solar corona, but are released at larger distances from the Sun and then carried by the solar wind. With Ulysses solar wind measurements, an inner source of pickup ions was discovered that originates near the Sun (Geiss et al. 1996). A summarizing discussion including recent data is given by Gloeckler et al. (2010). The inner source pickup ions can also be generated by the different dust interactions listed at the beginning of this section. Grün et al. (2010) claim that the "fluxes of inner-source pickup ions also point to the existence of a much enhanced dust population in the nanometer size range" and refer to a work by Schwadron et al. (2000). This claim is neither substantiated by the given reference, nor by other recent studies of the pickup ions that we are aware of. Schwadron et al. (2000) estimate the geometric cross-sectional area of dust inside 1 AU that is needed in order to explain the measured inner source pickup ions by charge recombination of solar wind ions on the surface of dust particles and find the condition for the total cross sectional area $G_{total} \geq 1.3 \times 10^{-15}\,m^{-1}$ of dust near 1 AU. The IFM, in contrast, generates over the entire mass interval $G_{total} = 4.6 \times 10^{-19}\,m^{-1}$. The nanodust flux observed with STEREO roughly lies on the IFM curve for which, as seen in Fig. 4 of the Grün et al. (1985) paper, the vast majority of the geometric cross-sectional area comes from the mass interval $10^{-12}\,kg \leq m \leq 10^{-6}\,kg$.[6]

Another model that Wimmer-Schweingruber and Bochsler (2003) suggest to explain the inner source also collapses because of the required large number density of nanodust. The authors suggest that pickup ions are solar wind ions that acquired single charge state after passing the nanodust. Another shortcoming of this model is that it does not take into account the velocity distribution of the ions after charge

---

[6]This distribution of geometric cross section is derived from the flux assuming dust relative velocities of 20 km/s. Since the small particles have larger velocities, their number densities are overestimated when applying this model.

exchange in the dust. Solar wind ions that pass the dust have radial velocities ranging from solar wind speed and zero (Minato et al. 2006). The pickup process of these ions should be quite different from those of the ions with initial azimuthal velocity assumed by Schwadron et al. (2000) in their study of the velocity distribution of the pickup ions in the solar wind.

5.5 *The Fate of Nanodust*

Once the nanodust that is in majority generated inside 1 AU has reached high speed it moves outward without being strongly deflected.

Trajectory considerations suggest that the nanodust with $Q/m = 10^{-4}$–10^{-5} e/m$_p$ that forms at larger distances from the Sun is also accelerated and in certain cases suggest that the speed of nanodust can reach even 700 km/s (Czechowski and Mann 2010). Larger particles ($Q/m \leq 10^{-6}$ e/m$_p$) are less efficiently accelerated and their acceleration depends on the field configuration. The total momentum that is carried with the nanodust flux is small, i.e., $\leq 10^{-6}$ of that of the solar wind near 1 AU (Mann et al. 2011). The trajectory calculations suggest, that the nanodust passes the heliopause without significant interactions. One might expect that the particles are destroyed by sputtering before reaching the heliopause. The time to reach a distance of 100 AU is of the order of 1 year, which is short compared to the 30-year sputtering lifetime of a 10-nm magnetite grain at 0.5 AU given by Mukai and Schwehm (1981). The sputtering rates further decrease with distance from the Sun but vary with the dust composition as well as with the energy and the composition of the solar wind and the sputtering of nanodust still needs to be studied in detail.

6 Summary

While it is plausible that the observed nanodust forms as a component of the interplanetary dust cloud of the solar system, our current knowledge does not much constrain the composition nor the formation process. Different from the other nanodust populations observed in the solar system, it most likely forms by collisional fragmentation. Collision velocities are comparable to those in the interstellar medium where collisional fragmentation may also produce nanodust. As far as the interactions of nanodust with the solar wind are concerned, the nanodust plays a role, for instance, for the pickup ions and neutrals. The present observational data, however, of these solar wind constituents cannot be traced back convincingly to the influence of nanodust. Estimating the flux of nanodust in interplanetary space is hampered by our lack of knowledge of the dust mass distribution and spatial distribution inside 1 AU and by our lack of knowledge of the fragmentation process during dust collisions.

We summarize the current results concerning nanodust in the interplanetary space as follows:

- Trajectory calculations show that a fraction of the nanodust that forms in the inner solar system is ejected outward and crosses Earth orbit with large speed that allows detection near 1 AU by means of impact ionization.
- The measured mass flux in the nanodust is a tiny fraction of the dust mass that fragments by collision inside 1 AU and it is plausible that it forms by fragmentation of larger dust.
- The cross-sectional area is small compared to that of the entire dust cloud and the momentum of the accelerated nanodust is small compared to that of the solar wind.
- At this point, we are unable to say what causes the observed flux variations of the nanodust. Dynamical effects, as well as source variations, may play a role.
- The different physics of the nanodust suggests that the used conventional semi-empirical fragmentation laws possibly do not apply for describing its formation.

At this point observations, rather than further model calculations, may guide the investigations. We suggest the following questions be addressed when analyzing observations of nanodust in interplanetary space:

- Study the flux variations of the nanodust and possible correlations with measured variables of the solar wind, the solar corona and the meteoroid streams.
- Determine lower size limit of nanodust in the solar wind and check for variations and their correlation with other variables.
- Derive the size distribution of nanodust and its time variation.
- Derive the average composition information and single particle composition information.

The observations of nanodust with plasma wave instruments provide already valuable information to address some of these questions. They are particularly suitable for studying the flux variations and possibly the size distribution of the nanodust. Based on more observational data, it may become worthwhile to re-consider the dust fragmentation laws and to further study other potential processes for the formation of the nanodust.

Acknowledgments This work was partly supported by the Belgium Solar Terrestrial Center of Excellence and the Polish Ministry of Science (grant No N203 4159 33). Parts of this chapter were prepared while IM stayed as Visiting Astronomer at LESIA, CNRS, Observatoire de Paris Meudon and as Visiting Professor at Universite Pierre et Marie Curie in Paris and this support is greatly acknowledged. This work was initiated and partly carried out with support from the International Space Science Institute (ISSI) in the framework of an International Team entitled "Nano Dust in the Solar System: Formation, Interactions, and Detections."

References

Bernatowicz, T., Bradley, J., Amari, S., Messenger, S., and Lewis, R.: 1999, *Lunar and Planetary Institute Science Conference Abstracts* **30**, 1392.

Burns, J.A., Lamy, P.L., and Soter, S.: 1979, *Icarus* **40**, 1.

Carpenter, J.D., Stevenson, T.J., Fraser, G.W., Bridges, J.C., Kearsley, A.T., Chater, R.J., and Hainsworth, S.V.: 2007, *J.Geophys. Res. E* **112**, 8008.

Collier, M.R., Moore, T.E., Ogilvie, K., Chornay, D.J., Keller, J., Fuselier, S., Quinn, J., Wurz, P., Wüest, M., and Hsieh, K.C.: 2003, *Solar Wind Ten* **679**, 790.

Czechowski, A. and Mann, I.: 2010, *Astrophys. J.* **714**, 89.

Czechowski, A. and Mann, I.: 2012, Nanodust Dynamics in Interplanetary Space, In: Mann, I., Meyer-Vernet, N., Czechowski, A., (eds.) *Nanodust in the Solar System: Discoveries and Interpretations*, 76.

Dennerl, K.: 2010, *Space Science Reviews* **157**, 57.

Dohnanyi, J.S.: 1969, *Journal of Geophysical Research* **74**, 2531.

Fechtig, H., Hartung, J.B., Nagel, K., Neukum, G., and Storzer, D.: 1974, *Lunar and Planetary Science Conference Proceedings* **5**, 2463.

Fomenkova, M.N. and Mendis, D.A.: 1992, *Astrophysics and Space Science* **189**, 327.

Fujii, M., Hayashi, S., and Yamamoto, K.: 1998, *Journal of Applied Physics* **83**, 7953.

Fujiwara, A., Kamimoto, G., and Tsukamoto, A.: 1977, *Icarus* **31**, 277.

Geiss, J., Gloeckler, G., and von Steiger, R.: 1996, *Space Science Reviews* **78**, 43.

Gloeckler, G., Fisk, L.A., and Geiss, J.: 2010, *Twelfth International Solar Wind Conference* **1216**, 514.

Grün, G. and Zook, H.A.: 1980, *Solid Particles in the Solar System* **90**, 293.

Grün, E., Zook, H.A., Fechtig, H., and Giese, R.H.: 1985, *Icarus* **62**, 244.

Grün, E., Baguhl, M., Divine, N., et al.: 1995, *Planetary and Space Science* **43**, 971.

Grün, E., Horanyi, M., Moebius, E., Sternovsky, Z., Auer, S., Srama, R., and Juhasz, A.: 2010, *AGU Fall Meeting Abstracts*, 1673.

Gumbel, J.: 2001, *J. Geophys. Res.* **A 106(A6)**, 10553.

Habbal, S.R., Arndt, M.B., Nayfeh, M.H., Arnaud, J., Johnson, J., Hegwer, S., Woo, R., Ene, A., and Habbal, F.: 2003, *Astrophys. J.* **592**, L87.

Havnes, O., Trøim, J., Blix, T., Mortensen, W., Næsheim, L.I., Thrane, E., and Tønnesen, T.: 1996, *Journal of Geophysical Research* **101**, 10839.

Horanyi, M., Houpis, H.L.F., and Mendis, D.A.: 1988, *Astrophysics and Space Science* **144**, 215.

Hornung, K. and Kissel, J.: 1994, *Astron. Astroph.* **291**, 324.

Hornung, K., Malama, Y.G., and Kestenboim, K.S.: 2000, *Astrophysics and Space Science* **274**, 355.

Hsu, S, Krüger, H. and Postberg, F.: 2012, Nanodust Measurements by the Cassini Plasma Spectrometer, In: Mann, I., Meyer-Vernet, N., Czechowski, A., (eds.) *Nanodust in the Solar System: Discoveries and Interpretations*, 118.

Hunten, D.M., Turco, R.P., and Toon, O.B.: 1980, *Journal of Atmospheric Sciences* **37**, 1342.

Ip, W.-H. and Chow, V.W.: 1997, *Icarus* **130**, 217.

Ip, W.-H. and Jorda, L.: 1998, *Astrophys. J.* **496**, L47.

Ishimoto, H.: 2000, *Astron. Astroph.* **362**, 1158.

Jones, A.P., Tielens, A.G.G.M., and Hollenbach, D.J.: 1996, *Astrophys. J.* **469**, 740.

Kalashnikova, O., Horányi, M., Thomas, G.E., and Toon, O.B.: 2000, *Geophys. Res. Lett.* **27**, 3293.

Kamitsuji, K., Ueno, S., Suzuki, H., et al.: 2004, *Astron. Astroph.* **422**, 975.

Kharchenko, V. and Lewkow, N.: 2012, Charge-Exchange and X-ray Processes with Nanodust Particles, In: Mann, I., Meyer-Vernet, N., Czechowski, A., (eds.) *Nanodust in the Solar System: Discoveries and Interpretations*, 194.

Kimura, Y.: 2012, Phenomena of Nanoparticles in Relation to the Solar System. In: Mann, I., Meyer-Vernet, N., Czechowski, A., (eds.) *Nanodust in the Solar System: Discoveries and Interpretations*, 46.

Kimura, H., Ishimoto, H., and Mukai, T.: 1997, *Astron. Astroph.* **326**, 263.
Kirsch, E., McKenna-Lawlor, S., Korth, A., Schwenn, R., O Sullivan, D., and Thompson, A.: 1991, *International Cosmic Ray Conference* **3**, 370.
Krüger, H., Graps, A.L., Hamilton, D.P., et al.: 2006, *Planetary and Space Science* **54**, 919.
Kuhn, J.R., Penn, M.J., and Mann, I.: 1996, *Astrophys. J.* **456**, L67.
Le Sergeant D'Hendecourt, L.B. and Lamy, P.L.: 1978, *Nature* **276**, 800.
Le Sergeant D'Hendecourt, L.B. and Lamy, P.L.: 1980, *Icarus* **43**, 350.
Le Sergeant D'Hendecourt, L.B. and Lamy, P.L.: 1981, *Icarus* **47**, 270.
Leinert, C., Roser, S., and Buitrago, J.: 1983, *Astron. Astroph.* **118**, 345.
Leinert, C., Bowyer, S., Haikala, L.K., et al.: 1998, *Astronomy and Astrophysics Supplement Series* **127**, 1.
Li A., and Mann I.: 2012, Nanodust in the Interstellar Medium in Comparison to the Solar System. In: Mann, I., Meyer-Vernet, N., Czechowski, A., (eds.) *Nanodust in the Solar System: Discoveries and Interpretations*, 30.
Mann, I.: 2010, *Annual Review of Astronomy and Astrophysics* **48**, 173.
Mann, I. and Czechowski, A.: 2005, *Astrophys. J.* **621**, L73.
Mann, I. and Murad, E.: 2005, *Astrophys. J.* **624**, L125.
Mann, I., Murad, E., and Czechowski, A.: 2007, *Planetary and Space Science* **55**, 1000.
Mann, I., Czechowski, A., and Meyer-Vernet, N.: 2010, *Twelfth International Solar Wind Conference* **1216**, 491.
Mann, I., Czechowski, A., Meyer-Vernet, N., Zaslavsky, A., and Lamy, H.: 2010, *Plasma Physics and Controlled Fusion* **52**, 124012.
Mann, I., Kimura, H., Biesecker, D.A., et al.: 2004, *Space Science Reviews* **110**, 269.
Mann, I., Pellinen-Wannberg, A., Murad, E., Popova, O., Meyer-Vernet, N., Rosenberg, M., Mukai, T., Czechowski, A., Mukai, S., Safrankova, J., and Nemecek, Z.: 2011, *Space Science Reviews* **161**, 1
McBride, N. and McDonnell, J.A.M.: 1999, *Planetary and Space Science* **47**, 1005.
Megner, L., Siskind, D.E., Rapp, M., and Gumbel, J.: 2008, *J.Geophys. Res. D* **113**, 3202.
Mendis, D.A.: 2001, *Physica Scripta* **89**, 173.
Meyer-Vernet N., and Zaslavsky A.: 2012, In Situ Detection of Interplanetary and Jovian Nanodust with Radio and Plasma Wave Instruments. In: Mann, I., Meyer-Vernet, N., Czechowski, A., (eds.) *Nanodust in the Solar System: Discoveries and Interpretations*, 160.
Meyer-Vernet, N., Czechowski, A., Mann, I., Maksimovic, M., Lecacheux, A., Goetz, K., Kaiser, M.L., Cyr, O.C.S., Bale, S.D., and Le Chat, G.: 2010, *Twelfth International Solar Wind Conference* **1216**, 502.
Meyer-Vernet, N., Maksimovic, M., Czechowski, A., Mann, I., Zouganelis, I., Goetz, K., Kaiser, M.L., St. Cyr, O.C., Bougeret, J.-L., and Bale, S.D.: 2009, *Solar Phys.* **256**, 463.
Minato, T. and Mann, I.: 2006, *36th COSPAR Scientific Assembly* **36**, 3046.
Minato, T., Köhler, M., Kimura, H., Mann, I., and Yamamoto, T.: 2004, *Astron. Astroph.* **424**, L13.
Minato, T., Köhler, M., Kimura, H., Mann, I., and Yamamoto, T.: 2006, *Astron. Astroph.* **452**, 701.
Morlok, A., Koike, C., Tomioka, N., Mann, I., and Tomeoka, K.: 2010, *Icarus* **207**, 45.
Mukai, T., Yamamoto, T., Hasegawa, H., Fujiwara, A., and Koike, C.: 1974, *Publications of the Astronomical Society of Japan* **26**, 445.
Nakamura, A.M., and Michel, P. In:Mann, I., Nakamura, A.M., and Mukai, T.: 2009, *Lecture Notes in Physics, Berlin Springer Verlag* **758**.
Mukai, T. and Schwehm, G.: 1981, *Astron. Astroph.* **95**, 373.
Nuth, J.A., III, Rietmeijer, F.J.M., and Hill, H.G.M.: 2002, *Meteoritics and Planetary Science* **37**, 1579.
Rapp, M., Strelnikov, B., Wilms, S., Luebken, F.-J., Gumbel, J., and Henkel, H.: 2003, *Sixteenth ESA Symposium on Rocket and Balloon Programmes and Related Research, ESA-SP* **530**, 379.
Rapp, M., Strelnikova, I., Strelnikov, B., Hoffmann, P., Friedrich, M., Gumbel, J., Megner, L., Hoppe, U.-P., Robertson, S., Knappmiller, S., Wolff, M., and Marsh, D.R.: 2010, *J.Geophys. Res. D* **115**, 0.
Rietmeijer, F.J.M.: 2000, *Meteoritics and Planetary Science* **35**, 1025.

Rietmeijer, F.J.M.: 2002, In Meteors in the Earths Atmosphere, ed. by E. Murad, I.P. Williams (Cambridge University Press, Cambridge, 2002), 215.
Rosinski, J. and Snow, R.H.: 1961, *Journal of Atmospheric Sciences* **18**, 736.
Schwadron, N.A., Geiss, J., Fisk, L.A., Gloeckler, G., Zurbuchen, T.H., and von Steiger, R.: 2000, *Journal of Geophysical Research* **105**, 7465.
Singh, J., Sakurai, T., Ichimoto, K., Hagino, M., and Yamamoto, T.T.: 2004, *Astrophys. J.* **608**, L69.
Sugita, S., Schultz, P.H., and Adams, M.A.: 1998, *Journal of Geophysical Research* **103**, 19427.
Tielens, A.G.G.M., McKee, C.F., Seab, C.G., and Hollenbach, D.J.: 1994, *Astrophys. J.* **431**, 321.
Tielens, A.G.G.M., Seab, C.G., Hollenbach, D.J., and McKee, C.F.: 1987, *Astrophys. J.* **319**, L109.
Tuzzolino, A.J., Economou, T.E., Clark, B.C., Tsou, P., Brownlee, D.E., Green, S.F., McDonnell, J.A.M., McBride, N., and Colwell, M.T.S.H.: 2004, *Science* **304**, 1776.
Utterback, N.G. and Kissel, J.: 1995, *Astrophysics and Space Science* **225**, 327.
Utterback, N.G. and Kissel, J.: 1990, *The Astronomical Journal* **100**, 1315.
Simpson, J.A., Rabinowitz, D., Tuzzolino, A.J., Ksanfomaliti, L.V., and Sagdeev, R.Z.: 1987, *Astron. Astroph.* **187**, 742.
Wehry, A. and Mann, I.: 1999, *Astron. Astroph.* **341**, 296.
Wimmer-Schweingruber, R.F. and Bochsler, P.: 2003, *Geophys. Res. Lett.* **30**, 020000.
Wurz, P.: 2012, Erosion Processes Affecting Interplanetary Dust Grains, In: Mann, I., Meyer-Vernet, N., Czechowski, A., (eds.) *Nanodust in the Solar System: Discoveries and Interpretations*, 178.
Zaslavsky, A., Meyer-Vernet, N., Mann, I., Czechowski, A., Issautier, K., Le Chat, G., Pantellini, F., Goetz, K., Maksimovic, M., Bale, S.D., Kasper, J.C.: 2012, *Journal of Geophysical Research*, in press.
Zel'dovich, Y.B. and Raizer, Y.P.: 1967, *New York: Academic Press, 1966/1967, edited by Hayes, W.D.; Probstein, Ronald F.*.
Zook, H.A., Grun, E., Baguhl, M., Hamilton, D.P., Linkert, G., Liou, J.-C., Forsyth, R., and Phillips, J.L.: 1996, *Science* **274**, 1501.

Index

I. Mann et al. (eds.), *Nanodust in the Solar System: Discoveries and Interpretations*,
Astrophysics and Space Science Library 385, DOI 10.1007/978-3-642-27543-2,

MIX
Papier aus verantwortungsvollen Quellen
Paper from responsible sources
FSC® C105338

If you have any concerns about our products,
you can contact us on
ProductSafety@springernature.com

In case Publisher is established outside the EU,
the EU authorized representative is:
Springer Nature Customer Service Center GmbH
Europaplatz 3, 69115 Heidelberg, Germany

Printed by Libri Plureos GmbH
in Hamburg, Germany